Agroforestry Systems

NIPA® GENX ELECTRONIC RESOURCES & SOLUTIONS P. LTD.
New Delhi-110 034

About the Authors

Dr. Alok Kumar Patra graduated in Agriculture from Chiplima campus of Odisha University of Agriculture & Technology in 1988, obtained his M. Sc. (Agriculture) in Agronomy from Institute of Agriculture, Visva Bharati, Santiniketan in 1990 and Ph. D. from Bidhan Chandra Krishi Viswavidyalaya, Mohanpur, West Bengal in 1996. He has an MBA degree from IGNOU, New Delhi.

Dr. Patra has published more than 90 research papers, 125 popular scientific articles and bulletins, ten books and 35 book chapters. He has 30 years of experience in research and teaching in UG and PG level on agroforestry and agronomy. Presently he is working as Professor of Agronomy in the All India Coordinated Research Project on Integrated Farming Systems, OUAT, Bhubaneswar 751003.

Dr. Manas Ranjan Nayak graduated in Forestry from College of Forestry under Odisha University of Agriculture & Technology, Bhubaneswar in 2010, obtained his M. Sc. (Forestry) in Silviculture & Agroforestry from the same institute in 2012 and Ph. D. from ASPEE College of Horticulture & Forestry, Navsari Agricultural University, Navsari, Gujarat in 2015.

Dr. Nayak has published more than 40 research papers, 45 popular scientific articles and bulletins, seven books and ten book chapters. He has eight years of experience in research, extension and teaching on agroforestry. Presently he is working as Assistant Professor of Forestry in the All India Coordinated Research Project on Agroforestry, OUAT, Bhubaneswar 751003.

Agroforestry Systems

Alok Kumar Patra
Professor of Agronomy
All India Coordinated Research Project on Integrated Farming Systems
Odisha University of Agriculture & Technology
Bhubaneswar-751003, Odisha

Manas Ranjan Nayak
Assistant Professor of Silviculture & Agroforestry
All India Coordinated Research Project on Agroforesty
Odisha University of Agriculture & Technology
Bhubaneswar-751003, Odisha

NIPA® GENX ELECTRONIC RESOURCES & SOLUTIONS P. LTD.
New Delhi-110 034

NIPA® GENX ELECTRONIC RESOURCES & SOLUTIONS P. LTD.

101,103, Vikas Surya Plaza, CU Block
L.S.C. Market, Pitam Pura, New Delhi-110 034
Ph : +91-11-43860225, Mob.: +91 9717133558, 9540816132
E-mail: newindiapublishingagency@gmail.com
Website: www.nipaersources.com

Print ISBN: 978-93-58871-86-9
ebook ISBN: 978-93-58873-70-2

Composed and Designed by NIPA®.

Preface

The practice of agroforestry is not new; it is as old as the agricultural practices that have existed for millennia. Realising the full potential of integrating commercially managed trees into farming systems is, however, still a long way off. Agroforestry systems can be advantageous over conventional agricultural and forest production systems through increased productivity, economic benefits, social outcomes and the ecological goods and services.

Implementation of the National Agroforestry Policy (2014), the Sub-Mission on Agroforestry (2016), and the National Mission on Bamboo (2018) have further accelerated the momentum of adoption and expansion of agroforestry in India. Inclusion of agroforestry as one of the eligible activities for receiving funds through Corporate Social Responsibility Funding (2014), and consideration of 14th Finance Commission of India to include green cover (over forest cover) to provide additional funding to State Governments is expected to further promote agroforestry in the country.

During the last two decades the science of agroforestry has progressed fast, both in its techniques and practices. New areas of studies have also emerged. Today, agroforestry has established itself as a viable approach of integrated land management system not only for meeting the deficits of food, fodder, firewood and timber but also for ecological considerations like soil conservation, watershed protection, wasteland management, biodiversity conservation, carbon sequestration and mitigation of climate change effects.

In this backdrop, agroforestry has been recommended as a core subject in the curriculum of the state agricultural universities. Keeping this in view, an effort has been made to write a textbook on agroforestry systems which would be useful to the students as well as the teachers. The common people, who love trees, would also find it worth reading.

This book is primarily based on the syllabus of the course 'Agroforestry Systems' taught to under-graduate forestry students. The book has been divided into fifteen chapters covering all aspects of agroforestry including concepts, definition, history, benefits and limitations, systems classifications, soil productivity, tree-crop interaction, multipurpose trees and their propagation, agroforestry management, watershed and wasteland management through agroforestry, climate change adaptation and mitigation, agroforestry design & diagnosis. The glossary and relevant bibliography including the web links will also help for easy access to the literature on the subject. Any suggestion to improve the contents of the book will be highly appreciated. The authors take full responsibility for any errors in this book. Any shortcomings may be intimated so that it will be taken care of in the next edition.

In writing this book, the literature on agroforestry developed by various organizations and agencies like ICRAF, ICFRE, FAO, AFNETA, ICAR-CAFRI, ICAR-CAZRI, etc. is freely used. The authors extend their sincere thanks to the authors and editors of various books, journals and periodicals which have been used as reference material in this book. Every care has been taken to cite the bibliographic references. However, any omissions, misrepresentations, incorrect citations or other mistakes that may have occurred are regretted.

The authors are very much thankful to their family members for constant support and encouragement for writing this book. They also express their gratitude to the New India Publishing Agency, New Delhi for bringing out the book timely and nicely.

15 August, 2024
OUAT, Bhubaneswar

Alok Kumar Patra
Manas Ranjan Nayak

Contents

Preface ... *v*

Glossary ... *xiii*

1. Agroforestry: A Sustainable Land Use System ... 1

Definitions of Agroforestry ... 2

Objectives of Agroforestry ... 3

Relevance of Agroforestry ... 4

Potential of Agroforestry in India ... 7

Implementation Constraints ... 12

Agroforestry and the Future ... 14

2. Land Capability Classification ... 17

Usda Land Classification Method ... 18

Land Evaluation ... 21

3. Classification of Agroforestry Systems ... 25

Basis of the Classification of Agroforestry Systems ... 25

Structural Classification of Agroforestry Systems ... 27

Agrisilvicultural Systems ... 27

Silvipastoral Systems ... 46

Agrisilvipastoral Systems ... 50

Other Agroforestry Systems ... 53

Classification Based on Arrangement of Components ... 55

Functional Classification of Agroforestry Systems ... 56

Socioeconomic Classification of Agroforestry Systems ... 57

Ecological Classification of Agroforestry Systems ... 59

Physiognomic Classification of Agroforestry Systems ... 60

Historical Classification of Agroforestry Systems ... 61

Classification Based on Utilization of Land ... 61

Floristic Classification of Agroforestry Systems ... 62

4. Selection of Multipurpose Trees for Agroforestry 63
Benefits from Multipurpose Trees 63
Selection of Appropriate Species 65
Matching Species to Requirements 65
Characteristics of MPTs Suitable for Agroforestry 66
Role of MPTs in Agroforestry 67
Suitable MPTs for Agroforestry Under Different Climatic Conditions 73
Suitable MPTs for Agroforestry Under Different Soil Types 74
Sitable MPTs for Agroforestry for Different Regions 75
Improving Mpts for Agroforestry Systems 76

5. Social Forestry 79
Definitionof Social Forestry 79
Characteristics of Social Forestry 81
Objectives of Social Forestry 81
Scope of Social Forestry 82
Components of Social Forestry 83
Benefits from Social Forestry 86
Strategies for Implementation of Social Forestry 87
Steps for Establishment of A Social Forestry Plantation 89
Roadside Plantation 92
Railway Line Plantation 93
Canal Banks Plantation 94
Difference Between Social Forestry and Agroforestry 94

6. Role of Agroforestry in Soil Productivity and Conservation 97
Trees Improve Soil 98
Beneficial Effects of Trees on Soil 99
Adverse Effects of Trees on Soil 102
Nutrient Cycling in Agroforestry Systems 103
Soil Organic Matter in Agroforestry Systems 105
Biological Nitrogen Fixation in Agroforestry Systems 106
Effect of Agroforestry on Soil Erosion 111

7. Tree Crop Interaction in Agroforestry 115
Nature of Interactions 116
Factors Affecting Interactions 117
Types of Interactions 118
Management Options to Neutralize Negative Interactions 124
Method for Quantifying Interactions 125

8. Structural and Functional Attributes of Agroforestry Systems 129
Key Traits of Agroforestry Practices 129
Premises of Agroforestry 130
Shifting Cultivation 131
Taungya System 132
Alley Cropping 134
Homegardens 136
Windbreaks 140
Shelterbelt 143
Multiple Cropping 144
Intercropping 146
Mixed Cropping 150
Sequential Cropping 152
Crop Rotation 153
Energy Plantations 154
Silvipastoral Systems 160

9. Agroforestry Planning and Management 167
Factors Affecting the Selection of Tree Species 167
Tree Propagation Methods 169
Nursery Management 173
Tree Management 177
Agricultural Crop Management 185
Management of Fruit Trees in Agroforestry Systems 187

10. Agroforestry for Different Agroecological Zones 189
Agroforestry for Humid and Subhumid Tropics 189
Agroforestry for Arid and Semiarid Tropics 191

Agroforestry for Tropical Wetlands 195
Agroforestry for Hill Lands 199

11. Agroforestry for Watershed and Wasteland Management 203
Principles of Watershed Management 203
Objectives of Watershed Management 204
Agroforestry- A Sustainable Option for Watershed Management..... 204
Contribution of Trees Towards Improvement of Soil Water Regime 205
Biological Measures for Watershed Management 207
Conservation Measures for Watershed Management 209
Management of Wastelands 212

12. Agroforestry Diagnosis and Design 219
Key Features of D & D 219
Agroforestry Systems Research Process 220
Macro D & D 220
Micro D & D 223
Methodological Considerations in D & D 226
Design and Analysis of Agroforestry Experiments 228
Design Requirements 229
Principles of Field Experimentation 230
Types of Agroforestry Field Experiments 231
Plot Size and Arrangement 232
Experimental Designs 233
On-Farm Research in Agroforestry 235

13. Agroforestry Systems Research and Development 239
History of Agroforestry Research 239
History of Agroforestry Research in India 242
Forest Policies in India 244
National Forest Policy 1894 245
National Forest Policy 1952 245
National Commission on Agriculture 1976 248
National Forest Policy 1988 249
National Agroforestry Policy 2014 253

Latest Developments in Agroforestry Research 257
International Institutes Working in Forest Conservation and Research 268

14. Agroforestry for Climate Change Adaptation & Mitigation........275
Effects of Climate Change 275
Agroforestry for Adaptation to Climate Change..............................277
Agroforestry in Mitigating Climate Change 278
Carbon Sequestration Potential of Agroforestry Systems..............................281
Carbon Sequestration Potential of Agroforestry Systems in India 281
Factors Affecting Carbon Sequestration in Agroforestry Systems.............................. 286
Agroforestry as a Tool to Mitigate and Adapt to Climate Under LULUCF Accounting..............................289

15. Management Practices of Multipurpose Trees 295
Mangium (*Acacia mangium*)..............................295
Sissoo (*Dalbergia sissoo*) 298
Khaira (*Acacia catechu*)..............................300
Eucalyptus (*Eucalyptus* spp.)..............................302
Casuarina (*Casuarina equisetifolia*) 305
Teak (*Tectona grandis*)..............................308
Neem (*Azadirachta indica*)..............................311
Tamarind (*Tamarindus indica*) 314
Phalsa (*Grewia* spp.) 316
Mulberry (*Morus alba*) 319
Poplar (*Populus deltoides*)..............................321
Oak (*Quercus* spp.) 326
Bamboo (*Dendrocalamus strictus*) 328

Selected References 333
Web Links 347
Index 349

Glossary

Agrisilviaqu aculture: A form of agroforestry consisting of tree (woody perennial), agricultural crop and freshwater aquatic animal components.

Agrisilviculture: A form of agroforestry consisting of tree and crop components.

Agrisilvipasture: A form of agroforestry consisting of tree, crop and pasture/animal components.

Agroforestry: A collective name for land use systems and technologies where woody perennials (trees, shrubs, palms, bamboos, etc.) are deliberately used on the same land management units as agricultural crops and/or animals, in some form of spatial arrangement or temporal sequence.

Alkali soil: A soil that contains sufficient alkali (sodium) to interfere with the growth of most crop plants (*p*H 8.5 or higher).

Allelopathy: Any direct or indirect harmful effect that one plant has on another or mutually on each other through the production of chemical compounds that escape into the environment.

Alley cropping: A farming system in which arable crops are grown in alleys formed by trees or shrubs, established mainly to hasten soil fertility restoration and enhance soil productivity.

Annual plant: A plant that grows for only one season (or year) before dying, in contrast to a perennial, which grows for more than one season.

Aquasilviculture: A form of agroforestry consisting of tree and aquatic animals.

Arid climate: Climate in regions that lack sufficient moisture for crop production without irrigation. In cool regions annual precipitation is usually less than 25 cm. It may be as high as 50 cm in tropical regions. Natural vegetation is desert shrubs.

Bench terrace: Conversion of a steep slope into a series of steps with near-horizontal hedges and near-vertical walls between hedges, using retaining walls (rock) or steep banks (soil) for intensive cropping.

Biomass: The weight of material produced by a living organism or collection of organisms. The term is usually applied to plants to include the entire plant, or it may be qualified to include only certain parts of the plant, e.g., above-ground or leafy biomass. Biomass is expressed in terms of fresh weight or dry weight.

Breast height: Universally adopted standard height for measuring girth or diameter of standing trees. It is 1.37 m (4 feet 6 inches).

Browse: The buds, shoots, leaves and flowers of woody plants which are eaten by livestock or wild animals.

Bund: An artificial earthen embankment made across sloping agricultural land to cut short lengthy soil slopes and reduce water runoff and soil erosion. It demarcates plot boundary or used for other purposes.

Bush fallow: A fallow of usually 3-10 years in which the natural vegetation regenerates which is composed principally of shrubs and young trees.

Buttress: An outgrowth from the base of a tree connecting it with the roots, especially common in tropical rainforest species.

C_3 plants: Species with the photosynthetic pathway in which the first product of CO_2 fixation is the 3-carbon molecule, phosphoglyceric acid and comparatively photosynthetically less efficient than C_4 plants.

C_4 plants: Species with the photosynthetic pathway in which the first product of CO_2 fixation is the 4-carbon molecule, oxaloacetic acid and photosynthetically more efficient than C_3 plants.

CAM (Crassulacean Acid Metabolism) plants: Species with stomata that open primarily at night, and have organic acids, especially malic acid, as the primary CO_2 fixation products.

Carbon/nitrogen ratio: The ratio of the weight of organic carbon (C) to the weight of total nitrogen (N) in a soil or organic material.

Carbon-stock: Carbon stored in vegetation or soil.

Catchment area: An independent unit of treated or untreated land area contributing runoff water to a reservoir.

Clean (or clear) bole: The part of a bole or stem which is free from branches.

Clone: A group of plants originating from a single plant by vegetative propagation and therefore all having the same genetic make-up.

Community forestry: A form of social forestry, where tree planting is undertaken by community people on common lands for fuelwood, timber, leaf fodder and other products that results in the improvement of community life.

Contour: An imaginary line connecting points of equal elevation on the surface of the soil.

Contour hedgerow: Dense, narrow hedges of woody perennials, planted along the contour of a slope to prevent soil erosion.

Contour ridge: A series of parallel ridges on the contour of cultivated land which has grass or shrubs planted on them to control soil erosion and improve water management.

Coppicing: Cutting of certain tree species close to ground level to produce new shoots from the stump. Also occurs naturally in some species if the trees are damaged.

Cover crop: Crops grown primarily to cover the soil and to reduce the loss of moisture due to leaching and erosion by wind and water.

Crop rotation: The growing of different crops on a piece of land in a preplanned succession, usually to prevent soil exhaustion from a demanding crop or to control pests and diseases.

Cropping pattern: The yearly sequence and spatial arrangement of crops or of crops and fallow on a given area.

Cropping system: The cropping patterns taken up for a given piece of land, or order in which the crops are cultivated on a piece of land over a fixed period of time and their interaction with farm resources, other farm enterprises and available technology which determine their make-up.

Crown: The canopy or top of a single tree or other woody plant that carries its main branches and leaves at the top of a fairly clean stem.

Cut-and-carry: Fodder or other plant products which are harvested and carried to a different location to be used or consumed.

Deciduous plant: A plant that sheds all or most of its leaves every year during a certain season.

Deforestation: Disturbance, conversion or wasteful destruction of forest lands.

Denitrification: The biochemical reduction of nitrate or nitrite to gaseous nitrogen, either as molecular nitrogen or as an oxide of nitrogen.

Diagnosis and Design (D & D): A systematic and objective methodology developed by International Centre for Research in Agroforestry to initiate, monitor and evaluate agroforestry programmes.

Dormancy of seed: The inability of seed to germinate, even under favourable growing conditions.

Drought: The absence of precipitation for a period long enough to cause depletion of soil moisture and damage to plants.

Ecosystem: All the plants and animals in a given area and their physical environment, including the interactions among them.

Erosion: The wearing away of the land surface by running water, wind, ice or other geological agents, including such processes as gravitational creep.

Evaluation: A process of assessing and monitoring results for conformity with agreed objectives in order to improve ongoing activities and plan better for future ones.

Evaporation: Loss of moisture from surfaces other than plants.

Evapotranspiration: The combined loss of water from a given area and during a specified period of time, by evaporation from the soil surface and by transpiration from plants.

Evergreen: Plants which retain their leaves and remain green throughout the year.

Exotic: A plant or animal species which has been introduced outside its natural range.

Extension forestry: The practice of forestry in areas devoid of tree growth and other vegetation situated in places away from the conventional forest areas with the objective of increasing the area under tree growth.

Fallow: Period of one-to-many years during which a field is not cropped, to allow soil fertility to be restored.

Farm forestry: Commercial tree growing by farmers on their own land.

Farming system: An appropriate combination of farm enterprises, viz., crops, livestock, forestry, fishery, apiary, poultry, and the means available to the farmer to raise them for profitability. It interacts adequately with the environment without dislocating the ecological and socioeconomic balance.

Fixed-lift pruning: Complete pruning of all branches below a prescribed point on the stem of the tree. The lift is specified as height from the ground and the choice of height is based on the shape of development attained by the stand.

Fluting: Stem shows irregular involutions and swellings. It is considered a serious defect for timber production.

Fodder: Parts of plants which are eaten by domestic animals, these may include leaves, stems, fruit, pods, flowers or pollen.

Foliage: The mass of leaves of plants, usually of trees or bushes.

Forage: Vegetative material in a fresh, dried, or ensiled state which is fed to livestock (hay, pasture, silage).

Forest fallow: A longer fallow, usually of several years, in which the naturally regenerated vegetation is composed principally of larger or mature trees.

Forest garden: A land-use form on private lands outside the village in which planted trees and sometimes additional perennial crops occur.

Grafting: The practice of propagating plants by taking a small shoot from one and attaching it to another so that the cambium layers from both are in contact and the transferred shoot grows as part of the main plant.

Grassland: A plant community in which grasses predominate; trees are either altogether absent or present in relatively small number.

Green manure: Green plant material incorporated into the soil to improve its fertility.

Groundcover: Living or non-living material which covers the soil surface.

Gully: A deep, narrow channel cut into the soil by erosion.

Hardening off: Measures taken to prepare plants raised in a nursery for planting out in the field.

Hedgerow (or hedge): A closely planted line of shrubs or trees, often forming a boundary or fence.

Herbaceous: A plant that is not woody and does not persist above ground beyond one season.

Homegarden: A land use form on private lands surrounding individual houses with a definite fence, in which several tree species are cultivated together with annual and perennial crops, often with inclusion of livestock.

Indigenous: Native to a specific area; not introduced.

Infiltration: The downward movement of water into the soil.

Infiltration rate: A soil characteristic determining or describing the maximum rate at which water can enter the soil under specified conditions, including the presence of an excess of water.

Inoculation: The process of introducing pure or mixed cultures of microorganisms into natural or artificial culture media.

Intercropping: Growing two or more crops in the same field at the same time in a mixture.

Interface: The area where there is positive or negative interaction between two entities, such as between rows of trees and rows of crops.

Internal rate of return (IRR): The maximum rate of interest that a project can repay on loans while still recovering all investment and opportunity costs; or, the earning power of money invested in a particular venture.

Land equivalent ratio (LER): The relative land area under sole crop required to produce the same yield as obtained under a mixed or an intercropping system at the same management level.

Land-use system: The way in which land is used by a particular group of people within a specified area.

Layering: Method to stimulate growth of roots on shoots or twigs in order to propagate the layered plant true to type.

Leaching (of nutrients): The dissolving and washing away of nutrients down the soil profile by the action of rain water.

Ley farming: Rotation of arable crops requiring annual cultivation and artificial pastures occupying field for two years or longer.

Lignin: The complex organic constituent of woody fibers in plant tissue that, along with cellulose, cements the cells together and makes substances resistant to decomposition and after some modification may become part of the soil organic matter.

Litter: Layer of decomposing plant material (leaves, branches, etc.) covering the ground, especially under trees.

Lopping: Cutting one or more branches of a standing tree or shrub.

Microclimate: The specific local climatic conditions near the ground or area around plants up to 2 m height, resulting from the modifications of the general climatic conditions by local differences in relief, exposure and cover, etc.

Mineralization: The conversion of an element from an organic form to an inorganic state as a result of microbial decomposition.

Mixed farming: Cropping systems which involve the raising of crops, animals and/or trees.

Mixed intercropping: Growing two or more crops simultaneously with no distinct row arrangement.

Monoculture: The repetitive growing of the same sole crop on the same land.

Mother tree: Tree from which seeds or vegetative parts are collected to be propagated.

Mulch: A layer of loose material on the soil to reduce moisture loss, moderate soil temperature and inhibit weed growth.

Multipurpose trees: Trees yielding one or more products and offering environmental benefits as well.

Multistoried (or multistoreyed)**:** Relating to a vertical arrangement of plants so that they form distinct layers, from the lower (usually herbaceous) layer to the uppermost tree canopy.

Net present value (NPV): An indicator of a project's long-term value as estimated at the time of implementation. It is calculated by summing all the annual net costs or benefits over the prescribed life span of a project, discounted at a pre-selected rate.

Nitrogen cycle: The sequence of chemical and biological changes undergone by nitrogen as it moves from the atmosphere into water, soil and living organisms, and upon death of these organisms (plants and animals) is recycled through a part or all of the entire process.

Nitrogen fixation: The biological conversion of elemental nitrogen (N_2) to organic combinations or to forms readily utilized in biological processes, by nitrogen-fixing microorganisms. When brought about by bacteria in the root nodules of leguminous plants, it is referred to as symbiotic; if by free living microorganisms acting independently, it is referred to as non-symbiotic or free fixation.

Nutrient pump: A deep tree root system, that takes up nutrients from deep soil layers and brings them to the surface in the tree and its litter fall.

Opportunity cost: The true sacrifice incurred by the choice of a given action.

Overstory (or overstorey): The highest layer of vegetation, often the tree canopy, which grows over lower shrub or plant layers.

Perennial plant: A plant that grows for more than one year, in contrast to an annual, which grows for only one year (or season) before dying.

Pollarding: Cutting back the crown of a tree in order to harvest wood and browse to produce regrowth beyond the reach of animals and/or to reduce the shade cast by the crown.

Polypot: Polythene bag used as a pot to raise young plants.

Prick out: Transplant seedlings from seed trays or seedbeds into nursery beds or pots.

Propagule: A part of a plant that can give rise to a new plant.

Protective plant: Plants grown to protect crops, soil or land from adverse environmental factors.

Provenance: Germplasm from a single place of origin. Germplasm from different provenances of the same species can differ in ways such as growth habit, biomass production or drought hardiness.

Pruning: Cutting back plant growth, including side branches or roots.

Recalcitrant seed: Seed that has to be sown fresh because it loses its viability rapidly; e.g., that of many large seeded tropical fruits.

Recreation forestry: The practice of forestry with the objectives of raising flowering trees and shrubs mainly to serve as recreation forests for the urban and rural population.

Reforestation: Establishment of trees on ground that has been recently cleared of trees.

Regeneration: The renewal of a forest crop by natural or artificial means

Resource capture: The processes by which plants obtain light, water and nutrients.

Rhizosphere: Portion of the soil in the immediate vicinity of plant roots in which the abundance and composition of the microbial population are influenced by the presence of roots.

Root ball: The network of roots and the soil clinging to them when a plant is lifted from the soil or removed from a container.

Root sucker: A shoot arising from the root of a plant.

Rotation (in forestry): The length of time between establishment and harvesting of a tree.

Runoff: The portion of the precipitation on an area which is not absorbed by the soil but finds its way into the streams after meeting the persisting demands of evapotranspiration and other losses.

Selective pruning: Removal of some of the branches on the stem at various levels above the ground according to a particular prescription.

Semiarid: Term applied to regions or climates where moisture is more plentiful than in arid regions but still definitely limits the growth of most crop plants. Natural vegetation in uncultivated lands in these areas is short grasses, shrubs and small trees.

Sequential cropping: Growing two or more crops in sequence on the same field. The succeeding crop is planted after the preceding crop has been harvested.

Shelterbelts: A wind barrier of living trees and shrubs established and maintained for protection of crop fields.

Shifting cultivation: System of land use in which the site of cultivation is regularly changed, with older sites reverting to forest or bush fallow.

Shrub: A woody plant that remains less than 10 meters tall and produces shoots or stems from its base.

Silvipastoral system: A form of agroforestry systems consisting of the trees (woody perennial) and pasture/animal components.

Slash-and-burn system: A technique for land clearing to prepare for planting of annual food crops or trees.

Social forestry: The practice of using trees and/or tree planting specifically to pursue social objectives, usually betterment of the poor, through delivery of the benefits to the local people.

Soil conservation: A combination of all management and land-use methods that safeguard the soil against depletion or deterioration caused by nature and/or humans.

Soil organic matter: The organic fraction of the soil that includes plant and animal residues at various stages of decomposition, cells and tissues of soil organisms, and substances synthesized by the soil population.

Sole cropping: One crop variety grown alone in pure stand at normal density.

Species: A taxonomic category below genus.

Stress: Any factor that disturbs the normal functioning of an organism.

Suckers: A side shoot from the roots of a plant; a side growth arising from an auxiliary bud.

Sustainable land use: Land use that maintains productivity of the land while conserving or enhancing the resources on which future production depends.

Taungya: Method of raising forest crops in combination with an agricultural crop; the agricultural use to which the land is put does not generally continue throughout the rotation of the forest crop but is confined to that period which ends with the closing of the canopy of the forest crop.

Thinning: Intermediate cuttings that are primarily aimed at controlling the growth of stands by adjusting stand density.

Tip pruning: Pruning of a branch at a point other than at its junction with the stem.

Topography: The physical description of land; changes in elevation due to hills, valleys and other features.

Transpiration: The loss of moisture from plants in the form of water vapour.

Tree garden: Multistoreyed agroforestry systems in which a mixture of several fruit and other trees are cultivated, sometimes with inclusion of annual crops.

Variable-lift pruning: Complete pruning of all branches below a prescribed variable point on the stem of the tree. This point may be specified either as a proportion of the height or as a diameter limit.

Watershed: A geohydrological unit or all the land and water area bounded by a divide which contributes runoff to a common point. Watershed is considered to be synonymous with 'catchment basin' and 'drainage basin'.

Wildings: Wildings are planting materials collected from natural vegetation,e.g., seedlings or cuttings.

Windbreak: A strip of trees or shrubs or crop plants serving to reduce the force of wind and provide a protective shelter against wind.

Woody: Plants which consist in part of wood; not herbaceous.

Zero-grazing: Livestock production systems in which the animals are fed in pens or other confined areas and are not permitted to graze.

1

Agroforestry A Sustainable Land Use System

Agroforestry is a very old practice, dating back many years. Tropical farmers have long been growing food crops, timber, and livestock in addition to taking advantage of the variety of products that may be obtained from natural wood lots. Indian culture is deeply rooted in its appreciation of trees and forests. The forests gave birth to the greatest aspects of Indian culture. The sages who developed Hindu philosophy resided in forests, living in perfect harmony with the natural world. In fact, trees have been discussed in our ancient literature so much that people were planting trees on their own in addition to crops for agriculture. Maharishi Kashyap's book 'Krishishukti' divides land into multiple categories and designates regions that are suitable for tree planting. Particular sites for planting trees include all wet and dry regions as well as the vicinity of homes, wells, and tanks. However, foresters and farmers, who have historically worked within very strict disciplinary boundaries and focused on producing their favoured commodities - crops, livestock, and trees - used to disregard such integrated production systems that incorporate many disciplines.

But in more recent times, resources have significantly decreased and the forest area has diminished. The inhabitants can no longer rely on the forest to provide them with firewood, fodder, lumber, bamboo, etc. As a result of the wood shortage, these commodities' prices have significantly increased. A lack of raw materials has been an issue for numerous sectors reliant on forests. To address these problems, a lot of farmers have just lately begun planting trees on their farmlands along with agricultural crops; so, it originated from the idea of agroforestry.

Agroforestry is the term used to describe a variety of land use practices when trees are coexisting with crops and/or animals on the same piece of land.

- It combines the preservation of the resource base with the development of several outputs.
- It emphasizes the utilization of a variety of native trees and shrubs.

- It is especially appropriate for contexts with little input and fragility.
- It involves sociocultural values in a more interactive way than the majority of other land use regimes.
- It is more intricate both architecturally and functionally than monoculture.

DEFINITIONS OF AGROFORESTRY

Agroforestry is the practice of growing both agricultural crops and forest trees on the same piece of land. However, several professionals in the field of agroforestry have defined agroforestry.

Agroforestry, according to Bene *et al.* (1977), is a sustainable land management system that boosts overall production, implements management techniques that are in line with local cultural norms, and combines agricultural crops, forest plants, tree crops, and/or animals simultaneously or sequentially.

According to King and Chandler (1978), agroforestry is a sustainable land management system that boosts the land's overall yield by producing crops, including tree crops, and forest plants and/or animals on the same plot of land concurrently or sequentially. It also applies management techniques that are in line with the customs of the local populace.

Agroforestry is defined by Nair (1979) as a land use system that integrates trees, crops, and animals in a way that is supported by science, desirable for the environment, practically achievable, and well-liked by the farmers.

Lundgren and Raintree (1982) define agroforestry as a term used to describe a variety of land use systems and technologies in which woody perennials (trees, shrubs, palm bamboos, etc.) are purposefully placed alongside agricultural crops and/or animals in a particular spatial arrangement or temporal sequence. There are interactions between the many components in agroforestry systems on both an ecological and economic level.

According to Leakey's (1996) concept, agroforestry is currently defined by the World Agroforestry Centre (ICRAF) as a dynamic, ecologically oriented method of natural resource management that, by integrating trees on farms and in the agricultural landscape diversify and maintain productivity, benefiting land users at all levels in terms of social, economic, and environmental aspects.

These definitions give rise to several fundamental concepts of agroforestry.

- Typically, agroforestry entails the utilization of two or more plant species, or plant and animal species, with a minimum of one woody species.
- Agroforestry systems consistently yield two or more outputs.
- The agroforestry system's cycle always lasts longer than a year.

- There are both positive and negative interactions between the components (tree, crop).
- An agroforestry system, even at its most basic, is more complex both economically and ecologically (both structurally and functionally) than a monocropping system.

OBJECTIVES OF AGROFORESTRY

The major objectives of agroforestry include:

1. To lessen the strain on natural forests
2. To manage the land most efficiently in order to boost and restore its production
3. To promote the planting of trees alongside livestock and agricultural crops in order to increase rural households' total productivity, income, and standard of living, particularly for marginal and small farmers
4. To make economical and efficient use of the resources at hand
5. To give rural residents with year-round work options
6. To supply raw materials for small cottage businesses in rural areas
7. To minimise risk during extreme weather events by promoting resilient agricultural and farming systems and safeguarding and stabilising ecosystems
8. To increase the fuel supply in remote areas so that consumers can easily access it;an excessive amount of pressure is placed on traditional forests to provide fire wood, and 70 million tonnes of dried cow dung are utilised annually in India while there is an opportunity to repurpose this waste for natural organic fertiliser
9. To increase the availability of small wood that is utilised by villages for building houses, tools for agriculture, and other household needs;agroforestry can help the rural populace meet this need while also easing the burden on the forest
10. To increase the yield of food crops, legumes, and tubers in order to supply the constantly expanding food needs of the Indian populace
11. To increase the production of meat, milk, vegetables, and legumes,as a result, it can improve food's nutritional worth, which humanity in our nation really needs;the average Indian consumes 2000 calories a day against the need of 3000
12. To maintain the ecological balance in rural areas
13. To control soil erosion and maintain humidity in cultivated areas so that the land's productivity is not reduces

14. To lessen agricultural insecurity in drought-prone areas; in these locations, the dual system of tree and grass production guarantees stability with land productivity
15. To supply forage for a huge livestock population in the country so that they are fed properly to produce a large quantity of milk and meat

RELEVANCE OF AGROFORESTRY

Planting trees on agricultural land may contribute to increased food output. Not only that, but it also contributes to greater food security and a wider variety of food options that improve global nutrition and health.

Improved nutrition and food security

In agricultural areas, planting trees can contribute to higher food production and greater food security. Furthermore, because agroforestry produces food with greater diversity and quality, it improves nutrition and overall health. As a component of agroforestry, planting trees can provide food, fuel, and non-wood goods that can be sold or consumed, providing more food security.

In general, homes can benefit from the nutrient-rich fruits, nuts, and leaves that trees can produce. While leaves can be utilised as cattle fodder, felled trees and their leftovers can be used as wood energy for cooking and heating. Agroforestry promotes the production of a wide range of goods in addition to food, including fibre, wood, fodder, and forage, as well as artisan goods, pharmaceuticals, hedge materials, gums, and resins.

More and more benefits from trees

As a component of an agroforestry approach, planting trees can be good for crops, animals, and human health and survival. Trees can obstruct high winds, preventing harm to crops. It is also well known that certain crops, such as wheat, barley and lucerne, do well in shelter. Increased yields may result from combining suitable tree species with adaptable crops.

Trees also provide shade on hot days, reducing animal stress, and shield animals from wind chill on chilly days.

In agroforestry, trees contribute to the effective fixation of nitrogen into the soil and the recycling of nutrients, both of which increase soil fertility and suitability for the growth of profitable crops.

Furthermore, establishing trees can aid in improving feeding efficiency because temperature variations might lead to erratic feeding patterns. In addition, trees can provide natural treatments and medicines.

Creation of resilient livelihoods

Planting trees alongside crops and livestock can lessen agricultural production's vulnerabilities and potentially accelerate the recovery from risks, natural catastrophes, and socioeconomic downturns. For instance, agroforestry economics lowers the chance of economic failure by increasing the diversity of production within the system.

Tree roots help stabilise the soil, reducing soil erosion, boosting soil fertility, and averting potential landslides. Trees can also lessen the effects of desertification on the environment, agriculture, and society. Therefore, higher levels of sustainable productivity translate into higher levels of farm income. It's also important to remember that agroforestry can result in longer-term jobs and greater revenue, which raises the standard of living in rural areas.

Climate change mitigation and adaptation

Reducing the impact of climate change on agriculture can be achieved through growing trees in agricultural systems. Purchasing tree products made from farm-grown trees lessens the need to clear forests, which in turn slows down the rate of deforestation, which is rapidly eliminating the planet's carbon sinks and releasing carbon that has been stored in the atmosphere. Therefore, through carbon sequestration, trees in agroforestry systems contribute to the mitigation of climate change.

Furthermore, by adjusting microclimatic factors like temperature, the amount of water vapour in the air, and wind speed, agroforestry significantly contributes to the improvement of climatic conditions and weather patterns, which benefits crop growth and animal welfare. This can generally help to regulate air quality, water concentration, rain cycles and patterns, wind erosion, and other factors that can mitigate the consequences of climate change and global warming on agriculture.

Environmental benefits

Environmental natural resource protection can be achieved through thoughtful agroforestry design and integration. For instance, by capturing and filtering water resources, planting trees can assist increase both the quantity and quality of available water.

By giving insects, animals, and plants an appropriate habitat, trees also promote biodiversity. Soil fertility can be restored by agroforestry if the system includes plants that are capable of fixing nitrogen.

Additional advantages of agroforestry systems for the environment include lessening the strain on natural forests, increasing soil nutrients, improving the microclimate, decreasing surface runoff, and more effectively recycling nutrients via deep-rooted trees on the property. Additionally, the systems contribute to enhanced soil structure and ecological system protection.

Support local communities and cultures

Agroforestry has a comprehensive effect, ensuring long-term sustainability in the area while fostering local communities' ability to prosper and preserve their customs and values. Agroforestry fosters the growth of rural livelihoods and cultural variety while lowering the demand for migration to other locations.

Agroforestry is intended to support the growth of regional cultures and communities. Indigenous people and local communities can preserve their customs and beliefs while guaranteeing the long-term viability of their traditional systems with the aid of agroforestry experts.

Agroforestry also contributes to the preservation of the agricultural legacy of humanity by conserving native species and methods of labour.

Agroforestry can support local spiritual beliefs, cultural variety, and respectable rural livelihoods. By removing the need to relocate agricultural locations, agroforestry also contributes to the stabilisation and enhancement of nearby communities.

Employment opportunities

Agroforestry provides occupations and creates employment opportunities throughout the year. Planting trees, gathering products from trees, growing crops, and managing animal components all require professional labour that is available at all times and has the expertise necessary to complete the necessary tasks. Farmers stand to gain much from this, and some may decide to relocate to an area where agroforestry is common in order to reap the rewards. Those who were unemployed before now have the opportunity to work, support their families, and lead more sustainable lives.

Poverty reduction

In developing countries, trees and tree products have economic value that can provide agroforesters with a means of subsistence and possibly lower their rates of poverty. For some, adding value to recently generated tree products might mean earning a living and creating jobs.

Additionally, farmers can use the financial incentives provided to support agroforestry as a means of making a living. Planting trees can lower production costs, so that the household income is increased.

Agroforestry systems not only increase the production of food, fodder, fuelwood, and timber, but they also lessen the likelihood of total crop failure, which is a regular occurrence in monoculture and single cropping systems.

POTENTIAL OF AGROFORESTRY IN INDIA

There are unlimited benefits of agroforestry to the food, fruit, livestock, fuel, wood, fodder, and bioenergy industries. However, proper research interventions, sufficient funding, appropriate extension methods, harvest process technology, new product development, value addition, and market infrastructure are necessary to fully realise the potential benefits of agroforestry. At present, total area under agroforestry in the country is about 28.42 million hectares that might be expanded further, which will contribute to increased and stable productivity and production, satisfy the fundamental requirement for food derived from trees, reduce ecological degradation, and guarantee sustainable management of land, water, and biodiversity. The conversion of underutilised areas, especially wastelands, can extend multiple benefits of agroforestry across vast areas in the country.

Fuelwood

In 1990, there were around 169 million tonnes of fuelwood produced from various sources; by 2001, it increased to 201 million tonnes. More than half of it was from the forests. The remaining portion was available from farmlands and communal lands. A large number of people in both rural and urban regions are involved in collection and market of fuelwood. In places without trees, the lack of fuelwood has been clearly felt. As a result, agroforestry has helped these communities meet their firewood needs. Additionally, as there is less and less opportunity to get firewood from the forests, the only option available to the public is to support agroforestry on agricultural and communal lands. Wastelands may provide more of a supply if silvipasture or pure wood lots were encouraged.

Wood

The following table provides a scenario of wood for both industrial and residential use. At the moment, imports bridge the gap between supply and demand. By encouraging around 3,38,068 ha of poplars (@ 160 t/ha productivity at 7 years rotation) or a mix of other species as *Eucalyptus*, subabul, *Acacia mangium*, etc. and prohibiting imports, the gap of 5 million cum (2.27 million tonnes) as of today might be satisfied. This additional contribution, at the low cost of Rs. 1000/t, will come to Rs. 7230 million. Other important contributions include the creation of jobs and the value addition in processing. More than half of the need for small timber is satisfied by agroforestry, which has been its main contribution.

Table 1.1: Demand and supply of timber for furniture, agriculture and industry (million cubic metres)

Particulars	2001	2006	2010*	2020*	2025*
Demand	73	82	89	107	116
Supply Scorn forests	12	12	12	12	12
Supply from plantation and Agroforestry	47	53	61	90	100
Deficit	14	17	16	5	4

1 ton timber = approximately 2.2 cubic metre (cum)

Source: Ganapathy, 1997; Saxena, 1990 *estimates of the authors

Fodder

In dry, semi-arid, and hilly areas, trees and shrubs sometimes provide a significant quantity of leaf fodder during lean times by tree lopping and trimming, a practice known as top feeding. By silvipasturing farm/community wastelands and cultivating fodder on farmlands, the scarcity of fodder might be lessened. Twelve million t of extra fodder would be produced annually at a rate of 2.0 t/ha, assuming that an additional 10% of community wastelands are available for silvipasture.

Table 1.2: Supply and demand of fodder (million tonnes)

Particulars	2005	2010	2020	2025
Green fodder				
Supply	385	3 95	406	426
Demand	98 8	1061	I 134	1 170
Deficit	603	666	728	744
Dry fodder				
Supply	428	451	473	484
Demand	549	590	630	650
Deficit	121	139	1 5 7	166

Source: https://www.downtoearth.org.in/coverage/why-is-fodder-so-expensive-802

Food

The agroforestry systems' diversity in terms of food grains, vegetables, and fruits allows for the guarantee of nutritional security for the populations. Production of milk, meat, and animal products will be ensured by introducing fodder growing under agroforestry land use. Numerous food crops, pulses, and oilseeds can satisfy the various demands of the community. The food crops cultivated under agroforestry systems alone are expected to provide around 25 million t/annum to the anticipated 320 million t of food that will be needed by 2025.

Table 1.3: Contributions of agroforestry produce to national demand by 2025

Product category	Total demand	Contribution from agroforestry			Contribution from agroforestry (%)
		Traditional	Improved	Total	
Timber (million cum)	116	52	48	100	86
Fuelwood (million t)	330	60	13	73	22
Fodder (million t dry)	1040	100	12	112	11
Fruits (million t)	43	1.5	4.4	5.9	14
TBOs (million of bio-diesel)	10	0.5	5.4	5.9	59
Food (million t)	308	7	15	12	7
Bio-energy (MW)	16000	4000	1000	5000	3
Forest cover (million ha)	100	10	15	25	25

Source: NRCAF, 2007

Fruits

Horticulture has grown by 7% throughout the previous several years, whereas the growth rate of agriculture has been staying at less than 4%. In 2002, the nation produced 48.57 million t of fruits, with the mere output of bananas and mangos accounting for about 54% of the total (Kalloo, 2003). Fruits are given more importance in an agroforestry system since they have a shorter juvenile stage than wood trees and can guarantee consistent output for up to 40 years. Experiences have shown that farmers choose agrihorticulture over agrisilviculture, perhaps because of the guaranteed returns and perceived security (Kareemulla *et al.*, 2002). Aonla (*Emblica officinalis*) has been cultivated under an agrihorticultural system in recent years, and improved cultivars of the species are currently planted across 50,000 hectares of land. Value addition and job creation are two benefits that are achieved from this practice. Furthermore, there is a lot of room to grow fruit trees on wastelands as hortipasture components in order to increase biomass yield and make money right away without compromising the production of feed. Horticultural land usage may be established on 14 million hectares of rainfed land.

Livestock

Promoting silvipasture helps in two ways. It provides more fodder, which guarantees that livestock are raised correctly, and it protects the environment. Under agroforestry, small-scale ruminant farming with a focus on sheep and goats is most suitable. Approximately Rs. 439 crores would be contributed annually if livestock production based on silvipasture were to account for 10% of the entire potential for silvipasture (Kareemulla *et al.*, 2002).

Bioenergy

The primary biomass energy sources utilised by rural families include agricultural biomass, cow dung, and wood from homesteads, croplands, and forests. One important benefit of agroforestry is the production of bioenergy from tree biomass. Certain species, such as *Prosopis julifiora*, are utilised in Andhra Pradesh and Karnataka to generate electricity. *Prosopis julifiora*is selected in these power plants because of its high calorific value, which exceeds 5000 Kcal. This species covers an estimated 0.2 million ha of land. The potential for bioenergy is 1000 MW, even if only 25% of the land is used for power generation and the remaining 75% is used for fuel and charcoal. Similar to this, many tree species may be employed to power biomass-based power plants. Therefore, under mild conditions, 5000 MW of electricity may be generated from biomass sources derived from trees, which is over one-third of the 16000 MW of total biomass potential. The construction of power plants with this kind of capacity costs Rs. 90,830 crores. Research and development can boost output and support power plants in managing captive plantations on degraded land. Encouraging bioenergy production using *Prosopis julifiora* also contributes to the creation of job opportunities.

Agroforestry for problematic soils and environment protection

Degradation of land is an international issue. India has a severe problem with about 100 million hectares of waterlogged regions, mining spoils, industrially contaminated areas, and wind/water erosion. The issues in these regions should receive more attention now that the National Rainfed Area Authority has been established. On such soils, agroforestry may be extremely important for both reclaiming and maintaining the resource status.

Employment opportunities

The importance of agroforestry products and environmental services in meeting low-income households' subsistence requirements and laying the groundwork for a larger and more stable standard of living for society has been emphasised by Dhyani *et al.* (2003). A notable surge in small-scale companies using wood and wood-derived goods has been observed recently due to an increase in the availability of wood on the market. These sectors have greatly increased the area under farm forestry and supported agroforestry. Several business corporations, limited companies, like ITC, WIMCO, West Coast Paper Mills Ltd., Hindustan Paper Mills Ltd., and financial institutions like IFFCO, have entered the market and started agroforestry activities in cooperation with farmers on a large scale after realising the potential of agroforestry. The livelihood impact of agroforestry, in terms of income and job creation, is one of its main economic benefits. In addition to the current agroforestry practices, the 25.4 million hectares of agroforestry land provide enormous potential for creating 943 million mandays of work yearly through better agroforestry systems. The most potential may be seen in 'tree borne oil seeds', with silvipasture coming in second. According to Dhyani and Sharda (2005), the Indian Himalayas alone contain 5.763 million mandays per year that might be generated through agroforestry for rural development and employment.

Table 1.4: Employment generation potential through agroforestry

AF Systems	Area (million ha)	Additional employment generation/ha/year (mandays)	Total employment/ year (million mandays)
Silviculture	1.8	30	53.3
Agrisilviculture (irrigated)	2.3	40	91.3
Agrisilviculture (rainfed)	1.3	30	38.0
Agrihorticulture (irrigated)	1.5	50	76.1
Agrihorticulture (rainfed)	0.5	40	20.3
Silvipasture	5.6	30	1 67.4
TBOs	1 2.4	40	497.1
Total	25.4		943.4

Source: NRCAF, 2007

Agroforestry for carbon trading

Because agroforestry is applicable in both agricultural and reforestation settings, and because it has the capacity to store carbon in its diverse plant species and soil, it is significant as a carbon sequestration approach. Since natural forests are the main sources of terrestrial carbon sequestration, agroforestry may also indirectly aid in carbon sequestration by relieving strain on these forests. Using agroforestry techniques for soil conservation might improve carbon storage in soils and trees, providing another indirect pathway for sequestering carbon. According to existing and forecast trends, India's emissions of greenhouse gases (GHGs) increased at a rate of 4% year between 1990 and 2000. These emissions are expected to continue growing to fulfil the country's target of 94.7 million hectares. Of total, 69.70 million hectares are covered in forests, while 25 million hectares are used for agroforestry. Therefore, over the course of the next 25 years, these places have a total capacity to sequester 118.37 million tonnes of carbon/ha/year. The rate of CO_2 equivalent emissions in the next 25 years, both current and projected, is 1227.68 million tonnes carbon/year. The biomass of trees produced by forestry and agroforestry would absorb 9.3% of the carbon dioxide in the atmosphere. The actions suggested for possible land use and terrestrial carbon management plans. Agroforestry systems and practices on pasture and agricultural land, enhanced tree management techniques to boost growth rates, and afforestation, reforestation, and restoration of degraded and wastelands are all effective methods for sequestering carbon. Regarding carbon substitution (replacing fossil fuels), there are two main approaches; first, increasing the use of biofuels and second, increasing the conversion of wood biomass into durable wood products for use in the production of energy-intensive materials. The former involves conserving biomass and soil carbon in existing sinks, while the latter involves improving wood processing and lopping and harvesting techniques, fire safety, and more efficient use of burning in both forest and agricultural systems. The agroforestry practices such as multistrata tree gardens, home gardens, agrisilvihorticulture and hortisilvipasture systems for food, fruits, and fodder especially in hill and mountain regions and coastal areas; boundary and contour planting for wind and soil protection; agrisilvipasture systems for fodder production as well as soil and water conservation; and biofuel plantations are suitable for sequestering atmospheric carbon.

Export

In the best-case scenario, the excess timber output may be exported. Exporting horticultural output from an agrihorticulture system is another possibility. Considering the growing global interest in herbal remedies, flowers, and organically grown goods, including woody perennials and herbaceous vegetation would boost the agricultural sector's export potential in addition to enhancing crop diversification. It is important to note that the growth of agroforestry is mostly focused on marginal areas, which are essentially chemical-free. It is projected that crop diversification and agroforestry would increase foreign exchange earnings by several folds. New opportunities will arise at the same time from the widespread use of agroforestry. By itself, it will lessen the annual burden of imports of forest products, which exceeds rupees 9000 crores.

IMPLEMENTATION CONSTRAINTS

Agroforestry has a lot to offer when conditions are right, but it also has drawbacks due to biophysical, socioeconomic, and political factors. A few of these restrictions are listed below in brief.

- Compared to monocropping, agroforestry is less understood and more challenging to implement due to its greater structural and functional complexity.

- Trees competing with crops for sunlight can lower crop productivity because they shade crops and alter the spectral quality of light, which can stunt crop development.
- Tree species that develop quickly may have high requirements for soil moisture, which might have a negative impact on crops, particularly in dry areas.
- Although litter fall, root biomass, and other factors may eventually somewhat offset this, the temporary loss of nutrients by trees may cause nearby crops to become nutrient-starved.
- Tree harvesting activities may cause damage to the food crops.
- Certain tree species can have detrimental chemical and biological impacts that reduce agricultural development and productivity and occasionally cause crop failure. These consequences include acidification, allelopathic, the build-up of toxic exudates, the sheltering of alternate hosts for pests and diseases, and more.
- Sometimes trees serve as hosts to insect pest and diseases that are harmful to agricultural crops. Increased susceptibility to pests and diseases often leads to dependence on potentially harmful pesticides.
- As with any plant, a poorly maintained tree stand can hasten erosion.
- Using farm machinery requires more precise and highly skilled management techniques, making it more challenging to operate.
- Because agroforestry systems need a lot of manpower, there may occasionally be a labour shortage for other farm operations.
- Farmers are usually unwilling to displace food crops with trees, especially where land is scarce. Rapid regeneration by prolific trees which may displace food crops and take over the entire field.
- Although tree farming requires high initial investment and return is usually delayed. There is no policy for financial support to the tree growers or agroforesters through nationalized banks.
- Because trees take a long time to provide benefits, farmers are hesitant to plant trees on their property.
- Insufficient availability of high-quality seed or planting materials for potential agroforestry species.
- Tight legal regulations that limit the sale, transportation, and harvesting of trees.
- Inadequate infrastructure and a lack of skilled labour for agroforestry.

- Generally speaking, no government organisation sets the minimum support price for tree products. As a result, the producers of the tree components of agroforestry systems do not receive a fair price for their output.

AGROFORESTRY AND THE FUTURE

Agroforestry is bound to play a major role in future, not only for its importance in food and livelihood security, but also for its role in combating the environmental challenges. It is estimated that by 2050, the demand for fodder, food grain, timber and fuelwood in the country would increase by 1.5, 2, 3 and 2 times, respectively (CAFRI, 2015). Agroforestry has the potential to meet this demand for food, fodder, firewood and timber. With the modern-day crisis of shortage of land for forestry and agriculture, agroforestry is well positioned to provide a perfect balance and a viable solution.

Table 1.5: Total domestic demands for various commodities and agroforestry contribution in 2050

Items	Projected for 2050	Contribution from agroforestry in 2050	Contribution (%) from agroforestry in 2050
Food grains (millions t)	457.1	41.14*	9.00
Fruits (millions t)	305.3	47.74*	15.64
Fodder (millions t)	1545	154.50	10.00
Fuelwood (millions t)	629	308.00	48.97
Timber (millions t)	347	295.00	85.01
Biodiesel (millions t) required for 20 % blending of diesel	37.92	30.34	80.01
Area (millions ha) required for tree-borne oilseeds	21.67	17.34	80.02
Agroforestry (millions ha)	53.32		

*Food-grains/fruits production from systematic agroforestry systems, viz., agri-silviculture/agri-horticulture only considered

Source: CAFRI, 2015

Agroforestry today has become a sustainable method to manage forest and agriculture together, while being economically and environmentally viable. This has the potential to reduce regional disparity, bring desirable peace, prosperity and happiness and ensure an optimistic future for the generations to come. Thus, the need of the hour is to invest in further research and development in this relatively new science. Agroforestry research should focus on development and demonstration for the adoption of various agroforestry models linked with market for enhancing productivity and profitability for small holding farmers. The state governments should evolve state agroforestry policy in consonance with the National Agroforestry policy 2014, to plan, prioritize and develop agroforestry action plan suitable for each agroclimatic zone and focus on strengthening research and activities through public-private partnership mode for developing profitable, ecologically and socioeconomically viable agroforestry models for all farmers. The promotion of sustainable agroforestry practices on a large scale in future only possible through amalgamation of proactive farmer policies of government, involvement of the industries, support services from non-government organisations and willingness of farmers (Verma *et al.*, 2017).

2

Land Capability Classification

A scientific evaluation of the physical characteristics of the land, the properties of the soil, and the management techniques is called land capacity categorization. Understanding potentiality, capacity, and appropriateness for the best possible use of land is the primary goal of land capability categorization. Measuring land capacity provides an objective assessment for land preservation in certain ecological scenarios.

On the one hand, land capacity aids in determining if a piece of land is efficient for a certain use, and on the other, it aids in preventing inappropriate use of the land, which poses a risk of erosion and degrades the quality of the land. Therefore, in order to maximise production, each piece of land's potential should be assessed taking into account both its pedo-geomorphic characteristics and any constraints resulting from environmental threats. Additionally, under the necessary management practices, capacity categorization allows farmers to use the land appropriately for sustainable output.

The land capacity classification method classifies soils mainly according to how well they can support typical farmed crops and pasture plants without degrading over an extended length of time. Capability class and capability subclass are the two divisions of land capability categorization. The following three elements are crucial in forming the basis of this categorization.

1. The soils are farmed in a mechanised manner and are well maintained.
2. Land that has to be improved, such as cleared, and that the farmer can do on his own is categorised based on its restrictions or potential risks associated with using it once the modifications are completed. Land that needs upgrades beyond what the farmer can afford is categorised based on its current state.
3. Other variables are not taken into account, such as the risk of crop damage from natural disasters like storms, the distance to markets, the kind of roads, the location, the size of farms, the ownership type, cultural norms, and the expertise or resources of individual operators.

The ability of the soils for trees, tree fruits, tiny fruits, decorative plants, recreation, or animals is not included in the categorization. Rather of being based on their limitations in terms of agriculture, the classifications are based on intensity. There are several types of soil in each class, and many of these soils need for distinct care and handling.

USDA LAND CLASSIFICATION METHOD

The Soil Conservation Service of the U.S. Department of Agriculture classified lands into eight capability classes mostly on the basis of topographic situation.

The following three sections make up the land capability classification.

1. Capability Classes
2. Capability Subclasses
3. Capability Units

Capability Classes

There are eight recognised capability classes. Class-I soils are the most capable of responding to treatment and have the fewest constraints, while Class-VIII soils are the least capable and have the most limitations.

- Class I-IV includes land suited for cultivation
- Class V-VIII land not suited for cultivation

Land suited for cultivation

Class I: Class I soils have no lasting limitations or very low chances of harm. They are excellent. They can be grown safely using standard, effective agricultural techniques. The soils are almost flat, deep, fertile, and easy to work. They are not vulnerable to harm from overflow or runoff. They are vulnerable to puddle erosion and fertility, though. This type of soils has very little or no risk of harm.

Management: In order to preserve the fertility and structure of Class I soils, which are utilised for crops, activities are required. Crop rotations, crop leftovers, cover crops, green manure crops, and fertiliser and lime usage are all part of these methods.

Class II: Class II comprises soils with mild usage restrictions. They have a moderate chance of becoming damaged. These soils are excellent. They are cultivable with readily implemented techniques. There are several ways in which Class II soils are different from Class I soils. Their key differences are that they have mild slopes, require drainage, are somewhat deep, and are prone to moderate erosion and overflows. Special consideration must be given to each of these elements.

Management: These soils could call for unique techniques like water control devices, soil-saving rotations, or customised tillage techniques. Often, they require a mix of techniques.

Class III: The use of Class III soils for crops is severely restricted. They are vulnerable to serious dangers or harm. These soils are passably decent. If they are treated properly and planted in excellent rotations, they can be utilised as crops on a regular basis. This type of soils has naturally poor fertility, relatively steep slopes, and more severe erosion.

Management: Cropping strategies that provide sufficient plant cover are necessary for these soils. In order to prevent soil erosion, the cover is necessary. Moreover, it maintains the structure of the soil. Cultivated row crops should be replaced with hay or other sod crops. It takes a mix of techniques to farm the land securely.

Class IV: If utilised as crop land, these soils present extremely serious, long-term restrictions or risks. The soils are not too bad. They might be grown on occasion if very carefully treated. They should be permanently housed on hay or sod for the most part. Class IV soils have unfavourable properties. They usually face significant erosion and are situated on steep hillsides. Their applicability for agricultural utilisation is limited. The soils have moderate slopes, poor fertility, and are shallow to fairly deep.

Management: For extended periods, these soils should be covered with hay or sod crops. They should only be sown among row crops on rare occasions.

Soils not suited for cultivation

Class V: Permanent vegetation should be maintained on Class V soils. They ought to be put to use for forestry or grazing. However, due to stoniness, dampness, or other issues, cultivation is not possible. The terrain is almost level. If handled properly, it is only slightly erodible by wind or water. They pose only minor risks and have few, if any, long-term restrictions.

Management: Restriction on grazing is necessary to preserve the vegetation cover.

Class VI: Grazing and forestry should be practised on these types of soils. They are not appropriate for cultivation and have moderate permanent limits. They are shallow or steep. Compared to Class IV terrain, Class VI land is either steeper or more vulnerable to wind erosion. Class VI ground is eroded by wind or is too steep.

Management: The plant cover should not be allowed to be destroyed by grazing. When maintained appropriately, Class VI land can provide timber products or fodder.

Class VII: When utilised for forestry or grazing, soils in Class VII are vulnerable to severe, long-term restrictions. They are harsh, shallow, dry, eroded, steep, and marshy. Regarding forestry or grazing, they are middling to bad.

Management: Land should be utilised for woods when rainfall is plentiful. It should be utilised for grazing in other locations. The latter situation calls for stringent management.

Class VIII: Even for woods or grazing, these soils are rather harsh. They are not appropriate for grazing or forestry. They ought to serve as a watershed, for pleasure, or for wildlife.

Land Capability Subclasses

The second category in the land capability classification system is capability subclass. It stands for the physical, chemical, or atmospheric constraints of the soil, which further limit the usage of the land. The following is a description of several subclasses of land capability.

'c' Adverse climate: This subclass indicates a strongly unfavourable environment for crop production, as characterised by a 'median' climate that has high enough growing-season temperatures to allow crops to reach maturity.

'd' Undesirable soil structure and/or low permeability: This subclass denotes soils that are hard to work, absorb water extremely slowly, or have circumstances other than a high water table or cemented bedrock limiting the depth of the rooting zone.

'e' Erosion: Soils in this category are those whose agricultural usage is restricted due to erosion damage. Damage is calculated based on production loss and the challenges of cultivating land that has gullies.

'f' Low fertility: This subclass comprises soils with low fertility that can be either easily corrected by using fertilisers and soil amendments under careful management, or that are hard to rectify practically speaking. Lack of plant nutrients, excessive levels of acidity or alkalinity, insufficient cation exchange capacity, high carbonate levels, or the presence of hazardous chemicals might all be contributing factors to the restrictions.

'I' Inundation by lakes or streams: This category comprises soils that are prone to flooding, which can harm crops or limit their use for agriculture.

'm' Moisture limitations: Because of the intrinsic properties of the soil, crops in this subclass are susceptible to drought. Typically, these soils are not very good at retaining water.

'n' Salinity: This subclass of soils has an abundance of soluble salts that negatively impact crop growth or limit the types of crops that can be cultivated.

'p' Stoniness: These soils are sufficiently stony to make planting, harvesting, and tillage difficult.

'r' Consolidated bedrock: Soils in this category are limited in their ability to be used for agriculture due to the presence of bedrock close to the surface.

Consolidated bedrock that is more than 90 cm below the surface is not regarded as a restriction, with the exception of irrigated fields where a deeper soil layer is preferred.

's' Two interpretations were accorded to subclass s: When creating maps that were typically created prior to 1969, subclass s was utilised instead of subclasses d, f, m, or n. Subclass s may also be used if two or more of subclasses d, f, m, or n apply to the same region. If a region is classed with a single category, the appropriate subclasses - d, f, m, or n - appear on the majority of maps created after 1969. Subclass s, which indicates a mixture of subclasses, will show for regions classed with two or more of the d, f, m, or n subclasses.

't' Topography: Soils in this subclass have topography as a constraint. The percentage of slope and the pattern or frequency of slopes pointing in different directions has an impact on crop maturity and growth uniformity, as well as creating an erosion danger.

'w' Excess water: Soils in this subclass have an agricultural use restriction due to excess water, other than flooding. Inadequate soil drainage, a high water table, seepage, or runoff from nearby places can all lead to excess water.

'x': Soils in this category have a restriction as a result of two or more unfavourable qualities acting cumulatively.

Capability Units

These are Capability Subclasses broken down even further. Soils that require fairly uniform conservation treatment and management approaches and are sufficiently uniform in their traits, potential, and constraints are included in a capability unit.

LAND EVALUATION

In general, land evaluation is the process of determining how well a piece of land performs for a certain purpose. Land can be assessed indirectly or directly, for example, by gathering and analysing crop-yield data. In the latter scenario, it is expected that land performance may be inferred from the observation of those attributes and that a variety of diagnostic criteria impact land performance in a sufficiently predictable manner. To identify and compare prospective land use types that are relevant to the evaluation's goals, land assessment in this context entails carrying out and interpreting surveys and studies of the land's landforms, soils, vegetation, climate, and other features. A parametric system or a categoric system can then be merged using the values assigned to those attributes.

Based on a more or less systematic examination of the physical characteristics of the land and their effects on both current and alternative land use systems, land evaluation evaluates the performance of the land. A tool or method for comparing the many uses and advantages that may be derived from land while taking the present and anticipated social and economic settings into account is land evaluation. In its most basic form, land assessment determines whether land is appropriate for a certain use and provides answers to the following queries.

- What are the characteristics of the land, and what kind of use provides the most advantages and/or sustainable production?
- What is the current land use, and what other uses are feasible in terms of both social and economic aspects?
- What would happen if current procedures continue as they are? And how may the current methods be made better?
- Which inputs are required to minimise risks and bad impacts while achieving the target sustainable production?
- What are the potential negative impacts on health, the economy, or society, as well as the advantages of a certain land use system or practice?

The review process offers information on which land use choices and options can be made, but it does not decide the land use or modifications to the land use. Stated differently, land evaluation facilitates the process of rational decision-making among land use possibilities for individual property owners, regional development organisations, or entire nations.

Need for Land Evaluation

There is a limited quantity of arable land on Earth, and degraded land is nearly lost to production forever. The populations that utilise this area may coexist peacefully with their natural surroundings as long as they don't become too numerous.

Nonetheless, the world's population has grown over the last 500 years, rising from an estimated 427 million in 1500 to 1650 million in 1900 and over 6000 million in 2000. According to recent estimates, there will be about 10 million people living there by 2050. As a result, the average amount of land accessible per person has decreased from 0.50 ha in 1950 to 0.25 ha currently. It is 0.15 hectares or less in a few East Asian nations.

Land is in high demand due to population growth and the pressing need for food and space. Even less appropriate ground, known as marginal zones, is rapidly being farmed since the finest suited land is already in limited supply. Fuelwood, income crops, building lumber, and animal grazing compete with food crops in many developing nations for available space, both on marginal and higher-quality land. When combined with improper land use techniques, this causes erosion, a build-up of salt, a loss of organic matter and plant nutrients, and harm to the structure of the soil. The need for more food in the industrialised world has intensified agriculture and increased the usage of agrochemicals. Over time, this circumstance has led to soil and groundwater contamination, which has raised concerns about environmental preservation.

Complete space reorganisation and more sensible use of the available land are required due to the need for more arable land, the risk of land degradation in developing nations, and the growing concern for environmental protection in industrialised nations, including the setting aside of agricultural land. This is only possible if the land's characteristics are understood, its natural resources are properly inventoried, and its relative appropriateness for various uses is appropriately evaluated. In this case, land appraisal offers a crucial method for presenting many situations and choosing suitable solutions for decision-making. Such land-use decisions must be decided at the plot or farm level as well as at the national, provincial, or district levels. The degree of detail and map sizes varies, but the fundamental procedure remains the same.

Currently, land is becoming a rare and non-renewable natural resource that is greatly sought after, for which there is increasing rivalry, and which, therefore, carries an appropriate exchange value. This makes land appraisal, and eventually land use planning, even more important. It is also connected to a scenario where judgments about altering the use of land may significantly impact societal views and society at large. For instance, agriculture is sometimes seen as a concern for the environment, but with proper management, it can also play a vital role in it. Decisions to alter land use may be obvious if high-quality indicators are available to track the effects of agriculture or any other land use activity.

3

Classification of Agroforestry Systems

It is vital to categorise agroforestry and its many systems in accordance with certain contemporary standards in order to have a thorough grasp of them and to enable future progress. Since each agroforestry system refers to a different set of agroforestry practices involving agriculture (crops), forestry (trees), and pasture (animals), the various uses of land are combined either temporally or spatially. Each agroforestry system has its own unique arrangement of various components, such as crops, animals, perennial trees, etc., and degree of interaction between the components. In terms of structure, composition, age intensity, technology, inputs, etc., one system varies from the other.

Creating a useful framework for the synthesis and analysis of data on current systems as well as the creation of new, promising ones should be the primary goal of classification. For this reason, every categorization system needs to have the accompanying the following standards.

- Provide a rational structure for organising the key elements that will determine how the system's output is produced.
- Describe the system's management approach.
- Provide flexibility in organising the data.
- Be simple to comprehend and manage.

BASIS OF THE CLASSIFICATION OF AGROFORESTRY SYSTEMS

Nair (1987) classified various agroforestry systems on four bases.

1. Structural basis (composition and arrangement of components)
2. Functional basis (component role and output)
3. Ecological basis (the ecological zones in which the system is found)
4. Socioeconomic basis (the socio-economic size and degree of management)

Subsequently, Dwivedi (1992) divides this base into seven categories, such as structural, functional, socioeconomic, ecological, physiognomic, floristic and historical.

Structural basis: The system's compositions as well as its temporal and spatial organisation are considered in this criterion. Incorporating woody species into a system has the potential to significantly modify both the vertical and horizontal plant arrangement, boost system diversity, and extend system life.

Functional basis: This describes how the woody components such as fruit, wood, fodder, etc. work. The addition of woody species typically results in an increase in the amount of products produced by the system, as well as other benefits like the ability to employ trees as windbreaks or to reduce soil erosion.

Socioeconomic basis: This describes the system's objective, which is often classified as intermediate, commercial, or subsistence. Moreover, agroforestry may be encouraged to achieve certain social objectives like reducing poverty (social forestry) or enhancing community access to resources on shared lands (community forestry).

Ecological basis: This describes how well an agroforestry system fits into a particular ecosystem. As a result, many forms of agroforestry exist for dry, tropical, and temperate climates that are tailored to the unique biological, ecological, and climatic circumstances of each region.

Physiognomic basis: This describes the xeromorphic, mesomorphic, and hydromorphic characteristics of vegetation.

Floristic basis: The species composition that is often used in various places can be taken into account when classifying plants.

Historical basis: This describes how agroforestry systems changed as a result of knowledge spreading throughout time.

Biological basis: Agroforestry techniques can improve biological production in a variety of ways. Planting nitrogen-fixing woody species in between crop rows can improve soil fertility. Numerous woody species' leaf may also be collected and used as green manure for agricultural plants, which not only supplies nutrients but also helps stop soil erosion and water loss and may even serve as a pest deterrent. Additionally, woody trees can be planted as windbreaks, to stop soil erosion, or to shade agricultural plants. All these methods, however, may be summed up by stating that the introduction of woody species into an agroecosystem can alter the ecosystem's flow and retention of nutrients as well as its physical structure.

Economic basis: Adding woody species to agroecosystems not only increases their biological production but also frequently boosts their economic benefits. Utilising woody foliage as fodder might save animal feeding expenses or shield the economy from suffering financial losses caused by livestock losses during a drought or famine. Woody plants can yield a variety of goods that can be sold or utilised on the farm, including fruit, fibre, nuts, construction and craft supplies, medicines, lumber, and charcoal. Farmers sometimes employ the longer-growing tree species used in agroforestry as a 'bank account' that can generate financial revenue in years when crops are poor. As a result, agroforestry techniques may help to improve economic stability in addition to diversifying the revenue streams accessible to farmers. Agroforestry has long been seen as an environmentally and socially friendly approach to sustainable development as its techniques increase economic returns.

STRUCTURAL CLASSIFICATION OF AGROFORESTRY SYSTEMS

The structural basis takes into account the time arrangement of various components as well as the spatial mixing of the woody components. The components of an agroforestry system and their intended roles or functions might be used to determine the system's structure, or the composition and dimensions of the crops. Equally crucial is the component's management. Therefore, agroforestry systems may be categorised based on their structural makeup such as nature of components and arrangement of components.

Classification based on Nature of Components

Agroforestry systems may be categorised into the following major systems according to the makeup of its components, and each major system can be further divided into smaller subsystems.

a) Agrisilvicultural systems
b) Silvipastoral systems
c) Agrisilvipastoral systems
d) Other systems

AGRISILVICULTURAL SYSTEMS

In agrisilvicultural systems, agricultural crops are interplanted in the spaces between the trees. In this approach, crops may be cultivated for up to two years under protected irrigation and for up to four years under rainfed farming. Grain crops are not economically viable to cultivate after the previously mentioned time, during which the crops can be grown commercially. It is possible to cultivate fodder crops, shade-tolerant crops, and shallow-rooted crops profitably, nevertheless. In order to provide easier cultural management and provide the intercrop with more sunshine, wider spacing is used without reducing the number of trees. In this arrangement, the tree crops perform better than in a monoculture. This system may be divided into the following forms according to the makeup of its parts.

- Improved fallow species in shifting cultivation
- Taungya system
- Multispecies/ multilayer tree gardens
- Alley cropping (hedgerow intercropping)
- Multipurpose trees and shrubs on farmlands
- Crop combination with plantation crops
- Agroforestry fuelwood production
- Shelterbelts
- Windbreaks
- Soil conservation hedges

Improved Fallow Species in Shifting Cultivation

Fallows are agricultural lands that are not planted for a season or several years. Recovering depleted soil nutrients is the aim of enhanced fallow species in shifting farming. Crops are reintroduced for one or more seasons when the soil has healed. As the name suggests, shifting cultivation refers to a pattern of land usage and a crop-production technique in which sections of land are cleared, and then left to rest for a longer amount of time than is necessary for cultivation. But the land returns to a modified version of its original cover during the rest phase. It is a production system that mostly relies on manual labour and nearly eliminates capital inputs. The north-eastern hill areas, which includes the states of Assam, Meghalaya, Manipur, Nagaland, Tripura, Arunachal Pradesh, and other states like Andhra Pradesh, Odisha, Karnataka, are home to a large population of practitioners of this system. This practice is known as 'podu' in the states of Andhra Pradesh and Odisha, and 'jhum' in the north-eastern states. It is thought to be especially harmful to forest regions. The primary characteristic of the agroforestry enhanced fallow system is the absence of co-growing trees and shrubs on the same plot as crops. While fallow times differ from place to place, they are currently getting shorter because of a severe lack of available land.

Species: Some plants can be introduced purely for their economic benefit, even if the fallow field's primary purpose is to preserve or restore soil fertility and decrease erosion. Plants with marketable products should be compatible with future crops, free of any detrimental physical or chemical impacts on the soil, and not in competition with the crop to be planted later on the same site. Species selection should not be limited to the soil improvers.

Establishment: Better fallows maybe established at different times and in different methods. The following techniques are usually considered for the establishment of the system.

- Planting seedlings in freshly ploughed and harvested areas
- Trimming brush selectively, and then planting tall seedlings for enrichment.
- Planting tall seedlings and cuttings on damaged ground to improve low-quality fallows
- Planting tree seedlings on blocks of cleared farmland in deeply spaced, precisely spaced planting holes or furrows

The precise methods differ depending on the land's past usage, its worth as fallow, its state as fallow vegetation, and how long the fallow is projected to last.

Demerits of the shifting cultivation

1. Shifting cultivation is an old farming technique that yields less yet demands arduous effort.
2. Altering the land's usage is a mistake.
3. Soil erosion and flooding in the surrounding region are caused by the excess runoff from the changing farmed areas.
4. Nutrients from the soil are lost by percolation, leaching, and runoff.
5. Cleaning and burning of plants reduces the number of soil microbes.
6. Soil's ability to exchange cations is diminished.
7. The forest's animals and plants are disturbed by shifting farming.
8. The depletion of hill slopes is caused by shifting farming.

Taungya System

Taungya is a Burmese term that literally translates to 'hill cultivation'. It was originally the name given to shifting farming in Burma. The taungya strategy basically involves planting forest tree species in the early years of the forest plantation's formation in addition to yearly agricultural crops. To put it simply, the taungya system is a method for establishing forest plantations where field or agricultural crops are temporarily cultivated alongside forest crops.

Thus, the taungya system is an additional phase in the transition from shifting agriculture to agroforestry. In the early phases of the construction of forest plantations, shifting cultivation involves growing woody species and agricultural crops in a sequential manner, whereas taungya involves combining the two components at the same time.

Similar to shifting cultivation and other smallholder systems, the primary incentive for using the taungya system is food production, even though wood production is the ultimate goal.

The taungya technique was originally applied by the British in 1850 to replant teak in Burma. It was later applied in Java in 1856. Dietrich Brandis brought the concept to India in 1856, and the first plantations were established in North Bengal in 1863. Kerala, West Bengal, Uttar Pradesh, Tamil Nadu, Andhra Pradesh, Odisha, Karnataka, and the north-eastern hill states all practise it. The term 'Kumri' refers to this method in south India. In many regions of India today, agroforestry practices like these are commonly used. The approach, which has been altered in several tropical regions, is essentially a modification of the classic shifting agriculture method, which allows farmers to cultivate food crops in forested areas, typically areas that are a part of forest reserves, for at least a year. Then, when the farmers gradually move their farming operations to new reforestation sites, the succeeding areas are transformed into plantations. In India, taungya systems have taken three different forms.

Departmental taungya: Under this arrangement, the forest department hires labourers on a daily pay basis to plant agricultural crops in addition to perennial tree crops. In the early years of a plantation, the primary purpose of the agricultural crops is to reduce undesired vegetation growing between the trees.

Leased taungya: The forest department leases the plantation land to interested cultivators so they can grow crops in between the tree rows for a set amount of time and maintain the tree plantation.

Village taungya: Under this system, individuals are permitted to establish themselves in a village within a forest with the aim of cultivating crops and tending to a tree plantation. Every family typically owns between 0.8 and 1.7 hectares of land where they may grow food and raise trees for three to four years.

Components of taungya system

A wide range of agricultural crops maybe cultivated in the taungy systems, suitable to the climate and soil. A variety of trees are cultivated in India. The following are the variations in these combinations based on the agroecological region.

Plains: *Tectona grandis*, *Lagerstroemia speciosa*, *Toona ciliate*, *Dalbergia sissoo*, *Acacia catechu*, *Anthocephalus chinensis*, *Shorea robusta*, *Michelia champaca*, *Gmalina arborea*, *Bomax ceiba*, etc.

Foot hills (up to 600 m): *Tectona grandis*, *Shorea robusta*, *Betula cylindrostachys*, *Bomax ceiba*, *Terminalia mayriocarpa*, *Chukrasia velutina*, *Schima wallichii* etc.

Middle hills (600–1800 m): *Alnus nepalensis*, *Symingtonia populanea*, *Betula alnoides*, *Cuppressus cashmeriana*, *Machilus edulis* etc.

Upper hills (1800–2400 m): *Pinus patula*, *Acer campbellii*, *Quercus* spp., *Juglans regia*, *Cryptomeria japonica*, *Tsuga dumosa* etc.

Benefits of the taungya system

- The process of creating artificial forest regeneration is less expensive.
- Providing job chances.
- Assists in optimising the amount of forest cover (better use of land).
- Food crops produced on forest land.
- The growth of weeds is eradicated.
- Extremely lucrative for the forest agencies in every instance.

Drawbacks of the taungya system

- Threat posed by epidemics
- Decrease in soil fertility and soil exposure
- The susceptibility of land to rapid erosion rises
- Generates certain legal issues

Multispecies/Multilayer Tree Gardens

In order to create the illusion of a controlled mixed forest, this technique combines commercial perennial tree crops, particularly tree species, with traditional forest species planted in plantations. These tree gardens are frequently located on jointly held ground around communities with dense populations of homes, as contrast to homegardens, which are situated around individual homes.

The multilayer, multispecies dense plant groupings lack structured planting configurations. The main component group consists of many woody elements with distinct shapes and development patterns. Although they are rarely seen, herbaceous plants can occasionally be seen. This agroforestry technique maybe applied in regions with significant human population pressure, good labour availability, and fertile soils.

Acacia catechu, *Areca catechu*, *Phoenix dactilifera*, *Artocarpus* spp., *Cocos nucifera*, *Mangifera indica*, *Syzygium aromaticum*, *Acacia mangium*, *Acacia auriculiformis*, *Gmelina arborea*, and other significant woody perennial plants are used in this system.

Alley Cropping (Hedgerow Intercropping)

Alley cropping is a technique where annual crops are cultivated in strips, or 'allees', between rows of trees or shrubs, giving the appearance of hedges. It is sometimes referred to as hedgerow intercropping for this reason. Researchers that specialise in agroforestry introduced the idea of alley cropping. It was believed that it may be simpler to maintain soil fertility by growing woody plants continuously in between the field crops rather than planting field crops after an excessively short fallow time, as land scarcity results in shorter fallow periods and diminishing soil fertility. Woody plants should be arranged in hedges so as to minimise obstructions to field cultivation. Furthermore, the benefits of woody plants to the environment, including decreased wind or water erosion (hedges arranged to stop the wind) or contour planting (hedges planted along the margins), would also help the crops. Alley cropping, as an agroforestry technique, falls in between parklands, improved fallows, and hedgerow barriers.

- A combination of field crops and woody plants is called alley cropping, whereas shifting farmers cultivate crops after the tree or bush fallow.
- In parklands, trees are big and dispersed, but in alley cropping, they are arranged in a row and controlled to maximise their advantages and reduce their competition with crop plants.
- While alley cropping is largely done to preserve soil fertility, hedgerow barriers are erected primarily to minimise erosion.

Fast-growing small trees and shrubs that can be routinely trimmed are utilised in the alley-cropping method. They also provide an abundance of twigs and branches that may be used as mulch and green manure for the crops growing in the alleys. As an alternative, the cuttings can provide a sizable amount of animal feed, for example, when manure is returned to the field during the dry season.

Planting and maintenance

In general, hedgerows are placed 4 to 8 metres apart, with a row spacing of 30 to 100 centimetres. They are positioned parallel to the field's longest side or along the contours to block the wind when it is at its fiercest. Cuttings or seedlings can be used, or direct planting can be used to establish the hedges. They need to be shielded from pests, animals that browse, and footfall. The young trees will benefit from manuring, weeding, and other treatments much like a crop would. They will not need as much care if they get a solid start in life.

If the field can be cultivated all year round in a humid area, the hedges should be trimmed at least once when a new crop or relay crop is planted. To give them time to form strong roots, even the fastest-growing trees, like leucaena should be given six to twelve months to mature before the initial cutting.

In monsoon climates, hedges are heavily clipped when they are planted early in the rainy season. In areas where heavy rainfall is not forecast throughout the growth season, the height of the hedges can range from 30 cm to 60 cm. In order to minimise competition for water and lessen shade of the field crops, hedges that have a lengthy growth season could require one lighter pruning before harvest. The trees are either let to grow or chopped down for feed during the dry season. When the trees are to be clipped again during planting season, they can have grown to an astonishing size if they are allowed to grow unchecked and the dry season is not too harsh. In that case, coppicing is used as hedge, and the wood can be used as stakes or fuel.

Fig. 3.1: Alley cropping
Source: World Agroforestry Centre

Selection of species

Calliandra calothyrsus, *Etrythrina subumbrans*, *Gliricidia sepium*, *Flemingia macrophylla*, *Sesbania* spp. and *Leucaena* species, *Albizia* spp., *Dalbergia sissoo*, *Tectona grandis*, *Gmelina arborea*, *Cassia siamea*, *Tephrosia* spp., *Cajanus cajan*, etc. are ideal plants for establishing hedgerows. The species should have the following qualities to be chosen for alley cropping.

1. Quick growth to guarantee a large amount of trash and prunings produced
2. A light, open crown that allows sunlight to penetrate through, such as feathery leaves
3. A root system that grows downward as opposed to sideways
4. Legume or other nitrogen-fixing species

5. Good response to regular pruning, with old wood growing swiftly and readily
6. Leaf litter that breaks down either slowly to provide a more lasting mulch or fast to release nutrients
7. Site adaptation, including tolerance to wind, flooding, acidic or saline soil, and pest tolerance

Benefits of alley cropping

The primary benefit of alley cropping over conventional shifting cultivation and bush fallow systems is the ability to simultaneously conduct the cropping and fallow phases on the same piece of land. This enables the farmer to harvest for a longer amount of time without having to convert the field back to bush fallow.

1. Giving companion food crops access to mulch or green manure. Plant nutrients are recycled from deeper soil layers in this fashion.
2. Providing shade during the fallow period and prunings to act as mulch to prevent weed growth.
3. Creating ideal circumstances for micro- and macro-organisms in the soil.
4. Provide a barrier to reduce soil erosion when planted along the contours of sloping terrain.
5. Providing firewood, staking supplies, and trimming for browse (to feed cattle).
6. Giving the companion crop nitrogen that has been organically fixed.
7. Increased crop performance as a result of the soil/plant system's increased nutrient and organic matter content.
8. A decrease in the application of synthetic fertilisers.
9. An enhancement of the soil environment's physical characteristics. Mulch can help with greater infiltration, less runoff, and more efficient use of water by lowering soil temperatures, decreasing evaporation, and enhancing soil fauna activity and structure.
10. The rows of trees physically block the flow of soil and water over sloping terrain, thus reducing erosion losses.
11. The provision of extra goods when using a multifunctional tree legume as the hedgerow, such as feed, fuel, or stakes.
12. While mulch may prevent weed germination and establishment during the cropping phase, shading the interspaces during the fallow period may lessen weed development.

Limitation of alley cropping

1. Water competition
2. High labour costs for establishment and trimming
3. Dramatic cropping system adjustment, needing several years to reach a stable stage that can be routinely handled by the farmer
4. Tree crop arrangement must be adjusted to farm circumstances. For example, hedges should have a variety of species, and appropriate pruning schedules (such as coppicing, replacement pruning, and shearing) should be determined based on the time of year and how the prunings will be used.

Multipurpose Trees and Shrubs on Farmlands

This strategy involves the random or planned scattering of different multifunctional tree species on bonds, terraces, field borders, agricultural areas, pastures and rangelands. Naturally, some of the trees sprout from seeds that birds and other creatures scatter. These trees are kept by farmers as they prepare their land for farming, and they are frequently scattered haphazardly over the area. Common agricultural crops, fruit trees, and multifunctional trees make up the majority of this system's constituents. This system's main goal is to produce different tree products; fence, social values, and plot delineation serve as protective measures.

Acacia albida, *Acacia tortilis*, *Vitellaria paradoxa*, *Parkia biglobosa*, *Adansonia digitata*, *Tamarindus indica*, *Azadirachta indica*, *Pongamia pinnata* and other tree species are the most significant and dominating ones in this system.

Advantages

1. Farmers believe that these trees do not pose a threat to food crops.
2. Throughout the lengthy dry season, trees offer shade to cattle amid the extreme heat (*Faidherbia albida*, for example).
3. The trees extend the time that the land is farmed without the need for fertiliser, boost agricultural productivity, and diversify the items that farmers sell.
4. Control of wind erosion occurs when wind-blown dirt in farmlands is intercepted by tree canopies.
5. The revenue generated by non-timber goods like fruits, gum, firewood, flowers, and charcoal is greatly increased.

Restrictions

1. Trees that are dispersed are frequently not at the ideal density that would provide the most advantages to the environment and agricultural output.
2. It is challenging to employ superior germplasm to enhance the system when it is dependent on naturally renewing trees.
3. Benefits from farmlands trees take a long time to materialise since they grow slowly.
4. Livestock that is let to graze agricultural leftovers during dry seasons frequently browses the trees.

Crop Combination with Plantation Crops

Large-scale monoculture is the practice of growing perennial trees and shrub crops such as tea, coffee, coconut, cashew, rubber and cocoa; however, on small holdings, these plants are frequently grown in combination into intercropping systems in a variety of ways.

- Integrated multi-storey of plantation crops
- Mixture of plantation crops in alternate or other regular arrangements
- Plantation crops with dispersed shade trees
- Intercropping with agricultural crops

These perennial crops can be interplanted on small holdings for the following reasons.

1. When tree crops are young, there are a lot of gaps between them
2. Especially in the early phases of tree growth, the intercrops lessen soil erosion between widely separated tree crops
3. Certain combinations of tree intercrops have complementary roots
4. Under some plantation crops, understorey crops can flourish in shade (for instance, *Aloevera*, coffee, pineapple, kalmegh, etc.)
5. Adequate use of light
6. Intercropping early growth perennial tree crops is a financially effective strategy while waiting for long-mature tree crop harvest
7. Different work calendars make some intercrops compatible
8. For certain tree crops, an intercrop is needed to act as a nurse crop in the early stages of development

Agroforestry Fuelwood Production

Due to the ongoing exploitation of natural forests and woods, there is a huge demand for fuelwood and poles in emerging nations. As a result, these vital resources are critically threatened, and the world's developing nations are facing a major firewood shortage on par with the petroleum crisis. Alternative fuel sources including kerosene, gas, and electricity are available today. These, however, are not accessible to everyone and need financial inputs. Approximately 3 billion cubic metres of wood are harvested and utilised globally each year. Slightly more than half of this quantity is burnt for cooking and heating, mostly directly but also in the form of charcoal.

The availability of fuelwood has decreased in many regions of the world throughout the past 20 years or more. To tackle this issue, a number of strategies have been put forth, the most important of which is encouraging the planting of trees for fuelwood production. It is suggested for rural residents to include fuelwood production from trees into their current farming methods. It is possible to interplant these fuelwood species on or near agricultural land. This system's main productive function is the production of firewood; its defensive functions include demarcating boundaries, providing fences, and acting as shelterbelts.

Acacia nilotica, Albizia lebbek, Cassia siamea, Casuarina equisetifolia, Prosopis juliflora, Eucalyptus tereticornis, Pterocarpus erinaceous, Daniella oliveri, Isoberlina doka, Dalbergia sissoo and other tree species are frequently used as fuel.

Shelterbelt

A stretch of trees and shrubs planted to protect residents and their property from high winds is known as a shelterbelt. Typically, shelterbelts are made up of three or more parallel rows of grasses, bushes, or trees. The evergreens or kinds of shrubs that do not lose all of their leaves in the windy season are the ones that are employed. Sometimes people grow other plants, like grass, to block the wind from removing the soil from around the base of their trees and shrubs.

A shelterbelt divides two areas, the leeward and the windward zones, and acts as a mechanical barrier against wind impact. The side that the wind blows from is known as the windward zone, while the side that the wind passes through is known as the leeward zone. As a general rule, a belt may shield an area up to 20 times its height on the leeward side and up to its height on the windward side.

Characteristics of shelterbelts

1. Shelterbelts have a typical triangular shape acquired by raising tall trees in the centre.

2. A certain degree of penetration by wind is planned because by raising a solid wall of trees, the protection decreases very fast on the leeward side.
3. The ratio of height and width of a shelterbelt should be roughly 1:10. Shelterbelts up to 50 m width are considered ideal under Indian conditions.
4. Orientation of shelterbelts depends upon the direction and velocity of the prevailing winds. Rows of trees are established at right angles to the prevailing wind.
5. The minimum length of a shelterbelt should be about 25 m.
6. Tree species selected for shelterbelt establishment should have ability for wider adaptability to climatic conditions (drought, frost, extreme temperatures, etc.), high growth rate, quick crown formation, developing a strong and deep root system (so that they do not compete for moisture and nutrients with agricultural crops), evergreen foliage, increased land productivity.

Features of the shelterbelt

Height's effect: The protected downwind zone's size is determined by the height of the shelterbelt. The shelterbelt's tallest tree or shrub row controls this. This changes as the shelterbelt ages and is more varied from one to the next. Wind speed decreases on the windward side of a shelterbelt are detectable upwind for two to five times the shelterbelt's height. Wind speeds are reduced on the leeward side up to a distance equal to 20 times the shelterbelt's height.

Density's effect: There will not be as much wind moving through a thick shelterbelt. Dense shelterbelts cause a decrease in air pressure on their leeward side. Turbulence is produced as air passing over the shelterbelt is drawn lower by this low pressure region. The protective area is comparatively reduced when dense shelterbelts are present. Less thick shelterbelts allow for more air to pass through, which decreases turbulence and lengthens the downwind protected area. The reduction in wind speed is not as considerable, despite the bigger sheltered area. Different flow patterns and zones of protection are generated by varying the density of the shelterbelt.

The variables governing shelterbelt density include species composition, number of rows, and spacing between trees. Density is increased and a stronger wind barrier is produced by reducing the spacing between trees or increasing the number of shelterbelt rows. The species selected for the shelterbelt will affect the sheltered area's length and height in addition to density.

Shelterbelts can be found in fields, on borders, beside roadsides, surrounding animal facilities, and around farmyards. A key factor is the density of the shelterbelt. Low density shelterbelts allow wind to flow through open sections, lengthening the sheltered area and reducing downward turbulence on the leeward side. The reduction in wind speed is not as considerable, despite the bigger sheltered area.

Low density shelterbelt designs are more suited for field shelterbelts when uniform snow distribution and soil erosion are necessary. On the other hand, high density causes a low pressure region to form on the leeward side and less wind to flow through the shelterbelt. This causes turbulence, which decreases protection downwind and boosts wind reductions near the shelterbelt. It also draws air downward.

To put it briefly, the downwind sheltered area and the quantity of wind that passes through the shelterbelt both rise as density falls.

Effect of orientation: When placed at a right angle to the direction of the predominant winds, shelterbelts work best. Shelterbelt orientation is determined by the design goals. Feedlots and farmyards typically require shelter from chilly winds and blowing dust or snow. Field crops typically require shelter from wind-blown soil particles, hot, dry summer breezes, or both.

The best protection is obtained when shelterbelts are oriented perpendicular to the direction of the predominant wind. The sheltered area shrinks when the wind shifts and stops blowing straight against the shelterbelt. It is uncommon for the wind to blow just in one direction, even though it occasionally does.

Length effect: The amount of the covered area downwind is determined by the height of the shelterbelt. The length of the shelterbelt establishes the region that is protected. A shelterbelt's unbroken length should be at least ten times longer than its height in order to achieve optimal effectiveness. The entire shielded region is less affected by end-turbulence when this ratio is used.

A shelterbelt's continuity is crucial because holes can turn into funnels that concentrate wind flow and increase wind speed. Lanes or field access through shelterbelts should be placed to prevent gaps from decreasing the efficiency of the shelterbelt.

Selecting trees and shrubs species for shelterbelt

The site's features and the major purposes are to be considered while choosing the trees and plants for a shelterbelt. Instead of planting two or more rows of the same tree or shrub, it is preferable to establish a multi-row shelterbelt utilising a range of species. More benefits come from having a diversity of trees and plants, one of which being a decreased chance of the shelterbelt being completely destroyed by calamities like drought, pests, or disease. Every species is unique in terms of height, breadth, density, lifespan, growth, capacity for invasiveness, and resistance to illness and insects.

Grasses: *Saccharum spontaneum, S. munja, Panicum antidotale, Cencnrus* sp. etc.

Shrubs: *Calotropis procera, Clerodendron phlomoides, Cassia auriculata, Dodonaia viscose* etc.

Tree: *Acacia Arabica, Acacia leucocephala, Dalbergia sissoo, Lannea coromandelica, Eucalyptus* spp., *Tamarix articulate, Prosopis juliflora, Prosopis spicigera, Casuarina equsetifolia* etc.

Benefits from shelterbelts

Microclimate modifications: The shielded zone's microclimate is altered as wind speed behind a shelterbelt decreases. Both the temperature and the humidity often rise. Plant water loss and soil moisture evaporation are reduced as a consequence. The height, density, direction, length, and time of day of the shelterbelt all affect the actual temperature changes. The leeward side daily air temperature is typically a few degrees warmer than the open air temperature when it is within ten times the height of the shelterbelt.

The air temperature near the ground is often colder during the day when it is farther away than ten times the height of the shelterbelt. In protected regions, the temperature near the earth is somewhat warmer at night than it is outside. On calm evenings, though, protected spots can be several degrees colder than the open air.

In general, summer soil temperatures in protected locations are greater than those in uncovered ones. Increased warmth might facilitate early sowing of crops and quicken germination in regions with brief growing seasons. Because the east-west shelterbelt reflects heat, the south side of the land it abuts has a greater soil temperature. Due to shadowing, soil temperatures near an east-west shelterbelt on its northern side may be lower.

Depending on the density of the shelterbelt, relative humidity in covered regions is frequently greater than in open areas. Increased humidity slows down the rate at which plants utilise water, allowing crops to use water more effectively than they would in open spaces. The leeward side of a shelterbelt reduces heat loss from wind chill. In farmyards and livestock shelterbelts, where humans, animals, and buildings gain from enhanced energy efficiency, the wind chill factor is most significantly moderated.

Because of modifications to the shielded zone's microclimate, the majority of the advantages of a shelterbelt are tangential. The immediate advantage of lowering wind speed to prevent soil erosion is one exception. Within ten times the height of the leeward side of the tree row, a shelterbelt can almost completely eliminate soil erosion on the leeward side.

Leeward and windward wind speeds are both decreased by shelterbelts. The microclimate in these sheltered zones moderates as a result of the ensuing decreases in wind speed. Microclimate adjustments can be used to provide ideal conditions for safeguarding living and working places, growing crops, and rearing livestock with proper planning.

Reduction in soil erosion by wind: The modified microclimate reduces wind speed and, therefore, soil erosion is reduced. A significant reduction of wind speed occurs on leeward side for a distance extending to approximately 20 times the height of the shelterbelt and also 3 to 5 times its height on the windward side. Therefore, a shelterbelt 5 m in height will provide a degree of protection for soils and crops for a total distance of up to 25 times its height, i.e., 125 m. Since the zone of protection provided by a single shelterbelt is limited, a series of shelterbelts is required to protect the whole field. The shelterbelts must be planted perpendicular to the direction of the prevailing wind to provide complete protection.

Increased moisture for crop growth: Shelterbelts reduces evaporation, thus providing more moisture for crop growth. It also protects the leeward areas from the desiccating effects of hot wind. Shelterbelts use moisture and nutrients from a greater depth than most annual crops.

Reduced wind damage to crops: Crops benefit from the reduced wind speed in the protected zone. The crop plants are less likely to be twisted by the wind.

Potential for increased farm productions: Most of the research conducted around the world reports crop yield increases due to shelterbelts. However, crops vary in their yield response to shelterbelt protection. Drought tolerant crops show the lowest response, forage crops are moderately responsive and weather-sensitive crops such as vegetables show the highest response. Besides, shelterbelts provide potential source of income for farmers (e.g., biomass, timber and non-timber products).

Windbreaks

Windbreaks are narrow strips of trees, shrubs and/or grasses planted to protect crop fields, homes, canals and areas from wind and blowing sand. Where wind is a major cause of soil erosion and moisture loss, windbreaks can make a significant contribution to sustainable production. From the farmers' point of view, gentle winds are advantageous as they can contribute to pollination of crops and seed dispersal. Strong winds, on the other hand, are damaging and could be detrimental to agricultural crops, human life and properties. Below are some of the effects of strong winds on crops.

1. Wind may increase transpiration rate, and this may lead to soil moisture deficits.
2. It can spread pests and diseases because disease causing spores and insects are dropped whenever wind speed is reduced.
3. It can cause crop lodging and deform plants.
4. It can increase loss of topsoil through wind erosion especially in semiarid areas.
5. It can cause drifts during herbicide and insecticide spray, hence, leading to wasteful application and ecotoxicity.

Design and management

The basic design of a windbreak should include permeability, the appropriate orientation, placement, length and height, number of rows, spacing, density and continuity to provide effective protection.

Permeability: The effectiveness of the windbreak is influenced by its permeability. If it is dense, like a solid wall, the airflow will pass over the top of it and cause turbulence on the leeward side due to the lower pressure on that side. This gives a comparatively limited zone of effective protection on the leeward side compared to the zone that a moderately permeable shelter creates. Windbreaks must be semi-permeable, ideally filtering 50-60 percent of the wind to reduce its strength. The desired permeability can be obtained by carefully selecting tree species. Species such as *Eucalyptus* and *Casuarina* will form effective windbreaks. Permeability of dense windbreaks can be improved by pruning lower branches at 0.5-0.8 m from the soil level.

Orientation: When designing a windbreak, the direction of the wind must be considered. A barrier should be established perpendicular to the direction of the prevailing wind for maximum effect. To protect large areas, a number of separate barriers can be created as parts of an overall system. When the prevailing winds are mainly in one direction, a series of parallel shelterbelts perpendicular to that direction should be established. Before establishing windbreaks or shelterbelts, it is important to make a thorough study of the local winds and to plot on a map the direction and strength of the winds.

Height and length: While selecting species for windbreak more importance is given to the height than the thickness. It is generally accepted that a windbreak protects an area over a distance up to 3-5 times its own height on the windward side and up to 20 times its height on the leeward side. Thus, the species selected should be erect and tall growing, hardy and drought resistant, and should occupy less space as far as possible. Windbreaks are more effective when they stretch without major gaps for distances exceeding 12 times the mature height of the trees.

Tree rows and spacing: In reducing wind speeds, narrow barriers can be as effective as wide ones. Although in theory, one-row barrier should suffice, in practice three to five rows are more effective. If a single row windbreak is to be planted, tree species that retain their foliage to the ground and give a fairly dense growth should be selected. *Eucalyptus* tree are generally unsuitable as single row windbreaks because of their habit of losing the lower branches. The main disadvantage of a single row is that if one tree is lost, gap is created which reduces the efficiency of the entire windbreak. Distance between trees varies with the relative importance of the protective versus productive purposes of the windbreak. Initial spacing of 3 meters between the rows, with trees 1-2 meters apart in the rows is desirable.

Species: In general, tree or shrub species selected for windbreaks should have the characteristics of rapid growth, straight stems, wind firmness, good crown formation, deep tap root system and resistance to drought. Some of the tree genera used for windbreaks are *Cassia*, *Acacia*, *Casuarina*, *Leucaena*, *Grevillea*, *Cupressus*, *Pinus*, *Dalbergia*, *Syzygium*, *Erythrina*, *Mangifera*, *Bambusa*, etc.

Windbreak for fruit orchards

Fruit orchards when exposed to strong wind usually incur heavy losses. Heavy wind increases the losses of moisture both by increasing transpiration and surface evaporation. The high winds also cause damage to fruit trees by breaking of branches, destruction of blooms and dropping of immature fruits. The growth and yield in protected orchard is definitely better than the exposed orchard. Hence, establishment of a tall growing windbreak is necessary to protect an orchard. Planting of windbreaks should be done at least two years after planting of fruit trees. The first row of it should be planted 10 m away from fruit plants at a close spacing to form a thick screen. The windbreak trees sometimes may compete with the fruit trees for water and nutrient. To prevent this competition, a trench may be dug about 1 m deep and 5 m away from the row of windbreak trees and all the roots of windbreak trees exposed in the trench are cut off periodically.

Multipurpose windbreaks

Although planting windbreaks is an investment that takes some land out of production, well designed windbreaks protect the health and productivity of crops enough to make the overall return positive. A multipurpose windbreak is designed to provide multiple functions and/or products, in addition to wind protection. Multiple products from a windbreak can include yields such as fruit, timber, animal fodder, mulch, wildlife habitat and other economic farm products. Adding multiple functions or products to a windbreak plan can make the installation and management more satisfying and economically viable for the farmer. Multipurpose windbreaks require special care in planning and management to maintain the primary function of wind protection while maximizing secondary yields. Once the form and position are carefully determined, then multiple functions or products can be added.

Since planting trees for a windbreak involves a long-term investment, trees having timber value may also be included. The main drawback of having timber as a secondary yield from a windbreak is that wind damage may produce timber of poor quality. Also, since windbreak trees require no or very little pruning, the lack of pruning may reduce timber quality and quantity in certain species that require a lot of pruning for optimal timber production. Planning for timber harvest requires the most careful effort as the entire trees are removed. The planting, harvesting and replanting must be coordinated to avoid the creation of gaps in the tree rows. Integrating timber trees with permanent rows of non-timber windbreak trees will help maintain the effectiveness of the windbreak. *Grevillea robusta*, *Pterocarpus indicus* and *Azadirachta indica* are some of the best suitable species for this purpose.

Nitrogen fixing trees (NFTs) can also be integrated in a multi-row windbreak and pruned regularly to provide a nutrient-rich mulch for crops, or a nutritious fodder to supplement the diet of farm animals. Although pruning should be avoided for most windbreak trees, the practice of cutting back NFTs and allowing them to resprout can be integrated with windbreak management. Pruned NFTs are much more susceptible to wind damage if they are allowed to regrow to a large size, but if they are cut regularly and the regrowth kept small, they will be effective as a short row. To maintain the windbreak's primary function with this practice, it is essential to prune the NFTs regularly. These species are planted on the most sheltered side of the windbreak. The important species for this purpose are *Leucaena leucocephala*, *Sesbania sesban*, *Calliandra calothyrsus*, and *Gliricidia sepium*.

Guidelines for multipurpose windbreak design

1. The species used should be selected first for their wind tolerance and appropriateness for the site (climate, soils, etc.). The products should be a secondary consideration in selecting species.
2. Windbreaks designed for multiple products should comprise of multiple rows. It enables more flexibility in management and harvest of products without compromising wind protection by creating gaps.

3. Trees yielding products such as fruit, food, fodder or mulch should ideally be located in the interior or wind-sheltered rows of the windbreak for maximum protection.
4. A diversity of species should be used to allow for greater flexibility in management and for better resistance to damage to trees caused by insects or diseases.

Benefits from windbreaks

1. It protects crops and pastures from hot, drying winds and cold wind, frost.
2. It reduces/prevents soil erosion.
3. It improves the microclimate by moderating temperature.
4. It reduces evaporation from farm lands, also checks transpiration.
5. It prevents grass fire.
6. It provides fencing and boundary demarcation.
7. It increases overall farm productivity

Soil Conservation Hedges

Planting rows of trees or bushes tightly along the slopes of a piece of land and then pruning them into hedges is known as a hedgerow barrier. They are planted to lessen erosion caused by water runoff and go by the name contour hedgerows as well. The soil fertility is preserved by the hedgerows by halting the erosion of topsoil. Hedgerow barriers have two functions.

1. The hedge functions as a physical, although permeable, barrier to runoff, delaying the water's movement and causing the majority of the soil particles to fall out.
2. The deep root system and litter fall serve to enhance the soil structure around the hedge. As a result, runoff water in the vicinity of the hedge infiltrates at a significantly greater rate, leaving the soil particles behind.

Overall, the result is less water and significantly less dirt accumulation than one might anticipate near the foot of the hill. Where low soil moisture affects crop growth, enhanced water retention on the slope through reduced flow and strong infiltration near the hedges is a significant gain.

Design and management

Less than 10 tonnes of soil are lost per hectare in a year when erosion control measures are successful. On slopes up to 20%, contour hedgerows may easily accomplish this degree of control; on steeper slopes, however, the results are more unpredictable, and slopes higher than 60% shouldnot be farmed at all.

Planting hedgerows on contours that are two metres or less below one another, i.e. a vertical decrease in height of no more than two meters, is advised. This translates to a spacing of around 20 metres between rows on a 10% slope. The distance should be around 10 metres on a 20% slope. These numbers are only approximations. Depending on the frequency of rainstorms and the soil's erodibility, the actual distance could be smaller. On uneven slopes when the contours do not run parallel, the spacing between hedgerows might vary greatly. As a result, the cropped strip's width fluctuates, making planting and ploughing more difficult.

Plants are placed in single or double rows, with three to four plants per metre, using seeds, seedlings or cuttings. After they become well-established, they can be trimmed to lessen any interference with the herbaceous crop. Until then, they are left to grow naturally. Since hedge height is not crucial for preventing erosion, they are typically trimmed to a height of 30 to 50 centimetres. The trimmings can be fed to cattle, utilised as mulch or green manure for the crop, or utilised to strengthen the barrier. Ploughed along the contour are the strips of ground in between the hedges; if the field crop is grown on ridges, these are likewise aligned along the contour. Soil is moved from the high side of the cropped strip to the hedgerow below by erosion and, more significantly, by hoeing or ploughing. After a longer length of time, the outcome is horizontal terraces divided by steep banks. Within two or three years, this soil displacement causes the creation of terraces with a visible drop in height behind the hedge. Significant variations in soil quality occur across each harvested strip as a result of the topsoil loss immediately below the hedge and the deposition of that soil above the subsequent hedge. This is evident in the much better crop stand above the hedge than below the bank. Applying the majority of the green manure immediately below the banks, where crops grow poorly, can help to greatly alleviate this issue.

Selection of species

Grevillea robusta, *Acacia catechu*, *Pinus roxburghii*, *Acacia modesta*, *Prosopis juliflora*, *Alnus nepalensis*, *Leucaena leucocephala*, and other tree species are utilised for soil conservation.

In order to generate fruit, fodder, or timber, trees can also be planted alongside staple food crops. Fruit trees, such as ber (*Zyziphus*) papaya, dwarf mango, guava, citrus, etc., are typically planted along the borders of terraces.

SILVIPASTORAL SYSTEMS (Trees + Pasture and/or animals)

The main feature of silvipastoral systems is the integration of tree production with livestock and fodder production. Silvipasture as an agroforestry practice is specifically designed and managed for the production of trees, tree products, forage and livestock. Silvipasture results when forage crops are deliberately introduced or enhanced in a timber production system, or timber crops are deliberately introduced or enhanced in a forage production system. Integrating livestock production in the farming system promotes farm product diversification, improves food security and offers a diversified marketing opportunity that can stimulate rural economic development. The interactions among timber, forage and livestock are managed intensively to simultaneously produce timber commodities, a high-quality forage resource and efficient livestock production. Overall silvipastoral systems can provide economic returns while creating a sustainable system with many environmental benefits.

Before new silvipastoral systems are established, implications of merging forestry and agricultural systems should be explored thoroughly for economic and environmental considerations. Factors like availability of potential markets, soil type, climatic conditions and species compatibility should be considered while choosing tree and forage crop species. The timber component should be marketable, of high quality, fast growing, deep rooted, drought tolerant, and capable of providing the desired products and environmental services. The forage component should be a perennial crop that is suitable for livestock grazing, compatible with the site (soil, temperature, precipitation), productive under partial shade and moisture stress, responsive to intensive management and tolerant of heavy utilization or grazing. A successful silvipastoral system requires understanding forage growth characteristics and managing the timing and duration of grazing to avoid browsing of young tree seedlings. Improper management of silvipastures can reduce desirable woody and herbaceous plants by overgrazing and soil compaction.

Benefits from silvipastoral systems

1. Economic risk is reduced because the system produces multiple products, most of which have an established market. Production costs are reduced and comprehensive land utilization in silvipastoral systems provides a relatively constant income from livestock sale and selective sale of trees and timber products. Well managed forage production provides improved nutrition for livestock growth and production.
2. Grazing can enhance tree growth by controlling competition of grass for moisture, nutrients and sunlight. Well managed grazing provides economical control of weeds. Fertilizer applied for forage is also used by trees. In addition, livestock manure recycles nutrients to trees and forage.

3. Some forage species tend to be lower in fibre and more digestible when grown in a tree-protected environment. Trees that provide shade or wind protection can have a climate-stabilizing effect to reduce heat stress on livestock. Protection by trees can cut the direct cold effect by 50 percent or more and reduce wind velocity by as much as 70 percent.
4. The forage protects the soil from water and wind erosion, while adding organic matter to improve soil properties. Some forage species fix nitrogen.
5. Silvipastures provide an attractive landscape with an aesthetic value. It can also increase wildlife diversity.

Within the broad category of silvipastoral system there are several types of practices which can be identified depending on the role of the tree/shrub (sometimes collectively called 'trub') component. These practices include

1. Protein banks (fodder tree banks)
2. Trees and shrubs on rangeland or pastures
3. Live fences of fodder trees and shrubs
4. Plantation crops with pastures and animals

Protein Banks

In hilly, semi-arid, and dry areas, protein banks are intentionally planted blocks of forage plants to help with fodder shortages, particularly during the dry seasons. Leguminous trees, bushes, and herbaceous legumes are examples of forage plants. They can be cultivated with appropriate grasses. In order to feed no grazed livestock (animals housed in stalls), the fodder trees pruned regularly. The fodder banks are called protein banks when they are based on legumes, which turn them into significant suppliers of protein. On or near farmlands, the following multifunctional trees that store protein are planted.

The important tree and shrub species for this system are *Acacia nilotica*, *Albizia lebbeck*, *Leucaena leucocephala*, *Gliricidia sepium*, *Sesbania grandiflora*, *Artocarpus* spp., *Bombax malabaricum*, *Cordia dichtotoma*, etc.

Trees and Shrubs on Rangeland or Pastures

In range and grazing areas, the main function of trees has been to provide shade. In addition to their essential role on rangelands, trees also yield additional goods like timber and fodder. To augment fodder production, different tree and shrub species are placed in this system either in an irregular manner or in accordance with a systematic pattern. The number of species with potential for this practice is great. They include *Acacia nilotica*, *Acacia mangium*, *Acacia auriculiformis*, *Gliricidia sepium*, *Bauhinia* spp., *Leucaena leucocephala*,

Derris indica, *Azadirachta indica*, *Emblica officinalis*, *Psidium guajava*, *Prosopis* spp. and *Tamarindus indica*.

Live Fences of Fodder Trees and Shrubs

Planting trees or bushes along farm boundaries, farmyard borders, grazing plot borders, animal enclosure borders and agricultural field borders are known as live fences. They primarily act as field borders, keeping animals contained inside the farm and out of nearby agricultural regions or crop fields. They might consist of one or several rows that are thickly planted. An alternative would be to plant a single row of live fence posts widely apart and fill the spaces between the trees with wire, sticks, or dead branches. Both varieties are composed of everlasting rows of trees or shrubs that are routinely pruned and pollarded. In addition to offering shelter and shade, the barriers have additional biotic effects. In addition to providing wood and foliage products, the trees can act as windbreaks. The foliage can be used as fodder. Legumes are especially valued as they have high protein content.

Controlling livestock movement is the major rationale for installing live fences and the fundamental goal of fence management, making the animal component of the practice the most essential part of the whole.

In arid and semiarid zones, live fences are often made of thorny species of the *Acacia* or *Prosopis* genera. In the humid and subhumid tropics, leguminous species such as *Gliricidia sepium*, *Sesbania grandiflora* and *Erythrina berteroana* are used.

Plantation Crops with Pastures and Animals

On grazing land, plantation plants including coconut, cashew, rubber, sapota, mango, litchi, and so on might be sporadically planted. Under such planted trees, cattle and small ruminants can graze in pastures that are open and slightly shaded. The animals gain from this activity in many ways.

1. Pastures may grow more as trees bring up nutrients from below pasture rooting depth.
2. They can also improve the quantity and quality of accessible feed, like as protein and minerals.
3. Light shade may help pasture growth and quality in dry areas by improving soil surface microclimate.
4. Trees may supply fodder when pasture species are infested by insects or diseases.
5. Fodder tree leaves are less trampled than those of creeping or erect pasture grasses and legumes.

6. By having shade, animals for meat grow better and dairy cows produce more milk in the humid tropics.
7. Wind and cyclone protection is provided by the trees.
8. Animals can eat tree fruits and pods which are rich in minerals and vitamins.

AGRISILVIPASTORAL SYSTEMS (Trees + Crops + Pasture/animals)

Agrisilvipastoral system is the combination of agricultural crops, woody perennials and livestock. Agroforestry practices under this system include

1. Agrisilvicultural system converted to silvipastoral systems
2. Multistorey system with free grazing
3. Woody hedgerows for browse, green manure, soil conservation, etc.
4. Alley cropping system using pasture grasses/fodder crops and agricultural crops
5. Homegardens (tree-livestock-crop mix around homesteads)

Agrisilvicultural System Converted to Silvipastoral Systems

In this system, the initial cropping combinations include tree seedlings and annual agricultural crops. As the trees grow and close their canopies, the yield recovery of annual agricultural crops gradually decreases and it will no longer be possible to grow these crops. Instead, shade tolerant grasses and vines will be more compatible to land use system where animals are allowed to graze freely, thus converting agrisilvicultural system to silvipastoral system.

Multistorey System with Free Grazing

This is similar to the multistorey under agrisilvicultural system, except that in this case, grazing animals are an added component.

Alley Cropping with Pasture Grasses and Agricultural Crops

This is similar to hedgerow cropping with seasonal agricultural crops. However, instead of all alleys planted with agricultural crops, some alleys in between the hedgerows are grown with improved pasture grasses and/or fodder trees or shrubs which are regularly cut and fed to livestock.

Woody Hedgerows

In this system, a variety of woody hedges, particularly fast growing and coppicing fodder shrubs and trees are planted for mulching, browsing, green manure, soil conservation, and other uses. This system's primary goals are soil conservation and the production of food, fuelwood, and fodder. Common plants in this system include *Erythrina* spp., *Leucaena leucocephala* and

Sesbania grandiflora, among others, in various forms, particularly in dry and semi-arid environments to act as mulch or litter to retain the available moisture or to offer browse.

Homegardens

This is one of the earliest forms of agroforestry practices found extensively in the high rainfall areas in tropical south and south-east Asia. The word 'homegarden' has been used to describe diverse practices from growing vegetables behind houses to complex multistorey systems. It is the intimate association of multipurpose trees and shrubs with seasonal and permanent arable crops and livestock within the compounds of individual houses.

Mostly the whole system, i.e., crop-tree-animal unit is managed by family labour. These systems are common in all ecological regions in the tropics and subtropics, especially in humid lowlands with high population density. The average size of a homegarden is usually much less than one hectare. In India, every homestead has around 0.2 to 0.5 ha land for personal production. In many parts of the world the fruit, vegetable and other foodstuff produced from homegardens provide a substantial part of the household food requirement. Food production is the primary function and role of most of the homegardens. The combination of crops with different production cycles results in a relatively uninterrupted supply of food products throughout the year. In most cases, animals and birds are kept in the homegardens. Fodder and legumes are widely grown to meet the daily fodder requirement of the animals. The waste materials from crops and homes are also used as fodder for animals and feed for birds. Cattle are mostly kept for dairy products and land cultivation. Goats, chickens and fish are kept for household consumption. Products from animals or the animals themselves can also be sold in the market.

Structure of homegardens

The most obvious features of all homegardens are their layered arrangement and appropriate species mixing. An herbaceous layer near the ground, a tree layer at higher altitudes, and an intermediate layer are often found in all home gardens. The lowermost layer, which is less than 1.0 m tall and is dominated by various vegetables, may be divided into two sections. The second layer, which is between 1.0 and 3.0 m tall, is made up of food crops like bananas and papayas, among others. The uppermost layer of the tree layer, which is 25 metres tall, is occupied by emergent, fully grown timber, and fruit trees, while the next lower layer is occupied by medium-sized trees, which are 10 to 20 metres tall. Fruit trees make up the majority of the intermediate stratum, which is between 5 and 10 metres high.

Woody species in the homegardens include *Cocos nucifera*, *Artocarpus heterophyllus*, *Mangifera indica*, *Psidium guajava*, *Anarcardium occidentale*, *Citrus* spp., *Carica papaya*, *Passiflora edulis*, *Annona squamosa*, *Tamarindus indica*, *Artocarpus altilis*, *Theobroma cacao*, *Calliandra calothyrsus*, *Moringa oleifera*, *Gliricidia sepium*, *Leucaena leucocephala*, *Sesbania grandiflora* etc.

Herbaceous species in the homegardens include grain crops, oilseed crops, sweet potato, cassava, yam, potato, carrot, peanut, garlic, pineapple, onion, turmeric, chilli, cabbage, okra, tomato, beans, banana, pumpkin etc.

Fig. 3.2: Vertical stratification in a typical homegarden
Source: Das & Das, 2005

Benefits from homegardens

1. Production is steady and varied, with less risk involved. Products like food, fuel, fodder, fruit, spices, and timber for building are produced on a regular basis.
2. Homegardens provide a multitude of protective purposes, including soil preservation, water retention, and creating a comfortable microclimate for people, animals, and certain plants.
3. The use of labour for the farm is very efficient due to proximity of the garden to the house, within the walking distances.
4. Due to their close proximity to the house, the farm family may provide additional care, particularly at night, for valuable crops or animals that require protection;
5. In addition to providing shade, wind protection, and seclusion, home gardens can also meet aesthetic standards. Open spaces in the garden are frequently used as hubs for family or community gatherings.

Constraints of homegardens

1. Due to the wide range of biological niches, a homegarden may serve as a habitat for species such as fungus, insects, and snakes that can spread illnesses or become pests.
2. When harvesting trees, other flora are frequently damaged.
3. Houses next to homesteads are more vulnerable to harm from large lateral roots and falling branches.

OTHER AGROFORESTRYSYSTEMS

Other agroforestry systems, sometimes known as minor agroforestry technologies, consist of aquaforestry, agrisilviaquaculture, apisilviculture, sericulture, mixed woodlots etc.

Aquaforestry

The aquaforestry system comprises of composite fish culture in farm ponds, and various trees and shrubs (*Leucaena leucocephala*, *Morus alba*, *Gliricidia sepium*, *Moringa olifera*, etc.), leaves of which are preferred by fish are planted on the boundary and around fish ponds. Leaves of these trees are used as feed for fish. Inland fish such as *catla*, *rohu*, *mrigal*, common carp, silver carp and grass carp can be grown in the ponds. In the coastal regions, farmers are cultivating fish and prawn in saline water and growing coconut and other trees on bunds of ponds. Now fish culture in the mangroves is also advocated, which form a rich source of nutrition to the aquatic life and breeding ground for fish and prawn. Other terms that have been used for this practice are silvipisciculture, agrisilviaquaculture and aquasilviculture. Some other modified aquaforestry systems are

1. A well-balanced system of animal husbandry including goatery, poultry, duck farming, turtles and fishes in the small ponds in homegardens make a balanced system of high moisture, energy and nutrient use efficiency per unit area.
2. In paddy field, fish can easily be reared by planting trees on field bunds or boundary to provide leaves used as fish feed. This system can be practised in high rainfall areas.
3. Coconut plants can also be successfully planted on raised paddy field bunds with an alley space of 5 m width depression which is utilized for pisciculture purpose.

Apisilviculture

This strategy involves planting a variety of honey (nector) producing trees on the edge of agricultural fields, which are often visited by honeybees.

Bees benefit the trees and trees in turn provide series of benefits to bees. The primary purpose of this system is to produce honey. The science of bees and beekeeping is known as apiculture. The raising and rearing of bees is called beekeeping. The growing value of bees and honey has led to a continuous focus on apiculture in many developing nations. Apisilviculture with *Eucalyptus*, *Gliricidia*, *Grevillea*, *Gmelina*, *Leuceana*, *Syzygium*, *Madhuca* and *Albizia* species were more remunerative and a good source of generating additional farm income in rural areas. Fruit trees can also be integrated with apisilviculture for more profits.

Sericulture

Silkworm culture is known as sericulture. Mostly, these silkworms are fed with leaves of mulberry tree (*Morus alba*). Silkworms produce silk and several by-products, which are used for many purposes like textile, fibre, soap, vitamins, medicine, etc. Silkworms depend upon mulberry tree leaves for their normal growth and development. The mulberry tree is a small deciduous tree which tolerates drought and heat but prefers moist climate. The mulberry tree also gives a fine wood. Its branches can be used to make farm tools and its bark to make high quality paper and artificial fibre. The fruit is edible and can be used to make wine. The tree can produce fuelwood and fodder. It also provides shade and can be used as ornamental plant. It is useful for windbreaks, soil conservation, soil stabilization, hedgerow planting, bee forage and live fence.

Mushroom in mixed tree species

Cultivation of paddy straw mushroom under the shade of high-density plantation of mixed multipurpose tree species is also remunerative. The tree species suitable for the system are *Acacia auriculiformis*, *Acacia mangium*, *Casuarina equisetifolia*, *Gmelin aarborea*, *Dalbergia sissoo*, etc.

Multipurpose woodlots

In this system, location specific multipurpose tree species are grown mixed or separately for various purposes such as wood, fodder, soil protection, soil reclamation, etc. Although not considered an agroforestry practice *perse*, short-rotation intensive silviculture offers definite potential for the reclamation of abandoned land and is therefore one of the technological options that should be considered on agricultural land in harmony with other types of agroforestry production systems.

CLASSIFICATION BASED ON ARRANGEMENT OF COMPONENTS

The arrangement of components in an agroforestry system gives first priority to the plants even though animals are included in the system. Component arrangement may take temporal and spatial factors into account. Thus, agroforestry systems may be divided into two categories based on how their components are arranged; spatial arrangement and temporal arrangement.

Spatial arrangement

Plant placement in an agroforestry combination can produce (i) dense mixed stands, similar to home gardens, (ii) sparse mix stands, such as those found in pastures and agricultural fields; and (iii) trees planted in row plantations or in strips along field boundaries that permit intercropping. The species might be arranged in strips or zones with different widths. These zones can take many different shapes, from macrozonal configurations to microzonal ones like alternating rows. Alley cropping, also known as hedgerow intercropping, is a typical illustration of the zonal pattern. Planting trees around the boundaries of fields and plots for a variety of uses, such as fruit, fodder, fuelwood, fencing, protection, soil conservation, windbreaks, and so forth, is an extreme type of zonal planting. It is also crucial to remember that severe macrozonal arrangements can result in single-cropping systems; nevertheless, the boundaries between zonal agroforestry and sole crop plots can be determined by the interactive relationship of various components.

Temporal arrangement

In agroforestry, plant temporal arrangements may also take many different shapes. The traditional shifting agriculture cycles, which involve 2-4 years of cropping followed by more than 15 years of fallow period during which certain woody species or combinations of species may be planted, are an extreme example. Comparably, in certain silvipastoral systems, grass leys may be in a state of rotation, with certain grass species persisting on the land for a number of years. In agroforestry, these temporal groupings of elements are referred to as coincident, contemporaneous, overlapping, distinct, and interpolated.

Coincident: When various crops coexist on the same plot of land, such as grazing beneath trees and coffee under shade trees.

Intermittent (space dominated): It occurs when seasonal crops are grown with perennial ones, i.e., turmeric or pineapple under *Gmelina arborea* plantation, seasonal grazing of cattle in pasture under trees.

Separate: It is when the different components occupy space at different times, i.e., improved fallow species in shifting cultivation.

Interpolated (space and time dominated): It is when different crops occupy the space during different time, i.e., homegardens.

Concomitant: It occurs when different crops stay together for certain period of time, i.e., *taungya* system.

An agroforestry system may also be a simultaneous or sequential system.

Simultaneous agroforestry system

In simultaneous agroforestry systems, the tree and crop components occupy the same land unit at the same time. There is significant overlap in the growth cycles of tree/crop components. As a result there is direct interaction between the two species, i.e. the tree and the crop species. Simultaneous agroforestry systems include alley cropping, parklands, fodder/protein banks, live fences, boundary markings and homegardens.

Sequential agroforestry system

Here crops and trees coexist on the same piece of ground at various seasons with little direct contact between them. Even if both components may have been planted at the same time, the crop and the tree components grow at distinct periods. There is a chance that one component species may develop quickly and the other slowly. The component species' nutrient absorption maxima may also happen in a sequence, which makes the species complimentary in how they utilise the resources in the soil. With successive agroforestry systems, interactions between crop and tree components become less frequent over time. Improved fallows and rotating woodlots are two instances of sequential agroforestry systems.

While farmers in sequential agroforestry systems make use of the plants' residual benefits, management in simultaneous agroforestry systems should strive to minimise interspecific competition. Because of this, farmers who practise shifting agriculture stack chopped trees in smaller spaces, burn them and then plant a new crop that is dependent on the ash that has gathered. Nitrogen-fixing trees are purposefully planted in improved fallows to enhance the physical characteristics of the soil and its overall fertility, which benefits crops that are produced after the trees are harvested.

FUNCTIONAL CLASSIFICATION OF AGROFORESTRY SYSTEMS

Agroforestry systems serve a variety of purposes and provide a wide range of goods. Two essential components of every agroforestry system are the production function and the protection function. This suggests that agroforestry systems serve both a service and a production role, producing one or more products that typically satisfy basic needs. In addition to producing fuel, fruit, food, fibre, and fodder, they also boost soil productivity, mitigate drought and floods, preserve biological variety, and moderate the climate. Because of this, agroforestry systems may be divided into the following three major classes.

1. Productive agroforestry systems
2. Protective agroforestry systems
3. Multipurpose agroforestry systems

Productive agroforestry systems

The creation of necessities, items needed to fulfil the basic requirements of society, is referred to in this system. Tree intercropping, homegardens, planting trees in and around agriculture fields, and raising fish and animals alongside trees are all included in this category. The goal of land usage is to optimise crop productivity, both annual and perennial. The various productive functions of agroforestry systems are food, fodder, fuelwood, timber, fibres, honey, animal products, fish, etc.

Protective agroforestry systems

In addition to protecting the land, protective agroforestry systems also improve the climate, lessen wind and water erosion, increases soil fertility, offer shelter, and have other advantages. Soil conservation plantings, windbreaks and shelterbelts, intercropping hedgerows, cover crop introduction, etc. come in this category. The goal of these systems is to safeguard the land. The protective functions of agroforestry systems are windbreak, shelterbelt, soil conservation, moisture conservation, soil improvement, biodiversity preservation and shade (for crops, animals and human beings).

Multipurpose agroforestry systems

By optimising both protective and productive roles, it guarantees multifunctional production. This class may cover silvipasture, multifunctional tree plantations, hedgerow intercropping, homegardens, and more. Its output should be sustained throughout time.

SOCIOECONOMIC CLASSIFICATION OF AGROFORESTRY SYSTEMS

Agroforestry systems have been divided into three groups according to socioeconomic factors including production volume and the degree of technological input and management. These groups include commercial agroforestry system, intermediate agroforestry system and subsistence agroforestry system.

Commercial agroforestry systems

The primary focus of these systems is the output production scale. The operations typically operate at a medium to large size, with land owned by either the government or the private sector. Labour is employed either on a contract basis or on a daily salary basis. The following practices come under this category.

1. The commercial cultivation of plantation crops like coconut, rubber, and oil palm, together with the ongoing underplanting of food crops, pasture, and livestock.
2. Growing shade-tolerant plantation crops commercially, such as cocoa, coffee, and tea, beneath overstory shade trees.
3. Rotational timber/food crop systems include using a brief period of food crop production as a silvicultural technique to guarantee the establishment of timber tree species such as taungya.
4. Commercial grazing and ranching under large-scale pulp and timber plantations, etc.

Subsistence agroforestry systems

Subsistence agroforestry systems are those wherein the use of land is directed towards satisfying the basic needs and is managed mostly by the owner and his family. Some marginally excess produce may be available for sale. In this system, level of technology input and management are usually low. Shifting cultivation, scattered trees in the farm, and homestead agroforestry are the examples of subsistence agroforestry systems.

Intermediate agroforestry systems

Intermediate systems are those between commercial and subsistence scale of production and management. On small to medium-sized farms, these are practised. The system strives to produce enough goods to suit the requirements of the family as well as generate revenue from surplus goods that can be sold. The following practices come under this category.

1. Perennial cash crops and subsistence crops are produced, with the cash crops meeting the family's financial demands and the food crops meeting their nutritional needs.
2. The intermediate agroforestry system includes agroforestry systems based on plantation crops like coffee, cocoa, and coconut as well as agroforestry systems based on a significant number of fruit trees.

There are more classifications for the socioeconomic agroforestry systems based on the technology and management employed.

On the basis of management

System under intense management: To increase production per unit area, agroforestry systems are under intensive management. Homegardens and trees bearing food require intensive care and management.

System with extensive management: productivity is lower here and it involves shifting cultivation, silvipasture, etc.

On the basis of technology

Low technology system: This system relies on antiquated technology, such as shifting cultivation, to produce agricultural and forest products.

High technology system: This system uses contemporary technology, such as tissue culture for forest management.

Systems with intermediate technology: These systems, including agrisilviculture, fall between those with low and high technology.

ECOLOGICAL CLASSIFICATION OF AGROFORESTRY SYSTEMS

Agroforestry systems are associated with a number of ecological elements. Important ecological factors (climatic, edaphic, and physiographic) can be used to classify it. Based on major agroecological zones, agroforestry systems have been grouped into the three types; humid/sub-humid lowlands, arid/ semiarid lands and highlands.

Humid/sub-humid lowlands

The humid and subhumid agroecological zone is characterized by hot humid climate for most part of the year with evergreen or semi-evergreen vegetation. It is the most important ecological region in terms of total human population it supports, extent of area and diversity of agroforestry and other land-use systems. Many types of agroforestry systems, such as different types of homegardens, plantation crops with various combinations, multilayer tree gardens, improved fallow in shifting cultivation and multipurpose woodlots are found in these areas with a high human population due to climatic conditions that favour the rapid growth of a large number of plant species. In areas with a low population density, improved fallow in shifting cultivation and trees on rangelands and pastures are the common agroforestry practices.

Agroforestry systems in arid/semiarid lowlands

The arid and semiarid region is characterized by a smaller number of rainy days (9-21 days during July to September) with 2-4 wet months, high solar radiation incidence, high wind velocity, high potential evapotranspiration and high mean aridity index. Various forms of silvipastoral systems, windbreaks and shelterbelts, multilayered multipurpose trees on croplands, *Prosopis* based agrisilvicultural systems are the major agroforestry systems in this agroecological zone.

Agroforestry systems in highlands

Variable rainfall, degraded and shallow lands at high altitude to deep rich soils in valleys, and great climatic variations are the main features of highlands. Steep sloping lands and terrain resulting in soil erosion is a major problem in this area. This area is storehouse of huge biological diversity. Production systems involving plantation crops such as coffee and tea, use of woody perennials in soil conservation and soil fertility maintenance, improved fallows and various silvipastoral systems are the major agroforestry systems practised in this zone. The following agroforestry systems are found in tropical highlands.

1. Production systems incorporating plantation crops, such as coffee and tea in commercial as well as small-holder systems
2. Use of woody perennials in soil conservation and soil fertility management
3. Improved fallows
4. Combinations of silvipastoral practices

PHYSIOGNOMIC CLASSIFICATION OF AGROFORESTRY SYSTEMS

This approach categorises vegetation based only on its outward characteristics, disregarding its scientific attributes. The classifications in physiognomic categorization are thoroughly examined in terms of the specific ecosystem's flora, landforms, climate, and soils. Thus, depending on the general characteristics and plant appearance, agroforestry systems are divided into three groups, such as xeromorphic, mesomorphic and hydromorphic

Xeromorphic agroforestry system

Xerophytic plants are flourished in deserts and arid areas with little or no water. In this agroforestry, water is a limiting element for the growth of agricultural products and trees. This kind of agroforestry system is typically found in semi-arid and desert areas. The agroforestry systems include various types of tree based farming systems.

Mesomorphic agroforestry system

Water is most abundantly accessible in the agroforestry system. Terrestrial plants known as mesophytes are not exclusive to either a dry or wet habitat.

This group includes the majority of agroforestry systems seen in temperate and tropical climates. This environment is favourable for plant development.

Hydromorphic agroforestry system

This is about agroforestry in locations that are flooded with water or have damp soil. The hydrophytes can only thrive in wet or watery environments. There is sufficient water for the growth of crops and trees. Wetland agroforestry, aquaforestry and aquasilviculture are the common practices in these areas.

HISTORICAL CLASSIFICATION OF AGROFORESTRY SYSTEMS

The several agroforestry systems used today have evolved throughout time. It is commonly acknowledged that agricultural crop production began in the form of shifting cultivation around a millennium ago. Sedentary agricultural techniques emerged throughout time as a result of the dissemination of knowledge, the growth of communities, and the construction of irrigation and drainage infrastructure. Perhaps the most modern agroforestry systems are some of the most intense, combining trees and agricultural crops. Classifying different systems based only on their history might be challenging, though, as several systems evolved concurrently based on local edapho-climatic and socioeconomic circumstances.

CLASSIFICATION BASED ON UTILIZATION OF LAND

1. Homestead agroforestry: Production of multipurpose tree species (MPTs) in homestead.
2. Forestland agroforestry: Production of crops in the vacant spaces of the forest.
3. Crop-farm forestry: Production of crops and trees in the crop land.
4. Fish-farm forestry: Production of fishes and trees in the farm
5. Animal-farm forestry: Production of animal with forest trees.
6. Poultry-farm forestry: Farming of poultry birds and trees.
7. Dairy-farm forestry: Farming of milch cattle and trees.
8. Beef cattle-farm forestry: Farming of beef cattle and trees.
9. Goat-farm forestry: Farming of goats and trees.
10. Integrated farm forestry: Production of crops, animals, fishes along with trees.
11. Road side agroforestry: Production of deep rooted tall trees with narrow canopiesand soil binding grasses or crops along the sides of roads, highways, railways and embankments.

12. Public place agroforestry: Production of suitable trees only or along with decorative herbs, vegetables in places like markets, premises of school, college, university etc.

FLORISTIC CLASSIFICATION OF AGROFORESTRY SYSTEMS

Agroforestry systems can also be categorised using floral characteristics. An inventory of all current systems and the plant species that make them up is required in order to define agroforestry systems based on species composition.

It is evident that there might be several methods for classifying agroforestry. On the other hand, a system that classifies agroforestry systems according to the primary functional properties of each component and their nature seems like a more sensible, straightforward, and practical approach with a clear objective. Once more, a system's selection may be influenced by a variety of interrelated factors, including social, ecological, and financial ones.

4

Selection of Multipurpose Trees for Agroforestry

Farmers have been growing trees for different purposes for thousands of years. Tree species that are grown to provide more than one significant function are called multipurpose trees. These functions may be productive such as producing fuelwood, timber, fibre, fodder, food, medicine, etc. and/or protective such as soil conservation, shade, shelterbelt, microclimate amelioration, land sustainability, biodiversity preservation, etc. All trees are multipurpose; some, however, are more multipurpose than others. Tree species can be multipurpose in two ways.

1. A single tree can provide more than one function. For example, *Gliricidia sepium* is grown as living fences that provide fuel, fodder and green manure for agricultural crops - all at the same time.
2. Trees of the same species, when managed differently, can provide different functions. For example, *Leucaena leucocephala* is managed so that some trees will mainly yield wood while others mainly produce leaf fodder.

Farmers can grow multipurpose trees (MPT) in various combinations with other crops, as in agroforestry, in block plantations of trees or in naturally regenerating tree farms. In certain cases, multipurpose trees are grown and managed for only one purpose. For example, *Gliricida sepium* is grown only to provide shade in coffee plantations. The same species may be planted in some other places and is managed differently for a very different use.

BENEFITS FROM MULTIPURPOSE TREES

The tree has both productive and protective functions, and it can play an important role in enhancing the productivity of the lands to meet the demand of ever-growing human and livestock population as discussed below.

Food

1. Food for man from trees in the form of fruits, nuts, cereal substitutes, etc.
2. Feed for livestock from trees

3. Enhanced food and feed production from crops associated with trees as in various agroforestry systems through biological nitrogen fixation, better access to soil nutrients brought to the surface through deep tree roots, improved availability of nutrients due to high cation exchange capacity of the soil and higher organic matter in the soil
4. Enhanced sustainability of cropping systems through soil and water conservation by arrangements of trees to control runoff and erosion
5. Indirect production benefits through microclimate amelioration

Water

1. Improvement of soil moisture retention in rainfed croplands and pastures through improved soil structure and microclimate effect of trees
2. Regulation of stream flow for reduction of flood hazards
3. Improvement in drainage from waterlogged or saline soils by trees with high water requirements
4. Increased biomass storage of water in fodder trees for animal consumption in dry season

Energy

1. Fuelwood for direct combustion
2. Gas from wood or charcoal
3. Ethanol produced through fermentation of high-carbohydrate fruits
4. Augmentation of wind power using appropriate arrangements of trees
5. Oils, latex and other combustible saps and resins

Shelter

1. Building materials for shelter construction
2. Shade trees for people, livestock and shade loving crops
3. Windbreaks and shelterbelts for protection of settlements, croplands, pastures and roadways
4. Live fences and fence posts

Raw material for industries

1. Raw material for pulp and paper industry
2. Tannins, essential oils and medicinal ingredients
3. Wood for agricultural implements
4. Fibre for weaving
5. Wood for a variety of craft purposes

6. Fruits, nuts, etc. for food processing industries

Cash

1. Direct cash benefits from sale of tree products
2. Indirect cash benefits from increased productivity through associated crops and/or livestock

Environment

1. Production of oxygen during photosynthesis
2. Absorption of dangerous chemicals and other pollutants from the soil
3. Carbon sequestration through carbon dioxide absorption during photosynthesis
4. Reduction of air pollution through absorption of air pollutants
5. Reduction of noise pollution

SELECTION OF APPROPRIATE SPECIES

It is important to select the most suitable tree species since it is not easy to replace them once they have been planted. The success of growing MPTs mainly depends on the choice of the species in an agroforestry system. The following factors should be kept in mind while selecting tree species.

Environmental adaptation: A multipurpose tree must be able to adapt to the climate, soil, topography, and plant and animal life of the area. This is especially important for exotic species.

Needs of farmers: The selected species should meet the needs of farm families. Farmers must be involved in selecting species. The cost of acquiring planting materials should also be kept in mind.

Ease of maintenance: Some species require more attention and management than others. Farmers should consider beforehand how much time they can spare for the trees. If they require additional skills and knowledge to grow a particular species, training or demonstration programmes should be organized.

Availability of genetic materials: Seeds or seedlings of the species being considered must be easily obtained. If vegetative propagation is required, farmers should receive training on how to do this.

MATCHING SPECIES TO REQUIREMENTS

Site characteristics: Although multipurpose trees are supposed to provide multiple benefits, they may not always yield advantages in all places. It is crucial to take geography, climate, and soil conditions into account while choosing a species. The farmer might choose to cultivate native or alien plants after determining the planting goal and knowing about the soil, rainfall, and other site factors.

Indigenous versus exotic species: Trees that thrive in adjacent plantations or in wild stands close by are excellent selections. It is possible that some of these species are native to the area and grow and breed there naturally. Some could be exotics; they were brought in from abroad. For instance, *Eucalyptus* species were brought from Australia to Asia. It is to ascertain whether the planting site's soil, rainfall, and other needs align with those of the foreign species before introducing it. Local exotic species are those that have adapted to the agro-ecological circumstances of the area and may be found in parks, farms, botanic gardens, and roadside vegetation.

Environmental matching: Environmental matching is the method most commonly employed to choose an alien species. According to the approach, the species must be studied in its natural habitat or its growth must be monitored in other nations where the environmental circumstances are comparable to those in the native country. Certain species can thrive at locations distant from their birthplace as long as the new location has similar environmental conditions.

Potential competition with other crops: Unwanted negative effects may arise from introducing an alien species. One potential issue is that the species may develop so used to its new surroundings that it may outcompete key crops for farmers' livelihoods and turn into a weed. *Prosopis juliflora*, for example, is a native of Central America and the West Indies that was brought to India as a fuelwood and fodder tree.

CHARACTERISTICS OF MPTS SUITABLE FOR AGROFORESTRY

In the agroforestry context, MPTs are those trees and shrubs which are deliberately grown and managed for more than one preferred product and/ or service. The cultivation of these trees is usually economically but also sometimes ecologically motivated in a multiple output land use system. The MPTs are the most distinctive component of agroforestry and the success of agroforestry as a viable land use option depends on exploitable potential of these multipurpose trees. MPTs for agroforestry systems should have the following characteristics.

1. They should be adaptable to local climatic conditions.
2. They should possess self-pruning properties.
3. If not self-pruning, they should be able to tolerate relatively high incidence of pruning. Their photosynthetic efficiency should not significantly decrease with heavy pruning.

4. They should be light-branching in habit.
5. They should be tolerant of side-shade in the early stages of growth.
6. Their canopy should permit the penetration of light to the ground.
7. Their phenology, particularly with respect to leaf flushing and leaf-fall, should be advantageous to the growth of the annual crop in conjunction with which they are being raised.
8. Their rate of litter fall and litter decomposition should have positive effects upon the soil.
9. Their 'above ground' changes over time in structure and morphology should be such that they retain or improve those characteristics which reduce competition for solar energy, nutrients and water.
10. Their root systems and root growth characteristics should ideally result in the exploration of soil layers that are different to those being tapped by the agricultural species.

ROLE OF MPTS IN AGROFORESTRY

Alley cropping

MPTs planted as hedgerows between rows of agricultural crops reduce soil erosion. When planted on slopes, alley crops slow down runoff rainwater and trap sediment, which can form natural terraces after several years. Farmers should prune hedgerows regularly to prevent them from competing with the crops for sunlight and water. Leaves from the pruned branches are used as animal fodder and branches are used as fuel. The tops of the hedgerows are also pruned every 6-8 weeks to obtain green leaf manure. Hedgerows may be planted in single lines or double lines depending on how much land is available and the slope of the land. Desirable characteristics of species for alley cropping are as follows.

- Easy to establish with minimum labour for planting and maintenance
- Fast growing so that benefits become available to the farm family as soon as possible
- Good sprouting habit after pruning
- Ability to fix nitrogen
- Ability to provide palatable and acceptable fodder with high nutritive value
- Ability to produce heavy foliage for green manure
- Deep root system to draw nutrients and water from lower soil layers
- Easy to propagate

Homegardens and other multistorey systems

Mixed plantings of annual agricultural and perennial tree crops around dwellings, often termed as homegardens, are a common type of multistorey agroforestry system. In homegardens, the lowest level often consists of vegetables or root crops; the second level includes fast growing trees or crops such as bananas and papaya; a third higher level may consist of large trees that provide fruit, timber and shade. Homegardens also provide a pleasant shaded living area. Many farmers grow multipurpose trees in their homegardens for flowers and fruits. Trees grown mainly for food should not be pruned regularly for fodder or fuel. Pruning can interfere with flowering and fruiting. Desirable characteristics of species for homegardens and other multistorey systems are as follows.

- Understorey crops require the right amount of light for optimum growth. Thus, the crown of trees should be high, small and open having sparse foliage.
- They should have deep root system to draw nutrients and water from deeper soil layers so that they would not compete with shallow rooted crops.
- Roots should not spread laterally too far from the trunk to minimize competition with nearby crops.
- They should fix nitrogen to grow well under adverse conditions and help improve soil fertility.

Fuelwood

Fuelwood tree species should be included in agroforestry systems to reduce pressure on natural forests and save forests from denudation. Species for fuelwood should have the following characteristics.

- Adapt quickly and yield a maximum volume of wood in a short time
- Stabilize the soil and maintain its fertility
- Require minimum time and effort for management
- Tolerate diseases and pests
- Tolerate drought and other climatic stresses
- Ability to good coppicing
- Have other uses that contribute to the farm enterprise
- Produce stems without thorns with branches having a diameter small enough to be cut with hand tools and be easily transported
- Produce wood that splits easily

- Have low moisture content or dry relatively quickly
- Produce minimal and non-toxic smoke (low ash and sulphur content) when burnt
- Not spark while burning, for safety reasons
- Produce dense wood, to burn longer

Species often used for fuelwood include *Casuarina equisetifolia*, *Acacia auriculiformis*, *Gliricidia sepium*, *Calliandra calothyrsus* and *Leucaena leucocephala.*

Fodder

Agroforestry practices that combine tree plantations with livestock are found generally in arid and semiarid regions, where natural grasslands and farm sizes are larger. In such systems, farmers allow their animals to graze on forage grasses in stands of trees. In humid and subhumid regions, which have smaller land holdings, alley cropping and other 'cut and carry' methods for procuring fodder are practised. Integrating tree growing with livestock production increases the production of meat, milk and other animal products without sacrificing agricultural land, reduces surface soil erosion, reduces reliance on inorganic fertilizer by using animal manure and provides additional income through sale of livestock. Desirable characteristics of species for fodder production are as follows.

- Produce leaves and/or pods that animals like to eat
- Improve fodder quality through high protein and nutrient content
- Withstand lopping, pruning, pollarding and coppicing
- Grow quickly, especially in early stages
- Adapt easily to different sites and environments
- Withstand intense shade if planted with other species
- Free from accumulation of toxic substances in leaves
- Fix nitrogen to enhance soil fertility
- Withstand damage by pests, diseases and browsing animals
- Have other uses

Some species often used for fodder include *Acacia tortillis*, *Faidherbia albida*, *Albizia lebbek*, *Leucaena leucocephala*, *Acacia nilotica*, *Calliandra calothyrsus*, *Artocarpus heterophyllus*, *Gliricidia sepium*, *Prosopis cineraria*, *Dalbergia sissoo* and *Sesbania grandiflora.*

Roundwood (poles and posts)

Eucalyptus spp., *Acacia auriculiformis*, *Acacia mangium*, *Albizia lebbek* and *Casuarina equisetifolia* are the important species often planted for roundwood to be used as poles and posts. Species grown for this purpose should have the following characteristics.

- Have a straight stem
- Have a single main stem rather than multiple stems
- Produce few and thin branches
- Be able for self-pruning
- Produce wood without knots
- Have little taper from bottom to top
- Have bark that strips easily
- Produce wood that is durable, light and capable of supporting heavy cross loads
- Resist termites and other wood borers
- Be able to absorb chemical preservatives easily

Live fences

In many places, farmers plant multipurpose trees in rows along farm boundaries as 'living fences'. In addition to providing fodder and fuelwood, living fences provide privacy and protection from browsing animals. Species grown for this purpose should have the following characteristics.

- Grow well under adverse conditions
- Grow well in close spacing
- Require minimal maintenance
- Withstand lopping and trimming
- Ensures fast establishment through vegetative propagation while reducing the chance of spreading to cultivated areas
- Grow fast to a medium height
- Be long lived
- Have prickles or stiff branches and leaves that animals do not like to eat
- Should not have adverse effects on other tree species or crops they are associated with

Species often used for live fences are *Cajanus cajan*, *Erythrina poeppigiana*, *Calliandra calothyrsus*, *Gliricidia sepium*, *Casuarina equisetifolia* and *Pithecellobium dulce*.

Windbreaks and shelterbelts

Windbreaks are strips of trees planted closely together along the edges of croplands perpendicular to prevailing winds. Especially in dry areas, windbreaks can provide protection to crops and soils from the detrimental effects of wind. Species grown for windbreaks and shelterbelts should have the following characteristics.

- Tolerate harsh environments
- Withstand strong winds
- Have a bushy, deep crown that still allows some wind penetration
- Keep lower limbs for a long time
- Have strong roots
- Grow quickly
- Live long
- Tolerate pests and diseases
- Should not have roots that compete excessively with nearby crops for water and nutrients

Species often used for windbreaks and shelterbelts are *Acacia nilotica, Casuarina equisetifolia, Azadirachta indica* and *Erythrina peoppigiana.*

Soil protection and rehabilitation

The important species often planted for soil protection and rehabilitation are *Acacia auriculiformis*, *Casuarina equisetifolia* and *Prosopis juliflora.* Species for this purpose should have the following characteristics.

- Have faster growth rate
- Grow well under adverse conditions
- Have a spreading crown
- Have a vigorous root system to bind the soil and a strong taproot especially in the areas prone to landslides
- Associate well with nitrogen fixing organisms and mycorrhizal fungi in the soil
- Reproduce vigorously, for example, from root suckers or through abundant natural seed fall and seedling development
- Tolerate fire

Industrial uses

Species grown for sawn timber, plywood, veneer, pulpwood or other industrial uses should have the following characteristics.

- Have faster growth rate
- Have a maximum rate of growth in early stages of development
- Have a straight stem, uniform size and small branches
- Show good natural pruning with rapid healing of wounds if used for producing sawn timber and plywood
- Have a good physical, mechanical seasoning, preserving and processing properties if used for producing sawn timber
- Have wood that is easy to peel, slice and free of knots if used for producing plywood and veneer
- Have wood with a high density and long fibres if used for producing pulp

The species suitable for industrial uses are *Pinus* spp., *Tectona grandis*, *Populus* spp. and *Terminalia* spp. for sawn timber; *Dipterocarpus* spp. and *Tectona grandis* for plywood and veneer; *Acacia* spp., *Leucaena* spp., *Pinus* spp. and *Eucalyptus* spp. for pulpwood.

Shade and nurse trees

Species often used as shade or nurse trees are *Erythrina poeppigiana*, *Gliricidia sepium*, *Leucaena leucocephala* and *Sesbania grandiflora*. Species grown for this purposes should have the following characteristics.

- Have fast early growth
- Establish easily, preferably by cuttings
- Be evergreen
- Live long
- Tolerate soil compaction by animals walking and sitting beneath them, as happens on pasture land
- Coppice well and withstand some lopping
- Fix nitrogen in the soil to enhance soil fertility
- Have a dense and spreading crown, if used to shade grazing animals
- Have a light crown, if used as nurse trees for other crops
- Should not compete with crops for soil nutrients if used as nurse trees

Nitrogen fixing trees

Nitrogen fixing trees (NFTs) are valuable in subtropical and tropical agroforestry. They can be integrated into an agroforestry system to restore nutrient cycling and fertility self-reliance. NFTs are often deep rooted, which allows them to draw nutrients from subsoil layers. Constant leaf drop from them nourishes soil life, which in turn, can support more plant life. The extensive root system stabilizes soil and adds organic matter to the soil. There are many species of NFTs that can also provide numerous useful products and functions, including food, wind protection, shade, animal fodder, fuelwood, living fence and timber, in addition to providing nitrogen to the system. Many fast-growing NFTs can be cut back regularly for mulch production. The NFTs may be integrated into an agroforestry system in many different ways including clump plantings, alley cropping, contour hedgerows or shelterbelts. They can serve many protective functions such as microclimate amelioration for shade loving crops like coffee; trellis for vine crops like vanilla, pepper and yam; and living fence. The major nitrogen fixing tree genera adopted in agroforestry systems are *Acacia*, *Albizia*, *Alnus*, *Casuarina*, *Dalbergia*, *Erithrina*, *Gliricidia*, *Leucaena*, *Prosopis*, *Sesbania*, etc. NFTs should have the following characteristics.

- Ability to fix nitrogen at higher rates
- Have faster growth rate
- Ability for coppicing
- Ability to survive in edaphic and climatic stress situations
- Produce more seeds and easy to propagate

Other uses of MPTs

Trees can help reduce soil erosion along streams and gullies. Their roots serve to hold the soil in place and reduce the impact of storm water. Some tree species suitable for stabilizing stream banks and gullies are *Paraserianthes falcataria*, *Gmelina arborea*, *Leucaena leucocephala*, *Sesbania grandiflora* and *Moringa oleifera*.

Some multipurpose tree species can also be used for decoration. These should have attractive flowers and fruits that can be eaten by birds and small animals. The species include *Acacia auriculiformis*, *Casuarina equisetifolia*, *Pterocarpus indicus*, *Albizia saman* and *Tamarindus indica*.

Trees can also be planted to produce other commodities such as *Eucalyptus* spp. for essential oils and honey, *Acacia* spp. for tannin for making dyes, *Azadirachta indica* for bio-pesticides and *Hevea brasiliensis* for latex.

SUITABLE MPTS FOR AGROFORESTRY UNDER DIFFERENT CLIMATIC CONDITIONS

Hot desert: *Acacia tortilis*, *Capparis* spp., *Prosopis chinensis*, *Prosopis cineraria*, *Tecomella undulate*

Cold desert: *Populus alba, Populus cilita, Populus euphretica, Populus nigra, Populus tremula, Salix alba, Salix fragilis*

Tropical semi-arid: *Acacia nilotica, Acacia senegal, Acacia tortilis, Albizia lebbeck, Azadirachta indica, Eucalyptus camaldulensis, Prosopis, Salvadora persica, Tamarix* spp.

Subtropical semi-arid: *Acacia modesta, Albizia procera, Bauhinia variegata, Ficus* spp., *Morus indica, Pinus roxburghii*

Temperate semi-arid: *Corylus colurna, Juniperus macropoda, Pinus gerardiana*

Humid tropical: *Areca catechu, Artocarpus heterophyllus, Chukrasia tabularis, Cocus nucifera, Gmelina arborea, Pterocarpus santalinus, Schima wallichi, Tectona grandis, Terminalia alata, Terminalia myriocarpa*

Humid subtropical: *Acer oblungum, Acrocarpus fraxinifolius, Aesculus indica, Eucalyptus globulus, Prunus* spp., *Quercus* spp.

Humid temperate: *Acer campbelii, Alnus nitida, Pinus alata, Pinus wallichiana, Populus ciliata, Robinia pseudacacia*

Tropical sub-humid: *Adina cardifolia, Bombax ceiba, Casuarina equisetifolia, Dalbergia latifolia, Dalbergia sissoo, Eucalyptus citridora, Eucalyptus teriticornis, Leucaena leucocephala, Moringa oleifera, Morus alba, Populus deltoides*

Subtropical semi-humid: *Albizia chinensis, Celtis australis, Eucalyptus globules, Eucalyptus grandis, Grewia optiva, Morus indica, Pinus roxburghii, Toona ciliata*

Temperate semi-humid: *Acacia mearnsii, Cedrus deodara, Celtis australis, Juglans regia, Quercus* spp.

SUITABLE MPTS FOR AGROFORESTRY UNDER DIFFERENT SOIL TYPES

Desert soil: *Acacia senegal, Acacia tortilis, Acacia nilotica, Prosopis chilensis, Prosopis cineraria*

Recent alluvium: *Acacia catechu, Bombax ceiba, Dalbergia sissoo*

Saline-alkali soils: *Acacia nilotica, Ailanthus* spp., *Azadirachta indica, Eucalyptus* spp., *Pongamia pinnata, Prosopis* spp., *Tamarix* spp.

Coastal and deltaic alluvium: *Areca catechu, Casuarina equisetifolia, Cocus nucifera*

Red soils: *Acacia nilotica, Azadirachta indica, Dalbergia sissoo, Dendrocalamus strictus, Eucalyptus hybrid, Leucaena leucocephala, Madhuca indica, Mangifera indica, Pterocarpus marsupium, Tectona grandis*

Black cotton soils: *Acacia nilotica, Adina cardifolia, Aegle marmelos, Bauhinia* spp., *Dalbergia latifolia, Hardwickia binnata, Tamarandius indica, Tectona grandis*

Laterite and lateritic soils: *Acacia auriculiformis, Azadirachta indica, Emblica officinalis, Eucalyptus* spp., *Tamarindus indica, Tectona grandis*

Peaty and organic soil: *Bischofia javanica, Ficus glomerata, Glircidia sepium, Syzygium cumini*

SUITABLE MPTS FOR AGROFORESTRY FOR DIFFERENT REGIONS

Jammu & Kashmir: Almond/apple in rows with saffron/vegetables; walnut with maize and beans; *Robinia pseudoacacia/ Celtis/ Ailanthus/ Morus/ Salix/* poplar with wheat; *Leucaena/* mulberry/ poplar with paddy-wheat system; *Acacia nilotica, Eucalyptus* with pulses-maize system

Himachal Pradesh and Uttarakhand: Mixed plantations of fodder, fuel, timber and forest trees with vegetables under irrigation, and with ragi-wheat system in drylands; *Kydia calycina/ Morus* with paddy-wheat on terraces; *Sapiuin insigne* with paddy-wheat; *Boehmeria rugulosa, Grewia optiva, Terminalia, Celtis caucasica* and *Parkia roxburghii* with paddy-wheat or maize-wheat on terraces

North-Eastern states: Arecanut/ citrus with black pepper, banana and papaya; Arecanut with pineapple and betel vine; *Grevillea robusta* with paddy; bamboo, *Parkia roxburghii, Mangifera indica, Syzygium cumini, Emblica officinalis* and *Morus alba* with vegetables; *Dipterocarpus macrocarpus* with coffee, cocoa; *Terminalia microcarpa, Shorea robusta/ Grevillea robusta/ Alnus nepalensis* with tea, cardamom, turmeric and ginger; *Indigofera tasmianii* with *Albizia lebbeck, Albizia procera, Albizia odoratissima, Albizia stipulata, Albizia mollucana*; *Albizia sumatrama* with tea; *Terminalia mircrocarpa, Bischoffiaja vanica* and *Alnus nepalensis* with cardamom

Tarai region: *Eucalyptus* with soybean/maize-wheat; Poplar with soybean/ maize-wheat/lentil; sugarcane with poplar up to three years followed by paddy-potato; *Leucaena/ Sesbania/* Poplar with rapeseed; *Dalbergia sissoo* and *Acacia nilotica* with sugarcane for three years followed by wheat

Alluvial plains: *Acacia nilotica/ Tamarix articulate/Agave sisalana/ Cajanus cajan* on bunds with paddy-wheat or maize-wheat; *Leucaena/ Prosopis cineraria/ Azadirachta indica/ Albizia lebbeck/ Toona ciliata* in mixture with crops; *Leucaena leucocephala, Shorea robusta, Tectona grandis* with ginger/ turmeric

IMPROVING MPTS FOR AGROFORESTRY SYSTEMS

The existing goals of tree improvement programmes are mainly oriented towards the increased wood production under forest plantation. However, the goals of improving MPTs for agroforestry systems should be broader and more diversified. Owino (1992) set two goals for improving multipurpose tree species for agroforestry systems.

1. To improve (through breeding) high priority species for agroforestry technologies taking into account farmers' needs for tree products and services, compatibility with companion crops, and sustainability and enhancement of land productivity capability.
2. To improve tree species which should be able to produce the maximum amount of tree product and/or services in the shortest possible time thus ensuring maximum economic returns.

Three primary considerations should guide the choice of the set of species for research concentration in any given situation.

1. Well-perceived potential of the species to solve identified land use constraints (e.g., fodder shortage, fuelwood shortage, soil fertility, etc., and combinations thereof).
2. Site adaptation and comparative growth performance.
3. Farmers' preference and/or adoption potential.

Criteria for selecting high priority tree species for agroforestry research and development may be summarized under three broad categories (Owino, 1992).

Adaptability considerations

1. Adaptability and good growth performance in broad agroecological zones
2. Adaptation to specific soil conditions (e.g., acidic soil)
3. Adaptation to constrained soil water regimes (e.g., drought resistance in arid lands)

Agroforestry technology considerations

1. Compatibility with alternate crops (both below and above the ground)
2. Special products and services (e.g., fodder, nitrogen fixation)
3. Tree management requirements in specific agroforestry technologies
4. Freedom from pests and diseases (including those of alternate crops)

Species-specific potential gains from improvement research

1. Existing genetic variation as basis for further improvement
2. Potential gains through improved silvicultural practices
3. Economic justification on long-term improvement programmes

Based upon these considerations a wide range of multipurpose tree species that could play a role in agroforestry technologies can be identified. Selection and breeding research should be concentrated on best provenances of three to five best species for each broad agroecological region.

5

Social Forestry

The idea of social forestry is not new in India. It may be found in Lord Buddha's teachings from around 2500 years ago. According to Buddha's teachings, a devout Buddhist should plant one tree and nurture it for five years so that it matures into a complete tree. By doing this, a person should plant around five trees over his lifetime. It is also believed that the Great Emperor Ashoka planted fruit trees and shade trees by the roadsides for the benefit of passing travellers.

The rural economy in India has been largely supported by forests. In the past, the local people kept an ecological balance while managing their woods to provide food, fuel, fibre, fodder, timber and herbal remedies. However, India's forests have rapidly disappeared due to biotic pressure from expanding human and animal populations, a lack of technical skills, insufficient investments, and ownership changes. In addition to biotic pressure, other factors that led to the depletion of forest resources includes inadequate scientific and technical inputs; staff lacking the necessary skills and training to assume their expected new roles; poor investment in forest development; and damage from mining, irrigation projects, industries, roads; and shifting cultivation. As a result, in the early 1970s, new ideas in forestry development were introduced. Therefore, a variety of activities were incorporated in social forestry initiatives, including planting trees on agricultural bunds and roadsides, creating woodlots on shared lands, and integrating local people in the gathering, processing, and management of forest products.

DEFINITION OF SOCIAL FORESTRY

The word 'social forestry' was coined by Jack Westoby and used in the ninth Commonwealth Forestry Conference at New Delhi in 1968.The phrase was not used or recognised until 1973, when the National Commission on Agriculture brought it back into the spotlight with its interim report on 'production forestry and manmade forestry'. As a result, social forestry, also known as 'forestry of the people, by the people, and for the people' documents a significant global movement. In order to reduce the pressure on the natural forests, the National Commission on Agriculture in India popularised the idea of social forestry to supply firewood, small timber and minor forest products to rural population. This involved encouraging people who relied on fuelwood, fodder, and other forest products to satisfy their own requirements through a variety of activities. The FAO expanded on this idea in 1978 when it defined community forestry as a scheme that actively involves locals in afforestation regardless of their pattern of land ownership.

Westoby (1968) defined social forestry as a forestry which aims at producing flow of protection and recreation benefits for the community. Prasad (1985) defined social forestry as forestry outside the conventional forests which primarily aims at providing continuous flow of goods and services for the benefit of people. This implies that the production of forest goods for fulfilling the needs of the local people is social forestry. Thus, conceptually it deals with the people to produce goods such as fuel, fodder, small timber, etc. to meet the needs of local community particularly the underprivileged section (Shah, 1988).Thus, social forestry is a concept that involves working with the impoverished to generate items that meet the needs of the local community, especially the disadvantaged.

In order to provide a wide range of goods and services to both individuals and society as a whole, it is the science and art of growing trees and/or other vegetation on all available land in and outside of forest areas, as well as managing the existing forest with close community involvement and integration into other operations. This results in balanced and complementary land use.

In general, social forestry refers to the self-help maintenance and development of trees and perennials with the purpose of improving the livelihoods of locals, particularly the impoverished, in a sustainable manner. Participatory approaches are used in this process. The term 'social' refers to the social configuration of social forestry, which is dynamic, adaptive, and sensitive to the social environment and context. People are involved individually or collectively in forestry activities for the establishment, management, maintenance, protection, and use of forests for their mutual benefit. These terms, 'social forestry', 'community forestry', and 'people's forestry' are synonymous. It is a land transformation programme designed to maximise land uses and restore the environment. The community actively participates as the main gainer, harvesting secondary benefits other than traditional industrial raw materials and ensuring improved land, landscape, environment, and overall quality of life in the process.

CHARACTERISTICS OF SOCIAL FORESTRY

1. Social forestry is the forestry for the people and by the people. Here, people are the direct beneficiaries and it cannot develop without their full participation.
2. It is the forestry on a small scale. Social forestry is developed by an individual or a household or a group of households or a community. Here, the productivity inputs are limited and if the resources are made available either through Government or community then it can be increased. Social forestry activities are often divided into two classes, community forestry and farm forestry depending on the benefits of activity diverted towards community or part of the community or individuals.
3. The multipurpose tree species selected for plantation in the social forestry are fast growing, early maturing with both productive and protective benefits.
4. The social forestry may range from monocropping multiple use goals (i.e., fast growing tree species for fuelwood, poles, small timber, fodder, etc.) to integrated cropping goals like agroforestry for getting wood, food, fodder and green manure, etc.

OBJECTIVES OF SOCIAL FORESTRY

The National Commission on Agriculture (1976) established the goals of social forestry, which were centred on the community's economic requirements and intended to enhance living circumstances.

1. To increase tree cover outside the forest areas
2. To encourage participation of people and institutions in plantation related activities
3. To increase the production of forest produce (small timber, firewood, fodder, fruits and other non-timber forest products) to meet the daily needs of the local population and thereby to reduce pressure on natural forests for supply of forest produce
4. To suitably use low productive and non-productive lands through tree plantation
5. To rehabilitate the degraded forests so as to optimize their productivity and restore their potential to provide ecosystem goods and services on sustainable basis
6. To generate marketable surplus of forest products to yield cash incomes and improve the consumption level of village people

7. To produce raw materials for village level cottage industries
8. To increase crop yields through appropriate agroforestry models
9. To create livelihood opportunities to economically backward sections of rural communities by collection of dry leaves, fuelwood, dropped down seeds, fruits, etc.
10. To improve carbon stock in the tree cover outside forest areas
11. To increase employment opportunities for rural poor throughout the year by collection of seeds, preparation of nursery beds and raising seedlings, etc.
12. To increase the natural beauty of the landscape and to create recreational forests for the benefit of rural and urban population
13. To improve the environment for protecting agriculture from adverse climatic factors
14. To conserve and improve ecology and environment in the region

SCOPE OF SOCIAL FORESTRY

Scope of social forestry is very wide in India. Various kinds of lands which are available or can be made available for social forestry plantation include wastelands, roadsides, canal banks, ravine lands, marginal and sub marginal agricultural lands, degraded forest lands, urban unutilized areas, industrial complexes, backyards, etc.

Wastelands: Approximately 63.85 million hectares of land is lying as wastelands in India. Out of these lands, about 50% are non-forest lands, which can be made fertile again if treated properly. At least one-third of such non-forest wastelands can conveniently be available for social forestry plantation.

Roadsides, canal banks and rail side lands: A large area of land is available for social forestry plantation along the roadside, canal bank and rail side.

Ravine lands: There are about 10 million hectares of ravine land, mainly confined to the states of Uttar Pradesh, Madhya Pradesh, Rajasthan and Gujarat. These lands can be managed through integrated approach of soil conservation and afforestation. The best use of ravine lands is social forestry.

Salt affected areas: Salt affected degraded lands can be reclaimed through social forestry plantation with suitable multipurpose tree species.

Marginal and sub marginal agricultural lands: These are the agricultural lands which are not physically suitable for growing of agricultural crops because of the low inherent fertility status of the soil. Mostly, these lands are owned by the economically poor people who are usually unable to use high input production system. Therefore, social forestry has a great potential in upgrading the fertility of such lands and uplifting the economic level of land owners.

Other agricultural lands: Arid and semiarid areas are frequently prone to moderate to severe droughts. Tree plantation or forestry in these areas has a great role in protecting agriculture, improving productivity, stabilizing production and ameliorating the environment.

Degraded forests: About 40 million hectares forest areas are degraded due to heavy pressure on them for fuelwood and overgrazing. These degraded forests can be revived through social forestry.

Urban areas: There is a scope for tree plantation in unutilized lands in urban areas such as compounds of public buildings, public parks, roadsides and even the private gardens.

Industrial complexes: In industrial complexes there is scope for tree planting. Normally, at the time of establishment of industries much more land is acquired than what is actually needed immediately. Thus, most of the industrial complexes have some land which is not put to any specific use. These unutilized lands can be put under social forestry plantation.

Backyard planting: Generally, in rural areas a majority of households grow trees in their backyards and still they have some unutilized space which can be brought under multipurpose tree plantation.

Temple groves: The temple compounds and premises of other places of worship can be utilized for tree plantation. Traditionally and culturally, temple groves are the best protected in our country.

School premises: The spare lands in school premises can be utilized for tree plantation. Creation of school forests has tremendous educational value for the students.

Graveyards: Graveyards or cremation grounds usually have some land which can be utilized for growing trees.

COMPONENTS OF SOCIAL FORESTRY

Social forestry can be categorized into four groups; farm forestry, extension forestry, reforestation of degraded forests and recreation forestry.

Farm forestry

Farm forestry is the deliberate attempt of farmers towards planting or keeping trees in their farmlands in order to supplement daily farm/household needs, meet the contingencies and take market opportunities created by the dwindling supply of tree/forest products from the common lands. There are reports on trees being increasingly used as contingencies and also contributing to the household food security. Farm forestry was defined by National Commission on Agriculture (1976) as the practice of forestry in all its aspects in and around the farms or village lands integrated with other farm operations. The Commission noted that any programme of planting of trees in farms should serve the following objectives.

1. To supplement production of fuelwood and small timber to meet the increasing requirements of local people
2. To release cow dung for use as manure
3. To increase production of leaf fodder
4. To create a diverse ecosystem by having trees interspersed with cultivation

At present both commercial and non-commercial farm forestry is being promoted in one form or the other. Individual farmers are being encouraged to plant trees on their own farmland to meet the domestic needs of the family. In many areas the tradition of growing trees on the farmland already exists. Non-commercial farm forestry is the main thrust of most of the social forestry projects in the country today. It is not always necessary that the farmer grows trees for fuelwood, but very often they are interested in growing trees without any economic motive. They may want it to provide shade for the agricultural crops; as wind shelters; soil conservation or to utilize wasteland.

Extension forestry

Extension forestry is the practice of forestry in areas devoid of tree growth and other vegetation situated in places away from the conventional forest areas with the objective of increasing the area under tree growth. Thus, planting of trees on the sides of roads, canals and railways, along with planting on non-forest wastelands to increase the boundaries of forest is known as extension forestry. Under this project there has been creation of woodlots in the village common lands, government wastelands and panchayat lands. This can further be classified as

Mixed forestry: It is the practice of forestry for raising fodder grass with scattered fodder trees, fruit trees and fuelwood trees on suitable wastelands, panchayat lands and village common lands. Mixed forestry in the village wastelands and panchayat lands should be acceptable to people. Only quick growing and those products of immediate concern such as fodder and grass should be taken up with optimum input and technology.

Shelterbelts: Shelterbelt is defined as a belt of trees and/or shrubs maintained for the purpose of shelter from wind, sun, snow drift, etc.

Linear strip plantations: These are the plantations of fast-growing tree species on linear strips of land on the sides of public roads, canals and railway lines. The National Commission on Agriculture recommended the plantations of fast-growing tree species on the sides of roads, canals and railway lines. Various states have taken up the linear strip plantation recommended by the NCA. Such plantations have the capability to generate wood and other products whose utilization may help the local people to meet some of their basic requirements.

Community forestry

Community forestry is a set of institutional arrangements in which communities are involved wholly or in part in decision-making and benefits and contribute knowledge and labour to achieve healthy forests and social well-being (Danks and Fortmann, 2004). Thus, it refers to forest and tree management activities undertaken either individually or cooperatively by the local people, either on their own or on leased private lands, on communal lands or on state lands (Wiersum, 2004). It involves the process of making and implementing decisions with regard to the use and conservation of forest resources within a local community, with the organization of the activities being based on shared norms and the interests of the people living in that local community.

According to Wiersum (2004) a community may be either a locality in the sense of a human settlement with a fixed and bounded local territory, a local social system involving interrelationships among people living in the same geographic area, or a type of relationship characterized by a sense of shared identity. Consequently, different community forestry arrangements are possible depending on the type of territory and the type of social relations being considered. In respect of such institutional arrangements, three main types of community forestry may be distinguished:

1. Management of any woody resources on lands which are located within a local territory, irrespective of whether these resources are privately, communally or state-owned
2. Management of common pool resources, such as communal forest or grazing lands, which are shared or held in common and jointly used by people who are formally or informally organized in a forest user group
3. Collaborative management of state forest lands under cooperative arrangements with a public forest administration

Rehabilitation of degraded forests

The degraded area under forests needs immediate attention for ecological restoration and for meeting the socioeconomic needs of the communities living in and around such areas. The NCA suggested reforestation of degraded forests to achieve the following objectives.

1. To grow short duration fuel and timber species for meeting the requirements of local people
2. To organize fuelwood supplies at reasonable rates which would prevent pilferage from neighbouring commercial forests
3. To tie up degraded forest areas with nearby rural and semi-urban entrées for their requirements of fuelwood
4. To provide employment
5. To rehabilitate the degraded forests

Recreation forestry

It is the practice of forestry with the objective of raising flowering trees and shrubs mainly to serve as recreation for the urban and rural population. This type of forestry is also known as'aesthetic forestry' which is defined as the practice of forestry with the objective of developing or maintaining a forest of high scenic value.

BENEFITS FROM SOCIAL FORESTRY

Improve food availability

By choosing the fruit and other food trees in the social forestry plantation programme food availability can be increased. This will provide extra food for tribal and other rural people at cheaper costs. This will lead to improved rural health, the development of fruit-based subsidiary industries, and assistance in resolving national food crises.

Fodder availability

Social forestry increases the supply of fodder by making it easier to obtain nutrient-rich tree fodder, increasing cattle wealth, improving human health by making animal-based food items more readily available, and improving soil health by adding more manure.

Expand the supply of fuelwood

Social forestry expand the supply of fuelwood by making it simpler to obtain fuel, making it a less expensive alternative to commercial fuel, substituting cow dung for manure, putting less pressure on government forests, and using renewable energy sources.

Support to small and cottage enterprises

Social forestry provides raw materials derived from trees to the cottage industries and cooperative societies established under the rural finance organisations.

Employment prospects in rural areas

Social forestry creates jobs, thus checks rural population migration. The plantation programme provides work options to landless people, engage in a variety of economic activities, provide the villagers more revenue. All these activities lead to increase in per-capita income in rural areas.

Ecological and environmental balance

Social forestry reduces the pressure on natural forests. This also manages land wisely; maximises the use of renewable resources; develops water resources; preserves soil and water; reduces air, water, and noise pollution; establishes an oxygen bank; improves bioaesthetic development and overall health.

STRATEGIES FOR IMPLEMENTATION OF SOCIAL FORESTRY

There must be an intimate interaction between the project personnel and villagers where the social forestry project work is to be undertaken. The relationship among the project personnel, villagers and other participants such as voluntary organisations, etc. must be regulated by the village forest rules and other pertinent legal documents related to forest management and product utilization. There must be a village approach ,i.e., villagers should participate and the resource generated should belong to the village and that should be shared, regenerated and expanded by everybody of his own. The extension and training activities of the village forest workers must be closely monitored as this is the most important part of the strategy.

Pre-requisites for implementation of social forestry

There are four important pre-requisites for implementation of the social forestry project - people's participation, physical and legal provision, decentralization and cooperation.

Peoples' participation: The participation of people is required because in a long run the afforestation would be undertaken by the village communities or individuals without any financial support from the Government agencies. Besides, it is the only viable option for ensuring protection of plantations. So without people's participation the social forestry project would not be successful.

Physical and legal provision: The project work cannot be undertaken if land is not made available and legal questions related to production, management, harvesting, etc. are not resolved. So before implementing the project these problems must be handled carefully.

Decentralization: The social forestry programme must be based on decentralized activities. The social forestry supervisors and village forest workers must undertake their work at village level. Planning, training and motivating must be a village activity. Besides, the employment and support for activities related to the development and use of forest produce would take place at the village level.

Cooperation: The project must have to establish a close cooperation with other concerned departments like revenue, environment, community development, tribal welfare, soil conservation, horticulture and agriculture. Through cooperation many problems related to procurement of land, contact with other agencies, implementing rural development projects, getting finance from financial institutions can be solved easily.

Besides, while formulating the objectives and strategies for achieving the overall aims of social forestry, the following points should also be taken into consideration.

1. The required land for plantation must have easy access to the people.
2. The very fast-growing tree species must be put into rotation.
3. Specific training and extension are required for success of the plantation programme.
4. Socioeconomic diversity and ecological conditions must be considered while implementing the project.
5. Specific preference for certain tree species by the local people must not be ignored.
6. If livestock farming is one of the major occupations of the local people, fodder tree species should be included in the plantation.
7. The afforestation in social forestry, though a community or individual activity, it needs a supporting organisation which can give technical guidance.
8. Periodical evaluation of the social forestry projects is essential at different stages of implementation so that constraints would be identified and solved before being complicated.

STEPS FOR ESTABLISHMENT OF A SOCIAL FORESTRY PLANTATION

Preparation of village resource plan

The village resource plan indicates whether the social forestry programme would be successful in a particular village. It includes some general information about the communication facility, human and cattle population, nearest township, etc. Besides, the information on physical resources like availability of land resources and their type, location and productivity is equally important for preparation of resource plan for success of the plantation programme. Biological resources include the information on the tree species being grown earlier by the villagers and if they have some specific demand for certain species, and availability of quality planting materials in adequate quantity. The resource plan should also show the availability of water for nursery and sometimes for plantation activity. But the most important single factor for success of any plantation programme is the human resource. It includes the availability of skilled and unskilled workers, the presence of different organizations and leadership pattern in the village.

Selection of villages

Basing on the village resource plan the extension worker prepares a list of the villages according to their need and scope to undertake the social forestry programme. From that list the villages are selected.

Formation of village forest committee

The village forest committee (VFC) is an elected body collaborating with the project officials at the village level whereas the joint management plan (JMP) is the document through which the relationship between the project and the village is manifested and regulated. The village forest worker (VFW), the social forestry supervisor (SFS) and the deputy director of social forestry are responsible for constituting the VFC and compiling the JMP with active participation and cooperation of the members of VFC.

The formation of VFC must take place several months before any practical work in the village is started. First of all, the VFW and SFS must call for a meeting of the villagers. The meeting should be attended by the male and female members of all groups and representatives from all households. Here, the SFS explains the project activities and its benefits to the villagers. He should also discuss about responsibility of villagers and their role in management and protection of village forest. The village forest rules in detail should be explained to the villagers. Before the next meeting and before the election of VFC members, the VFW should visit the village, identify the aspiring candidates for the

VFC and discuss with them individually regarding their roles, duties and responsibilities as committee members. The VFW should intimate the date of meeting for election of the members of the VFC to the villagers. Then the SFS assisted by the VFW would conduct the election of the VFC members on the scheduled date in the presence of other revenue and forest officials. After formation of VFC, its members elect its chairman, secretary and treasurer, etc. The resolution of the meeting should be recorded in a village forest committee register and signed by all the members of the village present in the meeting.

Joint Management Plan is an agreement between the social forestry project and the village. This is a legal document and defines the legal rights of the community. The preparation of JMP is important for achieving the objectives of the project. The JMP consists of four parts. The first part is the summary of the existing resources in the village. The second part is connected with the selection of plantation site, component and species. This part also deals with the issue of cattle grazing since in many cases the land selected for plantation may be a common grazing land. The third part deals with the technical aspects of nursery and plantation establishment and the issues related to employment of the workers and their working conditions. The last and fourth part deals with the administrative, legal and management aspects of the project.

Selection of beneficiaries

The selection of beneficiaries must be done as per the information of the village resource plan. Beneficiary may be an organisation, an institution, an individual or a group of farmers.

Land assessment, identification and demarcation

The land assessment, identification and demarcation should be made very carefully in cooperation with the villagers and the village forest committee. The interest of the villagers, total land area available for plantation, previous use of the land and protection of the plantation, etc. should be taken into consideration at the time of land assessment.

Choice of plant species

1. The tree species must fulfill the objectives of plantation.
2. They should be adaptable to local climate, soil, topography, and plant and animal life. .
3. They should be easy to propagate.
4. The trees should be established with minimum labour for planting and maintenance.

5. The species should be fast growing so that benefits become available as soon as possible.
6. The trees must produce a variety of products such as fuelwood, small timber, fruit, etc.
7. They should have the ability to provide heavy and palatable foliage for green manure and acceptable fodder.
8. Species having deep root system should be preferred to draw nutrients and water from lower soil layers.
9. Usually, nitrogen fixing tree species are chosen to maintain the soil fertility.
10. The species should be tolerant to diseases, pests, drought and other climatic stresses.
11. The species which are immune to grazing or easy to protect from cattle should be selected.

Raising the planting stock

The raising of good quality planting stock of different species is the key factor for successful plantation activity. So, it is necessary to raise seedlings in a nursery to obtain true-to-type seedlings. However, before going ahead with nursery plans, a thorough economic analysis is required since a good nursery can be an expensive undertaking. If the requirement of seedlings is not too high, commercial seedlings are the best and cheapest source of planting material and these may be procured from a known and reliable source. A successful nursery operation depends on many factors like selection of site, planning and supervision, timely operations, etc. The persons involved in raising seedlings should be well trained in all aspects of nursery techniques.

The planting site must be properly demarcated and if required fencing should be done to protect young plants from stray grazing. The land should be weed free and the soil is loosened to improve the uptake of water. The usual spacing should be at 2 m × 2 m; sometimes high-density planting may be done at 2 m × 1 m. The size of pits should be 45 cm × 45 cm × 45 cm, however, in highly degraded soil it should be 60 cm × 60 cm × 60 cm. The best period for planting is after the first shower of monsoon rains. The young trees are provided with extra nutrients by mixing compost, manure or fertilizer in the topsoil before or during planting. Care of the young trees in the field will improve establishment and growth.

Management of the plantation

Once a plantation has been established it has to be managed carefully up to the point of time when it is ready for harvesting. A management strategy is a set of prescribed methods of manipulating the growth of trees so as to satisfy the objectives of the grower. These strategies must be developed with clear objectives appropriate to the existing physical and biological conditions, resources and technologies available.

ROADSIDE PLANTATION

The main objectives of roadside or avenue plantation include provision of comfort to the travellers, aesthetics and landscape improvement, stabilization of roadsides, improvement of ecological conditions and maximization of the productivity of the site to meet the requirements of the local people. Effective roadside plantations also check the movement of sand and dust to the road from adjoining areas and restrict soil erosion along stream banks or near the bridges. Each 30 m width of a tree plantation can absorb about 6 to 8 decibels of sound intensity. Thus, to reduce the noise generated by high-speed traffic on national highways to tolerable limits, about 20 to 30 m wide belts of trees and shrubs should be planted. Similarly, in cities 7 to 15 m wide belts of trees and shrubs may be required for this purpose. Evergreen trees are better for effective noise abatement.

The roadside plantation may be done in a balanced line, unbalanced continuous or discontinuous line, or in sporadic or parking system. The balanced line system gives a continuous green wall of uniform size trees whereas unbalanced line systems give alternate avenues of different species interspersed with ornamental trees, either continuous or discontinuous. The parking system of planting may be adopted at some selected spots to develop picnic spots or resting places for travellers. Single row planting is normally done along village and district roads while more than one row can be planted in highway roads. In case of multiple rows planting, the first row should be of shade and/ or ornamental trees and the remaining rows may be planted with fast growing trees species. The row spacing may be adjusted with respect to the position of telephone or electric lines or drains or other structures along the road. Trees should not be planted under telephone or electric lines. Plant to plant spacing should be of 12 to 14 m for shade trees having broad crown while for other trees it may be 5 to 10 m. Planting at about half the desired spacing is usually practised and when the crowns of trees begin to touch one another, less promising ones are thinned to leave the remaining trees at desired spacing. Too close spacing along hill roads and on the curves should be avoided. The trees should not obstruct the view at places where the pedestrians or domestic animals might usually be crossing the road near the villages.

The preferred species for roadside plantation include *Acacia arabica*, *Acacia auriculiformis*, *Ailanthus excelsa*, *Albizia lebbek*, *Albizia procera*, *Anthocephalus chinensis*, *Azadirachata indica*, *Bauhuinia purpurea*, *Bauhinia variegate*, *Butea monosperma*, *Cassia fistula*, *Cassia siamea*, *Dalbergia sissoo*, *Delonix regia*, *Erythrina* sp., *Ficus glomerata*, *Gliricidia sepium*, *Madhuca longifolia*, *Mangifera indica*, *Pongamia pinnata*, *Polyalthia longifolia*, *Prosopis juliflora*, *Samanea saman*, *Saraca indica*, *Syzygium cumini*, *Tamarindus indica*, *Terminalia arjuna*, *Terminalia catappa*, etc.

RAILWAY LINE PLANTATION

The main objectives of railway line plantation include stabilizing the railway track and protecting it against erosion, checking illegal encroachment of railway land, optimum utilization of the surplus land for tree cultivation to generate products needed by the local people and checking the shifting of sand in desert areas getting on to the railway track. The first row of trees is recommended to be planted at a distance of about 7.5 m from the centre of the track. The trees in the first row should be such that it may not attain a height more than their distance from the railway track so that in the event of tree fall during cyclonic wind, the tree tops do not reach the railway line and create traffic hazard. The inner side of the curves should not be planted to ensure clear visibility of the track. On either side of an unmanned level crossing, a length of about 100 m should be left unplanted. No planting should be done under the telephone or electric lines. The spacing between the lines and between the trees in a line maybe 5 m, 3 m or 2 m depending upon the species to be planted, the fertility of the site and the objectives of the plantation. Commercial timber species of 30 years or more rotation may be planted at 5 m × 5 m spacing. Fast growing species for fuelwood, poles or fodder maybe planted at closer spacing of 2 m × 2 m or 2 m × 1 m. Ornamental trees should be planted near the towns, railway stations, road crossings and places of tourist interest. Trees with brittle stems or branches should not be planted especially in the first row. Wind firmness is a desirable character for the species for railway line plantation. The plantations along the railway lines are prone to accidental fires originating from engine spark especially in areas infested with tall grass or bush growth. The species to be planted in such areas should be able to resist such accidental fires. Hardy species like *Acacia catechu*, *Acacia arabica*, *Albizia procera*, *Azadirachata indica*, *Butea monosperma*, *Cassia fistula*, *Cassia javanica*, *Delonix regia*, *Ficus bengalensis*, *Madhuca indica*, *Mangifera indica*, *Prosopis juliflora*, *Saraca indica*, *Terminalia arjuna*, etc. should be given preference for railway line plantation.

CANAL BANKS PLANTATION

The main objectives of canal bank plantation include stabilization of canal bank soil against erosion, checking the shifting of sand getting into the canal, provision of comfort to travellers using canal side road, utilization of the available land for tree cultivation and harvest of tree products, particularly fuelwood and small timber for local people, checking water logging in strips along canals and in adjoining areas and improvement of aesthetics in the area. The number of rows to be planted and the position of the first row depend upon the width and nature of the land available for planting. The spacing between the trees in a line depends upon the species selected and the desired end product from the plantation. The first row of trees should preferably be of shade trees. Ornamental trees maybe planted in the first row or in groves near villages, towns or places of tourist interest and crossings of the roads. For planting in the first row, the trees with a strong taproot system should be preferred. The choice of species depends on climatic factors, objectives of planting, etc. The preferred species for canal banks plantation are *Acacia arabica*, *Cassia javanica*, *Dalbergia sissoo*, *Delonix regia*, *Eucalyptus* spp., *Ficus bengalensis*, *Jacaranda ovalifolia*, *Populus* spp., *Vateria indica*, etc.

Difference between Social Forestry and Agroforestry

Sl. No.	**Social Forestry**	**Agroforestry**
1.	Social forestry is a plantation made on lands outside conventional forest areas.	Agroforestry is a method of farming that allows trees and shrubs to grow along with crops and/or livestock, therefore blending agriculture and forestry in the same production system.
2.	It is the forestry of the people, by the people and for the people.	Agroforestry is a system which is rather localized in its concept for managing the unit of land for maximizes production of agricultural crop and forest trees complimentary with each other.
3.	The major objectives of social forestry include supply of fuelwood to rural people, fodder for cattle of the rural population, protection of agriculture by creation of diverse ecosystems, arresting wind and water erosion, etc.	The major objectives of agroforestry is to reduce pressure on natural forests and to encourage tree plantation with agricultural crops and livestock to improve overall productivity, income and livelihoods of rural households, especially the marginal and small farmers.

4.	Social forestry can be categorized into five groups such as farm forestry, extension forestry, reforestation of degraded forests, community woodlots and recreation forestry. Besides roadside, railway line and canal banks plantations may also be included under social forestry.	There may be several agroforestry practices like agrisilviculture, silvipastoral, hortisilviculture, hortisilvipastoral, etc. depending upon the components included in a system.
5.	Planting of trees on massive scale is done on vacant land, community land, roadside, railway track and even degraded reserve forest.	Agroforestry is practiced mostly in farmer's field/own land.
6.	Trees and shrubs are to be used to harvest multiple products.	It involves integration of two or more than two components in a cultivated land.
7.	Social forestry is short rotation forestry. Return from a social forest is relatively late.	From an agroforestry system the returns are quick from agricultural crop components.

6

Role of Agroforestry in Soil Productivity and Conservation

Management approaches to soil, including problems of soil degradation and low soil fertility, have recently undergone major changes. The earlier concept was to concentrate on achieving high levels of production from the more fertile areas, leaving the marginal lands for extensive use only. Steeply sloping and highly drought-prone areas were mostly left without any cultivation as production from these areas was hardly cost effective. Soil constraints were to be overcome by inputs like improved crop varieties, fertilizers, chemical control of pests and diseases, and the use of irrigation.

The approach of use of newly developed high yielding crop varieties, improved agronomic management practices, use of chemical fertilizers, efficient water management and timely plant protection measures has been successful in achieving large increases in crop productivity in recent years. On the other hand, continuous application of fertilizers at higher rates leads to environmental problems. Yield responses to fertilizers have declined because of soil physical degradation and micronutrient deficiencies. Above all, large numbers of poor farmers simply can neither afford high levels of fertilizers and other purchased inputs, nor do they have the capital to take on the risk involved in their use. Increasing the area under irrigation has also run into severe constraints in the form of limits to available freshwater resources.

A new set of approaches has been recommended by Young (1997) to overcome these soil related constraints and to maintain or enhance the productivity level of soil.

- To find ways for sustainable use of marginal lands
- To reclaim and restore degraded lands
- To improve germplasm to produce plant varieties suitably adapted to soil constraints
- To maintain soil organic matter and biological activity, with benefits both for soil physical conditions and balanced nutrient supplies
- To improve nutrient cycling and nutrient use efficiency in agroecosystems

- To use fertilizers and other external inputs at moderate levels, seeking strategic use to overcome deficiencies that cannot otherwise be remedied
- To improve water-use efficiency

Agroforestry can contribute to all these aspects and has a major role to play in some of them. The capacity of trees to grow under difficult climatic and soil conditions coupled with their ability for soil conservation and enhanced nutrient cycling gives agroforestry systems a potential in the marginal, semiarid and sloping soils. Tree litter and prunings can substantially help to maintain soil organic matter and improve soil physical properties resulting in a continuous supply of nutrients.

TREES IMPROVE SOIL

Agroforestry systems have the potential to make use of marginal and degraded lands through the soil improving effects of trees. Trees have a different impact on soil properties than annual crops, because of their longer residence time, larger biomass accumulation and more extensive root systems. 'Trees improve soils' - this hypothesis is substantiated by the following facts (Young, 2005).

- The soil under natural forest is fertile. It is well structured, has a good water holding capacity and has a store of nutrients bound up in the organic matter. From time immemorial, farmers have known that they will get a good crop by planting in forest clearings.
- In natural forests, nutrients are efficiently cycled with very small inputs and outputs from the system. A forest ecosystem is a relatively closed system in terms of nutrient transfer, storage and cycling in contrast to agricultural systems.
- Reclamation of degraded land through afforestation or a similar type of tree-based land-use system is found to be best to restore soil fertility in many areas.
- The practice of shifting cultivation demonstrated the power of trees to restore fertility lost during cropping.
- The conversion of natural forest ecosystems to arable farming systems leads to a decline in soil fertility and a degradation of other soil properties unless appropriate corrective measures are taken.

Nitrogen fixation and a high biomass production are the two most desirable characteristics of a tree. However, many properties are specific to particular objectives of systems in which the trees are used. The following properties of a tree make it suitable for soil fertility maintenance or improvement.

1. A high rate of production of leafy biomass
2. A high rate of nitrogen fixation
3. The existence of deep roots and a dense network of fine roots with a capacity for abundant mycorrhizal association
4. A high and balanced nutrient content in the foliage; litter with high nitrogen, low lignin and polyphenols
5. An appreciable nutrient content in the root system
6. Rapid litter decay, where nutrient release is desired, and a moderate rate of litter decay, where maintenance of a soil cover is required
7. Absence of toxic substances in the litter or root residues
8. A capacity to grow on poor soils for reclamation
9. Absence of severe competitive effects with crops, particularly for water
10. Low invasiveness
11. Productive and service functions other than soil improvement

The existence of productive functions is not directly concerned with soils but is of the highest importance if the tree is to be effective in fertility maintenance. A species needs to be acceptable and desirable in agroforestry systems from production points of view.

BENEFICIAL EFFECTS OF TREES ON SOIL

Trees improve soil fertility by processes of increasing additions to the soil, reduction of losses from soil, and improving soil physical, chemical and biological conditions. Most of the beneficial effect of trees on soils is primarily due to their root, crown, litter fall and N-fixing ability and the effects are directly proportional to the density of the trees. Understanding the magnitude and rate of beneficial as well as adverse effects of trees on soil is the key to successful design and management of agroforestry systems.

Additions to the soil

Maintenance or increase in soil organic matter: Trees are believed to increase, or at least maintain, the organic-matter levels of the soil. This is mainly through litter fall and continuous degeneration or sloughing-off of roots of standing trees. Pruning materials from the woody perennials used in agroforestry systems add a huge quantity of organic matter to the soil, and this is generally high in nitrogen. The root biomass of trees is usually 20-30% of total plant biomass (Armson, 1977), but it may be as low as 15% in some rain forests or as high as 50% or more in semi-arid and arid climate (Nair,1989). This biomass consists of structural roots, fine roots and associated mycorrhizae. The ability of the root system to improve soil organic matter is a crucial factor in low-input agricultural systems with low productivity levels. Roots also store considerable quantities of nutrients.

Nitrogen fixation: Biological nitrogen fixation takes place through symbiotic and non-symbiotic means. Symbiotic fixation occurs through the association of plant roots of many leguminous plant species with N_2-fixing microorganisms, *Rhizobium* or *Bradyrhizobium.* A few non-leguminous tree species also nodulate with a genus of actinomycetes, *Frankia.* About 200 non-leguminous plant species belonging to 19 genera nodulate with *Frankia. Alnus*, *Casuarina* and *Allocasuarina* are the main genera of nodulating non-leguminous plants (Bond, 1983). The N-fixing ability of trees and shrubs in agroforestry system varies from 50-500 kg/ha/annum depending upon the trees species and growing conditions (Jhon, 1988).

Nutrient uptake: Trees are more efficient than herbaceous plants in taking up nutrients released by the weathering of soil at deeper horizons. Potassium, phosphorus, calcium magnesium and micronutrients are released by rock-weathering, particularly in the B/C and C soil horizons which tree roots often penetrate. Thus, nutrients in deeper soil horizons, that are unavailable to shallow rooted crops, are taken up by deep rooted trees.

Atmospheric input: Atmospheric deposition makes a significant contribution to nutrient cycling. It consists of nutrients dissolved in rainfall (wet deposition) and those carried in dust (dry deposition). Trees reduce wind speed considerably and thus favour dry deposition.

Increased water infiltration: Trees facilitate water infiltration along the root channels due to higher organic matter content in the soil. The leaf litter fall from trees also absorbs a large quantity of water. This substantially reduces runoff and increases water input to the soil.

Water retrieval: The tap roots of trees draw water from soil well below the depth ordinarily reached by roots of crop plants. This process increases water input in plant-soil system.

Reduction of losses from the soil

Protection from erosion: Soil erosion causes loss of soil organic matter and nutrients, and thus results in reduction in crop yield. Agroforestry systems reduce erosion through ground litter cover by tree leaves and understorey vegetation. The tree canopy may also reduce the erosion to a little extent by slashing raindrops. Trees and shrubs through their proper planting arrangement and management can act as effective barriers to control soil erosion.

Nutrient retrieval and recycling: The tree-root systems in an agroforestry system intercept, absorb, and recycle nutrients in the soil that would otherwise be lost through leaching, thereby making a more closed nutrient cycle. The mycorrhizal systems associated with tree roots are an agent in the nutrient cycling process. They penetrate a large proportion of the soil, facilitating the uptake of nutrients, which can move only short distance by diffusion. Thus, the nutrient-use efficiency in an agroforestry system is greatly enhanced. The leaf litter after decomposition forms humus, releases nutrients and improves soil properties. The nutrient contents of leaf litter may vary from 150-300 kg N, 10-20 kg P, 75-150 kg K and 100-300 kg ca/ha/yr which means that nutrient cycling can substantially reduce the fertilizer requirement of the associated crops grown in agroforestry system (Negi and Sharma, 1984).

Reduction of water loss through evapotranspiration: Tree canopies lower temperature and increase relative humidity in an agroforestry system which leads to low transpiration rate. Tree litter and prunings act as mulch on ground surface that reduces the evaporative loss of soil moisture. The water storage capacity of soil is also increased due to improved soil organic matter.

Reduction in the decomposition rate of humus: The rate of decomposition of organic matter is comparatively low under trees than in agricultural use through lower topsoil temperature and less soil disturbance.

Effect on physical conditions of the soil

Maintenance or improvement of physical properties: The soil physical properties such as structure, porosity, moisture retention, and erosion resistance are much improved in an agroforestry system than that of an open area or area with seasonal field crops. There is much evidence of the influence of physical properties of tropical soils on crop growth, independent of nutrient or other effects (Lal and Greenland, 1979).

Modification of extremes of soil temperature: A cover of tree leaf litter and pruning greatly reduce the surface temperatures of ground in the tropics, which sometimes exceed 50° C and adversely affect crop growth. In temperate regions, there is some protection from ground frost by the 'blanketing' effect of litter.

Penetration of compact layers by roots: Tree roots are capable of penetrating the layers of laterite, calcrete or hard pans in the soil, so drawing nutrients and water beneath.

Effect on chemical conditions of the soil

Reduction of acidity: Trees tend to moderate the effects of leaching through the addition of bases to the soil surface. The bases released by litter decay can help to check acidification caused by the application of fertilizers to the crops in an agroforestry system.

Reduction of salinity and sodicity: Perennial woody species in an agroforestry system can reclaim saline and alkaline soils by lowering salinity and reducing sodium saturation of the exchange complex. Trees as such cannot contribute to reduction of salinity, but they assist in water management that leads to soil amelioration and leaching out salts. Afforestation with *Acacia nilotica* and *Eucalyptus tereticornis* in Karnal region successfully reduced the soil *p*H from 10.5 to 9.5 over five years (Dagar, 2009).

Reduction of soil toxicities caused by pollution: The trees do not extract pollutants from a chemically polluted soil, but some species are tolerant to them. The trees build up soil organic matter and ameliorate fertility which results in reducing pollution levels in the soil.

Effect on biological conditions of the soil

Improvement in the activity of soil fauna: The presence of tree litter in soil is frequently associated with higher rates of activity of soil fauna.

Effects of shading: Shade caused by the tree canopy in an agroforestry system lowers ground-surface temperatures, which reduce the rate of loss of soil organic matter by oxidation. Shading is allowed to a certain extent while trees are grown with crops. In most of the cases, the branches are pruned up to $1/3^{rd}$ height of the trees with advancement of age of the trees.

Root nodulation: Roots of some nitrogen fixing trees have more nodules where they are in close contact with roots of non-nitrogen fixing plants, the latter stimulating the former.

Exudation of growth-promoting substances into the rhizosphere: Root exudates by a few tree species are believed to work as growth-promoting substances for other plant species.

ADVERSE EFFECTS OF TREES ON SOIL

Removal of organic matter and nutrients through tree harvest: Trees accumulate large quantities of carbon and nutrients in their biomass, part of which is removed in harvest. The problem is greater where there is whole-tree harvesting, with gathering of fine branches and litter by local people after a timber harvest. From a soil management point of view, all branches and litter should be allowed to decay *in situ*, although such a practice appears unreasonable to the local people.

Competition between trees and crops for nutrients: Usually an established root system of trees or shrubs dominates that of newly planted annual crops.

Ideally, trees in agroforestry systems should have a deep penetrating tap root system with limited lateral spread. Whereas lateral spread of the canopy can be controlled by pruning, root pruning is generally too expensive to be practical.

Competition between trees and crops for moisture: In the semiarid and arid zones, this is possibly the most serious problem encountered in agroforestry.

Production of substances which inhibit germination or growth: Some *Eucalyptus* species produce toxins which can inhibit the germination or growth of some annual herbs. Some trees release organic compounds to the soil which are phytotoxic and may cause problems in agroforestry systems.

NUTRIENT CYCLING IN AGROFORESTRY SYSTEMS

A nutrient cycle is the movement and exchange of organic and inorganic matter back into the production of living matter. Nutrient cycles occur within ecosystems. The plants take up nutrients from the soil to grow. Soil nutrients mostly come from the breakdown of mineral-bearing rocks and decomposition of plant and animal residues incorporated into the soil. The nutrients that plants get from the soil are stored in all plant tissues. Some of the plant parts such as dead leaves and roots are returned to the soil during the plant's growth. In addition, depending upon the nature of plants and type of land use, some parts are added to the soil when the plants are harvested. The biomass so added decomposes through the activity of soil microorganisms and the nutrients that had been bound in the plant parts are released to the soil where they become once again available to the growing plants.

Nutrient cycling occurs to varying degrees in all land use systems. Agroforestry systems show a high potential to improve soil fertility with more efficient nutrient cycling because of the beneficial effects of woody perennials on the soil. The woody perennials have more extensive and deeper root systems than herbaceous plants and thus have a potential to capture and recycle a larger amount of nutrients. Their litter contribution to the soil is also more than that of herbaceous plants.

Forest ecosystems have high rates of turnover and low rates of losses from the system and these are usually self-sustaining through closed and efficient nutrient cycling systems. But agricultural systems are often open and the turnover within the system is relatively low and losses are comparatively high per unit area and unit time. Nutrient cycling in agroforestry systems falls between these two extremes and as compared to agricultural systems more nutrients in these systems are reused by plants before being lost from the system.

Table 6.1: Nutrient content (%) of biomass (leaves) in agroforestry systems

Tree	N	P	K	Ca
Alchornea cordifolia	3.29	0.23	1.74	0.46
Cassia siamea	2.52	0.27	1.35	-
Dactyladenia barteri	2.57	0.16	1.78	0.90
Erythrina poeppigiana	3.30	0.18	1.16	1.52
Gliricidia sepium	4.21	0.29	3.43	1.40
Inga edulis	3.10	0.20	0.90	0.70
Leucaena leucocephala	4.33	0.28	2.50	1.49

Source: Nair, 1993

Agroforestry systems provide an opportunity for modifying nutrient cycling through management which results in more efficient use of nutrients in the soil. There are several management options for exploiting the advantages of efficient nutrient cycling in agroforestry systems.

- The deep root systems of trees reach deeper soil horizons, which are not often accessed by roots of common agricultural crops. Thus, there is an enhanced nutrient uptake as a result of rock weathering or percolation past herbaceous plant roots. This process, which is commonly called nutrient pumping, is a significant factor of soil fertility improvement in agroforestry systems.
- Gains from symbiotic N_2-fixation by trees can be enhanced through selection of tree species and admixture.
- Different trees and shrubs used in alley cropping systems vary in quantity, quality and decomposition rate of leaf biomass. The timing of hedge pruning can be regulated in such a way that the nutrient release through decomposition of biomass is synchronized with the peak period of crop's nutrient demand.
- Better management practices which improve organic matter status of the soil in an agroforestry system will also improve nutrient cycling and, thus better soil and crop productivity.
- A major management consideration in agroforestry is to reduce nutrient loss through soil conservation.

Table 6.2: Dry matter and nutrient inputs via litter fall/prunings in various production systems in the humid tropics

Systems	Dry matter (t/ha/year)	Nutrient inputs (kg/ha/year)				
		N	P	K	Ca	Mg
Alley cropping						
Leucaena leucocephala	6.5	160	15	150	40	15
Gliricidia sepium	12.3	358	28	232	144	60
Sesbania	7.5	25-110	-	-	-	-
Erythrina poeppigiana	9.6	278	24	216	120	52
Erythrina spp.	8.1	198	25	147	111	26
Shade systems						
Coffee/*Erythrina*/*Cordia*	15.8	331	22	162	328	69
Coffee/*Erythrina* pruned	20	461	35	259	243	76
Coffee/*Erythrina* non-pruned	7.6	175	11	75	122	33
Cacao/*Erythrina*	6.5	116	6	40	116	41
Cacao/*Cordia*	5.8	95	11	57	108	43
Cacao/mixed shade	8.4	52	4	38	89	26

Source: Szott *et al.*, 1991

SOIL ORGANIC MATTER IN AGROFORESTRY SYSTEMS

Woody perennials differ from herbaceous crops in the quantity, quality and time of addition of organic matter to the soil. Trees usually add organic matter in a steady rate whereas herbaceous crops add biomass mostly at harvest. Biomass added to the soil from trees contains more lignin than herbaceous crops, which affect the rate of decomposition and humus formation. Majority of the organic materials are of plant origin, others include microbial tissue and dead biomass of soil fauna. Soil organic matter consists of two parts, fully decomposed organic matter or humus and plant and microbial remains that are in various stages of decomposition, commonly called litter. Organic matter decomposition refers mainly to the conversion of litter to humus through the activity of soil fauna and microflora.

Plant litter with high nitrogen is decomposed rapidly and is considered to be of high quality, whereas woody residues and other materials like cereal straw which have high lignin content are more resistant to decomposition and are considered to be of low quality. Materials with low carbon to nitrogen ratios (C: N) decompose rapidly and release relatively larger quantities of nitrogen. But materials with high C: N ratio provide a source of energy (C) for the microbes to multiply rapidly and utilize the N reserves from the soil. This causes a temporary unavailability of nitrogen for the plants as the added material is very low in N. Subsequently when the C (energy) source is depleted, the microbial population declines and the N that have been temporarily incorporated in microbial tissues are once again released to the soil and available for plant uptake.

Prunings of many of the woody perennials used in agroforestry systems are generally high in nitrogen. Fresh leaves of some leguminous trees and shrubs such as *Leuceana leucocephala, Gliricidia sepium,* and *Erythrina* spp. decompose relatively fast and release a major part of their nutrients from the applied mulch within about four weeks under humid tropical conditions. But nutrient-rich and easily decomposable material may not always be desirable because nutrient release may exceed plant nutrient demands resulting in asynchrony between supply and demands. Mulches of species such as *Cassia siamea, Flemingia macrophylla* and *Dactyladenia barteri* that are also commonly used in agroforestry systems as hedgerow planting are generally slow to decompose. Suppression of weed growth or reduction of soil moisture loss through evaporation may be sometimes more desirable than providing a quick supply of nitrogen to the crops. In such situations these slow decomposing mulches which provide a good ground cover for a longer period of time are preferred. Thus, better nutrient use efficiency in an agroforestry system can be achieved by selecting tree species with differing rates of litter decomposition, adjusting the timing of pruning to regulate the time of addition of the mulch and modifying the method of application of the mulch, i.e., surface addition or soil incorporation.

The roots of trees add a substantial amount of organic matter to the soil. The roots of crops function on a seasonal basis but tree roots function all year round. The root biomass of trees is usually 20-30% of total plant biomass, but it may be as low as 15% in some rain forests or as high as 50% or more in semiarid and arid climates. This biomass consists of structural roots, fine roots and associated mycorrhizae. Mycorrhizae absorb carbohydrates from the host plant and in turn, function as an expanded root system which increases nutrient absorption.

BIOLOGICAL NITROGEN FIXATION IN AGROFORESTRY SYSTEMS

Through a natural process known as 'Biological Nitrogen Fixation', some bacteria and trees with nodules in their root systems are able to transform nitrogen from gaseous form into a form that may be used by other plant species.

Biological nitrogen fixation (BNF) offers an excellent opportunity for drawing upon the vast reserve of atmospheric nitrogen in an inexpensive and environmentally sound manner for meeting the needs of nitrogen of plants. BNF takes place through symbiotic and non-symbiotic means. Symbiotic fixation occurs through the association of plant roots with two types of nitrogen-fixing bacteria; rapidly multiplying strains belonging to the genus *Rhizobium* and slowly multiplying strains belonging to the genus *Bradyrhizobium*. One group of leguminous trees nodulates only with *Rhizobium* and another group nodulates only with *Bradyrhizobium*. A third group is more promiscuous since it nodulates both with *Rhizobium* and *Bradyrhizobium*. Moreover, about 200 non-leguminous plant species nodulate with N_2-fixing actinomycetes, belonging to genus *Frankia* by symbiotic means. When the nodules, the site and storage organ of N_2 fixation and the roots degenerate, the nitrogen contained in them are released to the surrounding soil medium and become available after mineralization to the adjacent plants.

Nodules: the warehouses of nitrogen

The single most crucial component of nitrogen fixing trees (NFTs) is their nodes, which are home to bacteria. Tiny microscopic creatures work hard inside tiny knobs of various sizes, fixing nitrogen and producing ammonia. Nodules can occur on stems or leaves, although they are most frequently seen on root systems. Nodule form varies greatly depending on the properties of the host plant.

Both the number and size of nodules vary. Some are as big as a baseball, but the majority are tiny, often measuring less than 0.5 cm in length or diameter. More nodule-bearing plants and trees typically have smaller nodes and are less effective at fixing nitrogen. Bigger nodules that are more effective at fixing higher nitrogen concentrations are bigger in size.

The inside of young nodules is often white or grey. They are not able to fix nitrogen. Mature and bigger nodules contain leghemoglobin, a protein that carries oxygen and nitrogen; they may be identified by their pink or reddish interior colour. They are the best fixers of nitrogen.

Nodulation in legumes by *Rhizobium*

Nodule formation is known as nodulation. There are certain key differences between legumes and non-legumes in this aspect of the nitrogen fixation process. The following stages are observed in the process of nodulation.

1. Whenever legumes are nitrogen-deficient, they emit flavonoids. This indicates to the rhizobia that they are interested in forming a symbiotic relationship.
2. Soil is normally home to rhizobia. They produce a nodulation factor in response to flavonoids, which stimulates the growth of malformed root hairs in the plant.

3. Rhizobia subsequently create an 'infection thread' to pass through the malformed root hairs and reach the root cells.
4. Rhizobia enter the infection threads through the coiled root hairs and proliferate there.
5. After entering the cortical area, the infection thread splits into many branches.
6. The cortical area exhibits meristematic development and young nodules form when there is a rhizobial infection. After one week of infection, tiny nodules become apparent to the unaided eye.
7. Eventually, mature nodules - misshapen rhizobium bacteria - develop, brimming with both immature and matured bacteroids.

Nodulation in non-legumes by *Frankia*

The nitrogen-fixing actinobacterium *Frankia* forms a symbiotic relationship with actinorhizal plants, such as trees and shrubs, to develop root nodules. Actinorhizal plants can be beneficially infected by *Frankia* in two different ways, depending on the host plant; either intracellularly or intercellularly.

1. Frankia uses an unidentified signal to cause the root hairs to curl, which initiates an intracellular infection.
2. Frankia enters the cells through the root hairs and spreads throughout them. Prenodules, which are tiny exterior protuberances that arise when cell divisions take place inside the cortex.
3. Frankia hyphae move towards the juvenile nodule lobe to generate mature nodules once the nodule lobe premordium develops.
4. The process of intercellular infection results in nodule formation even in the absence of prenodule growth or distortion of the root hair.

NFTs add nitrogen to the soil

New nodules develop in the roots of NFTs when nitrogen builds up, increasing the amount of storage space available. Moreover, the NFTs enlarge the nodes by gathering nutrients from the air and storing them in nitrogen-friendly warehouses.

An estimated 100-300 kg of nitrogen may be accumulated per hectare annually by a wide range of legumes that fix nitrogen. Legumes with greater yields have the potential to provide 500 kg of nitrogen per hectare annually.

There are trade-offs associated with this technique for the NFTs. Through this symbiotic relationship, the trees give the rhizobia carbohydrates (malate and succinate), which allows the rhizobia to get the ammonia required to produce amino acids. The bulk of the nitrogen is stored in the nodules, which allows the host plants to develop. But too much gas builds up in plant tissues, eventually killing the trees. Nevertheless, the surrounding soil and nearby plants gain from this process. The decaying trees nourish the soil with important nitrogen, which is then made accessible for the development of new plants. Additionally, the integrity of the general health of the ecosystem is preserved by employing permaculture techniques as opposed to destructive commercial treatments.

Additional effects of NFTs

1. NFTs are dependable providers of food, yielding a substantial amount of vegetables and grains. They serve as residential gardens' mulch banks as well.
2. In hostile, dry, and infertile environments, NFTs are essential pioneer species. They make land arable and fruitful by balancing the harshness of the environment and replenishing its fertility.
3. Because they consistently provide organic matter to the soil, there is less need for synthetic fertilisers.
4. They minimise erosion and enhance soil structure with their large root system.
5. Roots that are both developing and atrophying provide many channels that improve the aeration of good soil.
6. NFTs offer timber, live fences, wind protection, fuelwood, and feed.
7. Certain NFTs provide microclimates to support crops that need shade.
8. NFTs can also be used to support vine crops as trellis.

The species or provenances chosen for introduction in any agroforestry system should have the highest possible nitrogen-fixing potential for a given set of climatic conditions. They should also be capable of tolerating environmental constraints, low levels of nutrients and pests. When trees are introduced to a site for the first time inoculating the host plant with *Rhizobium* or *Frankia* through soil or crushed nodules may be beneficial.

- Trees in an agroforestry system can be crucial for recycling nutrients that are leached down through the soil profile and minerals that are released from parent material that is weathering, including rocks and sediments.
- These nutrients are utilised by the tree for its growth and development; many of them end up in the top soil as dead leaves, twigs, blossoms, and seeds that either slowly break down on the surface or are consumed by animals.Every tree contributes in some way to recycling the soil's nutritional status.

- While most evergreen species retain a certain amount of litter fall throughout the year, deciduous trees lose the majority of their leaves in the fall, leaving behind a dense mat on the ground. These nitrogen-fixing trees often produce high-nitrogen litter, which improves the soil's nitrogen status.
- The capacity of several tree species to transform atmospheric nitrogen into organic nitrogen for their own use via intricate symbiotic relationships with Rhizobium bacteria and their fine roots is another significant element.
- The bacteria create nodules on the roots that have the ability to transform atmospheric nitrogen gas into nitrogen that the plant can use.

Mycorrhizal infections also increase the absorption of phosphate and other poorly mobile ions in soil such as Zn^{2+}, Cu^{2+}, Mo^{2+}, and K^{+}. Mycorrhizal fungi are most often associated with the roots of nitrogen-fixing trees. Nodulation and nitrogen fixation require a high P level in the host plant, which can be facilitated by the mycorrhizal symbiont. Mycorrhizal fungi also increase water uptake of the host plant.

Soil acidity influences nitrogen fixation by the direct or indirect effects they have on the host plant and the symbiotic microorganisms. The detrimental effects of soil acidity can be overcome by selecting acid-tolerant host plants. The soil acidity can also be corrected by application of proper amendments to the soil. The higher organic matter content of soil under nitrogen-fixing trees in an agroforestry system leads to satisfactory yields of annual crops even when the pH is lower than usually recommended in conventional cropping systems.

Table 6.3: Estimate of N_2 fixation by some woody species suitable for agroforestry systems

Species	**N_2 fixation (kg/ha/year)**
Acacia mearnsii	200
Casuarina equisetifolia	60-110
Erythrina poeppigiana	60
Faidherbia albida	20
Gliricidia sepium	13
Inga jinicuil	35-40
Leucaena leucocephala	100-500
Sesbania rostrata	83-109

Source: Nair, 2008

Table 6.4: List of nitrogen fixing tree species suitable for agroforestry systems

Sl. No.	Name of the species	Sl. No.	Name of the species
1.	*Acacia albida*	24.	*Cassia siamea*
2.	*Acacia auriculiformis*	25.	*Casuarina equisetifolia*
3.	*Acacia catechu*	26.	*Dalbergia latifolia*
4.	*Acacia aneura*	27.	*Dalbergia sissoo*
5.	*Acacia dealbata*	28.	*Delonix regia*
6.	*Acacia decurrens*	29.	*Gliricidia sepium*
7.	*Acacia farnesiana*	30.	*Hardwickia binata*
8.	*Acacia implexa*	31.	*Leucaena leucocephala*
9.	*Acacia leucophloea*	32.	*Moringa oleifera*
10.	*Acacia mearnsii*	33.	*Oogeinia oojeinensis*
11.	*Acacia melanoxylon*	34.	*Parkinsonia aculeate*
12.	*Acacia mollissima*	35.	*Peltophorum ferrugineum*
13.	*Acacia nilotica*	36.	*Pithecellobium dulce*
14.	*Acacia planifrons*	37.	*Prosopis alba*
15.	*Acacia Senegal*	38.	*Prosopis chilensis*
16.	*Albizia chinensis*	39.	*Prosopis cineraria*
17.	*Albizia lebbek*	40.	*Robinia pseudoacacia*
18.	*Albizia procera*	41.	*Samanea saman*
19.	*Alnus nepalensis*	42	*Saraca indica*
20.	*Alnus nitida*	43.	*Sesbania aegyptica*
21.	*Bauhinia variegata*	44.	*Sesbania bispinosa*
22.	*Butea monosperma*	45.	*Sesbania grandiflora*
23.	*Cassia fistula*	46.	*Tamarindus indica*

EFFECT OF AGROFORESTRY ON SOIL EROSION

Agroforestry improves or maintains soil productivity through its effect on soil conservation. Soil conservation addresses not only erosion, but also other forms of physical, chemical and biological deterioration of soil. The adverse effect of erosion is expressed in terms of loss of soil organic matter and plant nutrients which leads to reduction of crop yields.

While construction and maintenance of physical barriers like bunds and terraces to control runoff involves huge investment and causes irreparable damage to valuable topsoil, agroforestry provides continuous ground cover formed of living and dead plant materials, crop residues and tree prunings which effectively check raindrop impact and runoff and thus, conserves soil. The presence of multiple layers of canopy in agroforestry systems can considerably reduce the velocity of falling rain drops and thus reduce the severity of their impact. Barrier hedges such as in alley cropping are effective in reducing runoff. Woody perennials also act as runoff barriers.

Agroforestry practices influence the soil erodibility factor through their effect on soil physical properties and soil organic matter. Soil structure under agroforestry has higher stability, lower detachability and higher water infiltration capacity than that of pure cropping. Agroforestry systems can reduce the rainfall erosivity. However, it may not be true in all agroforestry situations. Low and dense canopy usually reduce erosivity. Thus, well managed agroforestry systems can reduce overall erosion losses.

Where wind is a major cause of soil erosion and moisture loss, windbreaks and shelterbelts can make a significant contribution to sustainable production. Trees and shrubs on the farmlands are traditionally used for soil erosion control in a number of ways. These include both direct use of these species for reducing erosion and their supplementary use for stabilizing physical structures that are created specifically for erosion control. Direct use involves use of trees and shrubs in agroforestry systems for increasing soil cover, providing live or dead barriers and enhancing soil's resistance to erosion through maintenance or build-up of organic matter and desirable physical properties. The supplementary use consists of use of trees and shrubs to stabilize physical structures such as bunds, risers or embankments.

Table 6.5: Summary of the beneficial effects of woody perennials in an agroforestry system on soil

Nature of processes	Processes/Avenues	Main effects on soil
INPUT (Augment/ additions to soil)	Biomass production (litter and root decay)	Improvement or maintenance of organic matter
	Nitrogen fixation	N-enhancement
	Effects of rainfall (quantity and distribution)	Influence on nutrient addition through rain/dust
OUTPUT (Reduces losses from soil)	Protection against water and wind erosion	Reduce loss of soil and nutrients
TURN-OVER	Nutrient retrieval/ cycling/release	Uptake from deeper layers and deposition on surface
		Withholding nutrients that can be lost by leaching
		Timing of nutrient release

CATALYTIC (Indirect influences)	Physical	Improvement of soil properties
	Chemical	Moderating effect on acidity, salinity and alkalinity
	Microclimatic	Ameliorative effect on extreme conditions
	Biological	Effect on soil microorganisms; improvement of litter quality through diversity

Source: Nair, 1989 and Young, 1997

7

Tree Crop Interaction in Agroforestry

Agroforestry systems are not simply systems where trees and crops or animals give useful products to the farmers, rather systems where trees and crops and/ or animals interact. Interaction literally means influence or mutual or reciprocal action. So, component interaction refers to the influence of one component of a system on the performance of other component as well as the system as a whole.

In agroforestry systems, trees are grown in close proximity to crops and pasture. Their performance would largely depend on their ability to share various growth resources in a given environmental situation. Various interactions take place between the woody trees and herbaceous plants (crops or pastures) which is referred to as tree-crop interface. These interactions take place through the media of soil and microclimate and may exert favourable or adverse effects on the crop. Study of interaction helps to know how the components of agroforestry utilize and share the resources of the environment, and how the growth and development of any of the components will influence the others. Interaction occurs both above and below the ground and includes a complex set of interactions relating to radiation exchange, the water balance, nutrient budget and cycling, shelter and other microclimatic modifications.

The success of an agroforestry system relies heavily on exploitation of the component interactions. In an ideal relationship, production of trees as well as crops or grasses in combination could be comparable to their sole performance. Agroforestry could be even more advantageous if the production of associated components is increased due to influence of trees. This is possible because trees are capable of improving productivity of soil in many ways. A large number of trees are known to fix nitrogen symbiotically. Nevertheless, instances of crop inhibition in association of trees are not uncommon. Such inhibitions are primarily caused by shade effect as well as competition for belowground resources such as nutrients and water. In some cases, inhibitory effect may also result from allelochemicals secreted by some of the tree species.

NATURE OF INTERACTIONS

The nature of interactions between two components can be described on the basis of observable net effect of one component on another in a system. The basic principle deciding the nature of interactions depends upon the ability of different components to capture and use the most limiting essential growth resources effectively. The capture of the limiting resource (e.g., light, water or nutrients) depends on the number, surface area, distribution and effectiveness of the individual elements within the canopy or root system of the species or mixture of species involved.

Complementary interaction

In a system if the tree and crop components help each other, by creating favourable conditions for their growth in such a way that the agroforestry system provides a greater yield than the yield of their corresponding sole crops, then the interaction between the components is said to be complementary in nature. The complementary interaction is of two types, viz., spatial and temporal. When complementarity is achieved simply by varying the proportion of the desired species, it is known as spatial complementary interaction. In this, the geometry of the system is altered by changing the spacing and density of the components. Therefore, under complementary association the proportion of each species can be modified to suit one's requirements without any sacrifice. When two species complement each other, the capture of the major limiting growth resources by the mixture always exceeds that by the corresponding sole stands. The two components also complement each other through more effective use of growth resources by utilizing them at specific times due to difference in growth pattern and resource requirement. Indeed, there are substantial opportunities for temporal complementarities if species make their major demands on available resources at different times, thereby reducing the possibility of competition.

Supplementary interaction

If the two components interact in such a way that the yield of one component exceeds the yield corresponding to its sole crop without affecting the yield of other component, the interaction is known to be supplementary in nature. Therefore, there will be an additional yield from one of the components; thus the overall yield of agroforestry system will be more than that of sole crops of both the components. In this interaction one component is maximizing its resource utilization without limiting it for the other component. This is an independent relationship.

Competitive interaction

In this system the tree and crop components interact in such a way that the increase in the yield of one component leads to decrease in the yield of other component due to competitive interaction. Although agroforestry is envisaged as a system of plant species that benefit each other mutually or unilaterally, it is too optimistic to assume that all types of competition can be eliminated in these systems, especially in areas with poor soils and scanty rainfall. The intensity of competition is the greatest when requirements are similar, and the growth and development proceed synchronously for both the components. Thus, there is extensive overlap between species in their resource requirements, resulting in a severe competition. To reduce the competition in an agroforestry system, the tree and crop components should be selected in such a manner that they can utilize the finite resources (water, nutrients, solar energy, etc.) at different times and show different growth and development phases from each other. Competition between trees and crops is a long-term problem in plantations when the crop species is perennial.

FACTORS AFFECTING INTERACTIONS

The interactive relations among the components of agroforestry are affected by many factors such as choice of species, density and age of trees, site factors, management practices, etc. Many of these factors can be manipulated for better production from the system.

Species of trees and crops: The growth pattern of trees varies from species to species and thus, affects the interactions with the associated components. Some crops may perform better in association with a particular tree component, whereas the yield of other crops may reduce with the same tree component, because different crops interact differently with the same tree species. It is not only that different crops exhibit differently with the same tree species but also the same crop may be affected differently by the different tree components.

Density of trees: Canopy cover of trees intercepts light depending upon the density of trees and consequently affects the performance of the undergrown crops. Generally, the yield of undergrown crop is decreased with increasing density of the trees. The characteristics of undergrown crop species will also determine the impact of canopy cover. Trees also modify microclimate of crops grown below and improve the physical conditions and fertility of the soil. Thus an increase in tree density should accrue some benefits while increasing the competition for other resources at the same time. Therefore, optimum density in agroforestry would vary with variation in the components.

Age of trees: The demand for various growth factors and therefore competition offered by the components would be affected by the age and the growth stage of the components. Trees take long time to attain full size and stature, and thus no or less competition is usually offered to intercrops during early years of plantation. As the trees grow, their effect on crop growth becomes apparent. Due to initial slow growth of trees, even crop or grass component may dominate and affect the tree growth adversely. For example, *Ziziphus* spp. in initial years cannot withstand competition from established pasture of *Cenchrus ciliaris*; however, once *Ziziphus* bushes attain good growth in 2 years, *Cenchrus* pasture does not affect their growth.

Site factors: Variations in climatic (rainfall, temperature, humidity, etc.), edaphic (physical, biological and chemical properties) and physiographic features of an area affect the plant growth which leads to variation in the interactive relationship of component species in an agroforestry system.

Management practices: The interactive relationship between the components in an agroforestry system also varies with management practices adopted for the system. Several characteristics could be identified as desirable attributes for trees in the agroforestry systems but often it is not possible to choose trees with all these characteristics due to various reasons. Through various management practices the interactions between the components in an agroforestry system can be manipulated. For fodder tree species, lopping done for harvesting forage periodically can at the same time increase light availability to the undergrown crops and improve their performance.

TYPES OF INTERACTIONS

Trees are the most important of all components in agroforestry systems. Based on components, interaction types may be 'tree-crop interactions' (TCI) and 'tree-animal interactions' (TAI). TCI and TAI may be positive (beneficial) or negative (harmful). These positive or negative interactions can be direct or indirect or both. Both positive and negative interactions may further be divided into above or below ground interaction in case of TCI.

The main effects of positive interactions in agroforestry systems are increased productivity, improved soil fertility, nutrient cycling, soil conservation, water conservation, weed control and microclimate improvements. Similarly, the effects of negative interactions include light competition, nutrient competition, water competition, pests and diseases and allelopathy. However, these interaction processes are interdependent and the manifestation of their effects is influenced by the environmental conditions. The relative importance of each effect depends on both the types of agroforestry systems and site factors.

Positive interactions

Increased productivity: The biomass productivity of an agroforestry system is generally greater than that of an annual system. The basis for the potentially higher productivity is due to the capture of more growth resources (e.g., light and water) or due to improved soil fertility. The presence of trees causes shade which in turn leads to complex interactions. These interactions result into microclimate amelioration which affect transpiration rate, photosynthesis and energy balance of the associated crops. All these factors may translate into increased productivity from the system.

Improved soil fertility: The potential for micro-site enrichment by some trees is an extremely important aspect of agroforestry. *Faidherbia albida* in West Africa, *Pinus caribea* in the savanna and *Prosopis cineraria* in Rajasthan are known to accumulate mineral nutrients due to the process of capture of nutrient rich litter. Alley cropping using fast growing, nitrogen-fixing trees (e.g., *Leucaena leucocephala* and *Gliricidia sepium*) in the humid tropics can substantially increase the soil fertility. A major feature of this system is the capacity of the trees to produce a large quantity of biomass for green manure and the need for regular pruning to prevent shading to the intercrops.

Nutrient cycling: Soil nutrients are considered among the least resilient components of sustainability. A fundamental principle of sustainability therefore is to return to the soil the nutrients removed from it through harvests, runoff, erosion, leaching and other loss pathways. In agroforestry systems, trees provide additional nutrient inputs through biological nitrogen fixation and deep nutrient capture. Deep nutrient capture is the uptake of nutrients by tree roots acting as a 'safety net' from depths where crop roots are not active. Similarly, taking up of nutrients from weathered minerals in deeper layer by tree roots is called as 'nutrient pumping' or 'nutrient mining'. These nutrients are considered as additional inputs in agroforestry systems, because such nutrients would otherwise leach down as far as the crop root zone is concerned. They become an input upon being transferred to the soil via tree litter (leaves, pruned materials, root debris, etc.) decomposition. The trees also improve soil health by providing huge quantity of organic matter through litter fall. Litters of high quality (low C:N ratio, low lignin and polyphenolic content) are decomposed rapidly and make nutrients available to the crop and the trees. In some cases, nitrogen is supplied by tree roots to crop roots, either due to root decay or root death following tree pruning or by direct transfer if nodulated roots are in close contact with crop roots.

Soil conservation: Contour hedgerows are highly effective in controlling soil erosion. The woody hedgerows provide a semi-permeable barrier to surface movement of water, whereas litter mulch from trees reduces the impact of rain drops on the soil and minimizes splash and sheet erosion. Tree species that provide a more effective physical barrier to erosion and produce mulch material that has a longer lasting protective role should be selected for agroforestry systems.

Water conservation: The presence of trees has positive effect on water budget of the soil and crops growing between or beneath them. A mulch or litter layer increases the infiltration of rain water while simultaneously reduces evaporation from the soil. Mulch is important, especially on sandy soils as it maintains soil moisture during the dry season where water supply for the crops is a problem.

Weed control: Another potentially positive interaction in agroforestry systems is related to weeds. Shade suppresses light-demanding weeds. Weed biomass yield is positively correlated with available solar energy. Apart from shading, weed suppression is also determined by factors such as land-use history, weather and mulch quality and crop competitiveness. Shade also reduces dry-season fire risks.

Microclimate improvement: In agroforestry systems, microclimate amelioration results primarily from the use of trees for shade, live fences or windbreaks and shelterbelts. The presence of trees reduces heat and light. Atmospheric temperature, humidity, movement of air, soil temperature and soil moisture are changed due to agroforestry systems. The nature of the understorey crops determines the extent of beneficial effects of shade trees. The smaller temperature fluctuations under shade are attributed to reduced radiation load on the crops during the day and to reduced heat loss during the night. A reduction in vapour pressure deficit causes a reduction in transpiration and hence reduced water stress for intercrops. This is especially beneficial during short periods of drought. Microclimate amelioration also reduces the disease and pest pressure by facilitating biological control agents.

Tree-animal interface (positive): The positive interactions at the TAI can affect overall system productivity in various ways. Some autotrophic production that is of no direct use to the farmer such as weeds or tree fodder can be transformed into animal biomass with high nutritional and monetary value. The productivity of the tree and herbaceous crop components can be increased through the addition of animal manure as a fertilizer source. Trees provide shade to the animals in the tropics. Heat stress on the animals in the tropics significantly reduces the time spent by them in grazing in open which leads to reduction in total feed intake. Shade reduces energy expenditure of the animals for thermoregulation and this is the main reason why animals in shade generally show higher feed conversion and ultimately higher weight gain or milk production. Shade may also have a beneficial effect on animal reproduction.

Negative interactions

As all the components of an agroforestry system utilize the same reserves of growth resources, negative interactions are likely to occur in every plant association. The major yield decreasing effects at the TCI arise from competition for light, water, and nutrients, as well as from interactions via allelopathy.

Competition for light: In agroforestry systems, shading by the trees reduces light intensity at the crop level and this is the most limiting factor in many situations, particularly those with relatively fertile soils and adequate water availability. The relative importance of light will decrease in semiarid conditions as well as on sites with low fertility soils. Since crops differ in their responses to poor nutrition, competition for light may either be reduced or amplified by a shortage of nutrients.

Competition for nutrients: An agroforestry system is assessed mainly by the yield of crop component and its reduction receives more attention than those of the associated tree species. The effect of nutrient competition is more severe for the crop components than for trees, because the crop root system is usually confined to the soil horizons that are also available to the roots of the trees but trees can exploit soil volume beyond the reach of the crop. Therefore, the effects of nutrient competition are more severe for the crop components.

Competition for water: Competition for water is likely to occur in most agroforestry systems at some period of time except in areas with well distributed rainfall or continuous supply of ground water. The competition depends on the severity of the drought and the drought tolerance of the plants. It plays a major role in the productivity of agroforestry systems despite the use of drought-tolerant and drought-adapted plants, especially in dry areas.

Microclimatic modification for pests/diseases: Bacterial and fungal diseases may increase in shaded, more humid environments in an agroforestry system. Reduced temperature and humidity fluctuations under shade can also have a favourable effect on spread of pests and diseases. Trees and crops can be a host of each other's insect pests and diseases.

Allelopathic interaction: Allelopathy refers to the inhibition of growth of one plant by chemical compounds (allelochemicals) that are released into the soil from neighbouring plants. Allelochemicals are reported to be present in practically all plant tissues, including leaves, flowers, fruits, stems, roots, rhizomes and seeds. The effects of these chemicals are dependent mainly upon the concentration as well as the combination in which one or more of these substances is released into the environment. The organic compounds released this way are often phytotoxins. Such toxic chemicals may be injurious to microbes and even to the seedlings of those plants releasing them. The effects of the chemicals may result in complete inhibition of growth or retarded growth. When the toxic exudates of the adult trees of a particular species suppress and eventually kill their seedlings, then the phenomenon is called autoallelopathy. Allelopathic compounds may be released into the environment by volatilization, leaching from living or dead tissues, exudation from roots and decay of plant tissues. The mechanism of action of allelochemicals is diverse and includes.

1. Inhibition of cell division and elongation
2. Inhibition of action of gibberellins or indole acetic acid that induces growth
3. Reduction of mineral uptake
4. Retardation of photosynthesis
5. Inhibition or stimulation of respiration
6. Inhibition or stimulation of stomatal opening
7. Inhibition of protein synthesis and changes in lipid and organic acid metabolism

Table 7.1: Agroforestry tree species having allelopathic effect on crop plants

Tree species	Annual crops whose growth is inhibited
Alnus nepalensis	Soybean
Casuarina equisetifolia	Cowpea, sorghum, sunflower
Eucalyptus tereticornis	Cowpea, sorghum, sunflower, potato
Gliricidia sepium	Maize, rice, tropical grasses
Leucaena leucocephala	Maize, rice, cowpea, sorghum, sunflower

Tree-animal interface (Negative): The most important negative interactions between animals and plants have direct effects. Low quality tree fodder with toxic compounds can adversely affect livestock production. Mechanical damage of trees or deterioration of soil properties through compaction by the animals may have a negative impact on the woody perennial component. Many tree fodder species in semiarid areas contain secondary compounds that reduce the feed value. The presence of high levels of phenolic compounds (tannins) or strong odours found in the leaves of species such as *Cassia siamea* and *Gliricidia sepium* may reduce palatability or acceptability of the fodder. Digestibility of some fodder may be low as the leaves contain toxins or toxic concentrations of certain micronutrients. Some of the harmful compounds reported from the tree fodders are mimosine in *Leucaena*, cyanogenic glucosides in *Acacia* species and robitin in *Robinia.*

Further subdivision of the interactions as suggested by Vandermeer (1989) is given in table 8.2. Possible tree-crop interactions and tree-animal interactions are summarized in table 8.3.

Table 7.2: Analysis of interactions between two populations 'A' and 'B'

Type of interaction	Effect of the interaction on the population		Nature of the interaction	Agroforestry example
	A	B		
Mutualism	+	+	Interaction favourable to the two populations	Mycorrhizae, Rhizobium - legume
Facilitation	+	0	Interaction favourable for A but not obligatory; B not affected	Windbreaks, shade trees Alley cropping (well managed)
Commensalism	+	0	Interaction obligatory for A; B not affected	Support trees for vines; Improved fallows
Neutralism	0	0	None of the populations affects the other in crop land	Scattered trees
Parasitism/predation	+	-	Interaction obligatory for A; B is inhibited	Pest and disease
Amensalism	-	0	A inhibited; B not affected	Allelopathy
Competition and interference	-	-	Each population is inhibited by the others use of (above or below ground) growth resources	Alley cropping (poorly managed)

0: No significant interaction; +: Advantage for the population in question (growth, survival, reproduction, etc.); -: disadvantage for the population in question

Table 7.3: Summary of tree-crop interactions and tree-animal interactions

Type	At tree-crop interaction (TCI)	At tree-animal interaction (TAI)
Positive	a) Shading trees (stress reduction) b) Efficient use of light or reduce waste of light resources c) Biomass contribution d) Microclimate amelioration e) Balanced utilization of nutrients f) Efficient use of aerial space g) Water conservation h) Weed suppression i) Soil conservation	a) Shading b) Manure decomposition
Negative	a) Competition for light b) Competition for nutrient c) Competition for water d) Weed growth increasing e) Allelopathy	a) Phytotoxins b) Browsing damage c) Trampling d) Disease or pest host

MANAGEMENT OPTIONS TO NEUTRALIZE NEGATIVE INTERACTIONS

The magnitude of interactive effects between trees and other components of agroforestry systems depends on the characteristics of the species, their planting density and spatial arrangement and management of the trees. The goals of management practices should be to increase the production of the desired products and to decrease competition of undesired components.

Management options to achieve increased growth of components in agroforestry systems are microclimate amelioration, fertilization, application of mulch or manure, irrigation, soil tillage, adapted species and supplemental feeding. Similarly, management options for decreased growth include pruning, pollarding, root pruning, trenching, excessive shading, application of herbicide and grazing or browsing. Manipulating densities and arrangements is probably the most powerful method for capitalizing on beneficial effects of trees while reducing negative ones. But when trees are used as supports for crop plants, the planting density of the trees is determined by the planting density of the crops. In such conditions, knowledge of the light transmission characteristics of the tree crowns and of the options for tree management becomes important.

In many cases one cultural treatment will accomplish both the goals simultaneously, e.g., pruning trees in alley cropping and applying the biomass to the soil. Though pruning will reduce the tree's competitive ability, it will also increase the growth of the associated intercrop by providing green manure and by allowing more light to penetrate to the crop.

Under conditions of severe below-ground competition, root pruning operations or trenching may eliminate or at least strongly reduce the negative effects of the trees on the intercrop. Trenching also lowers the moisture use by crops. In poorly fertile soils, addition of fertilizer enhances the productivity of components.

Fallowing the land can be beneficial where residual effects of the trees benefit the crops in subsequent years. Choice of tree species is crucial with regard to shading effects, root competition or provision of useful products for the farmer. As trees generally have a long lifetime, a good choice is a far-reaching decision, which has effects in the longer term.

METHOD FOR QUANTIFYING INTERACTIONS

Interactions between the components in an agroforestry system can be positive, neutral or negative. When the interaction is positive, there is complementarity between the components, while there is competition if interaction is negative. The balance between the positive and negative effects determines the overall effect of the interactions in a given agroforestry system.

ICRAF has developed an equation for quantifying tree-crop interaction (I), considering positive effects of tree and crop yield through soil fertility enrichment (F) and negative effects through competition (C) for growth resources between tree and crop. This equation is also called 'tree-soil-crop interaction'. The simple form of this is $I = F - C$. If $F > C$, interaction is positive,i.e., agroforestry system is beneficial or productive; if $F < C$, interaction is negative,i.e., agroforestry system is harmful or not productive; and if $F = C$, interaction is neutral.

Tree-crop interaction (TCI) equation

A simple approach is to quantify the effects of each factor in relation to the performance of the control or sole crop in the absence of any interference by trees, assuming that main factors are not confounded experimentally. For example, the overall tree-crop interaction (I) on crop yield can be expressed as $I = F - C \pm (M + P + L)$, where F is the benefit of tree pruning on soil fertility and microclimate of the soil surface; C is the reduction in crop yield due to interspecies competition between the tree and crop; M the consequence of above-ground changes in temperature, light, humidity and wind speed; P the consequences of change in soil properties; and L the reduction in losses of nutrient or water. The effect of each factor is expressed as a percentage of that with respect to the control crop.

Land equivalent ratio

The concept of land equivalent ratio (LER) is the most widely accepted index for evaluating the effectiveness of all forms of mixed cropping, which has been extended to agroforestry also. LER is the relative land area under sole crop required to produce the same yield as obtained under a mixed or an intercropping system at the same management level. It is calculated as sum total of the ratios of yield of each component crop in an intercropping or mixed cropping system to its corresponding yield when grown as a sole crop. Thus, LER can be expressed as, LER = (Xi / Xs) + (Yi / Ys), where X and Y are the component crops in either an intercrop 'i' or a sole crop 's' system. When LER = 1, there is no advantage to intercropping over sole cropping, when LER < 1, more land is needed to produce a given yield by each component as an intercrop. Intercropping is beneficial only when LER > 1.

Competition indices

The effect of interspecies and intraspecies competition can be analysed by varying the proportion of the species or the population of the mixture. This approach has been used for crop-weed competition and has been extended to agroforestry systems also.

Relative crowding coefficient is a measure of the relative dominance of one component crop over the other in an intercropping or mixed cropping system. The coefficient (K) is determined separately for each component crop. For crop 'a' in association with crop 'b', the relative crowding coefficient is, $K_{ab} = (Y_{ab} \times Z_{ba}) / [(Y_{aa} - Yab) \times Z_{ab}]$, where Y_{ab} is the yield of species 'a' in association with 'b', Y_{aa} is the pure stand yield of species 'a', Z_{ab} and Z_{ba} represent the sown proportion of 'a' (in association with 'b') and 'b' (in association with 'a'), respectively. Similarly, for crop 'b' in association with crop 'a', the relative crowding coefficient is, $K_{ba} = (Y_{ba} \times Z_{ab}) / [(Y_{bb} - Y_{ba}) \times Z_{ba}]$, where Y_{ba} is the yield of species 'b' in association with 'a', Y_{bb} is the pure stand yield of species 'b'. The product of the two coefficients, K_{ab} and K_{ba} designated as K, denotes yield advantage if K > 1, no advantage if K = 1 and disadvantage if K < 1.

Mulch + shade model

Fast growing trees produce a lot of mulch material, but also cast a lot of shade. VanNoordwijk and Hairiah (2000) presented explicit algebraic solutions for an agroforestry model, which links both the mulch production and its ensuing soil fertility effect and the shading which is assumed to have a negative effect on crop yields to the biomass production of the tree. The model leads to a simple mulch/shade ratio as a basis for comparing tree species. The model also predicts that at low soil fertility, where the soil fertility improvement due to mulch can be pronounced, there is more chance that an agroforestry system improves crop yields than at higher fertility where the negative effects of shading will dominate. The mulch + shade model, however, does not incorporate the dynamic interactions among water availability, N dynamics, and crop and tree growth. Incorporating these elements on the basis of a daily time step extends the model beyond what can be solved explicitly. It leads into the realm of dynamic simulation models, which keep track of resource stocks outside and inside the plants and use these to calculate daily resource flows and daily resource capture.

8

Structural and Functional Attributes of Agroforestry Systems

All agroforestry systems are characterized by three basic sets of attributes; productivity, sustainability, and adoptability (Nair, 2008).

Productivity: The majority of agroforestry systems, if not all of them, strive to maintain or raise both land productivity and output (of desired commodities). In several aspects, agroforestry may raise production. These include higher yields of related crops, less input use in the cropping system, enhanced labour efficiency, and higher production of tree products.

Sustainability: Agroforestry can achieve and indefinitely maintain conservation and fertility goals by conserving the production potential of the resource base, mainly through the beneficial effects of woody perennials on soils.

Adoptability: Improved or new agroforestry technologies that are introduced into new areas should fit the social as well as environmental characteristics of the land use system for which it is designed and should also confirm to local farming system.

KEY TRAITS OF AGROFORESTRY PRACTICES

Agroforestry practices are intentional combinations of trees with crops and/ or livestock which involve intensive management of the interactions between the components as an integrated agroecosystem. The four key characteristics - intentional, intensive, interactive and integrated are the essence of agroforestry which distinguish it from other agricultural or forestry practices. Thus, a land use practice to be called as agroforestry must satisfy all of these four criteria.

Intentional: Combinations of trees, crops and/or animals are intentionally designed and managed as whole units, rather than as individual elements which may occur in close proximity but are controlled separately.

Intensive: Agroforestry practices are intensively managed to maintain their productive and protective functions, and often involve annual operations such as cultivation, fertilization and irrigation.

Interactive: Agroforestry management seeks to actively manipulate the biological and physical interactions among the tree, crop and animal components. The goal is to enhance the production of more than one harvestable component at a time, while also providing conservation benefits such as control of soil erosion or protection of wildlife habitat.

Integrated: The tree, crop and/or animal components are structurally and functionally combined into a single, integrated management unit. Integration may be horizontal or vertical, and above- or below- ground. Such integration utilizes more of the productive capacity of the land and helps to balance economic production with resource conservation.

PREMISES OF AGROFORESTRY

The premises on which the concept of agroforestry is based are partly biological and partly socioeconomic.

Biological premises

Agroforestry has a beneficial effect on the soil through efficient nutrient cycling. The roots of trees take up nutrients from the soil, convert and utilize them for the production of plant material and then return them to the soil in the form of tree litter. This litter is transformed into humus and later incorporated into the soil. In a well-managed agroforestry system, the relatively more efficient nutrient cycle minimizes the leakages of nutrients from the system.

Trees are generally deep-rooted than agricultural crops and are often able to trap and utilize nutrients that have been leached from the upper layers of the soil. Some tree species have the capacity of 'pumping' nutrients from layers that are not normally tapped by agricultural crops. The compacting effects of falling rain on the soil are reduced in an agroforestry system decreasing soil erosion and thus another possible source of leakage of nutrients from the system is plugged.

Agroforestry is a system of land management in which tree crops are grown together with agricultural crops, one of its objectives being to optimize and sustain the total yields of the component crops. Competition among the different components of the system is not great enough to affect the total productivity of the system in an adverse manner. Water, nutrients and light are the limiting factors in an agroforestry system. The forest and agricultural species that are utilized in the system should be compatible and should complement each other during most stages of their lives. More specifically, with respect to water they should be unequal in competitive ability; with respect to nutrient, they should vary in ability to utilize the nutrients in different forms; and, with respect to light, those species should be selected which display growth patterns, rates of growth, phenology, and architecture permitting maximum interception of light by both the agricultural and forest crops at any one time.

Socioeconomic premises

Forests are being felled in by farmers who require the land to produce food for their very existence. These areas are basically unsuitable for arable agriculture, either because of the inherent infertility of the soils, or because the sites are prone to accelerated erosion, or because of a combination of these two factors. The people who clear the forests to produce food are often not unaware of the possible deleterious effects of their practices upon the ecosystem in terms of erosion hazards, the occurrence of droughts and floods, and the possibility of soil fertility decline. Yet, despite their knowledge of these adverse consequences, they persist with such practice due to lack of suitable alternatives for their survival.

The failure to develop the marginal lands often leads to retardation in the rate of improvement of the general economy. The developmental and technological options are fewer in marginal areas than in most other ecosystems. When the biological influences and services of forests are considered along with the specific socioeconomic problems of those who exist in marginal areas, the technological package should include agroforestry systems. If the economic returns from the agroforestry systems are significant and if these are designed to optimize the joint productivity of wood and food from the same unit of land with careful choice of agricultural and forest species and suitable management practices, the socioeconomic developmental problems of the area would be addressed adequately.

SHIFTING CULTIVATION

A pattern of land use and a crop production system known as 'shifting cultivation' involves clearing land (usually by fire), cultivating it for a few years to raise one, two, or three crops, and then abandoning the area to a new location until its fertility has naturally returned.

There are differences between the fire fallow cultivation method, which is slash and burn, and a much older shifting agriculture approach. The duration of the land's agricultural usage is the primary distinction between shifting cultivation and the slash and burn method. The conversion under the slash and burn method is long-term and frequently irreversible. A more transient use of the land for cultivation is shifting cultivation.

People have been using shifting cultivation for subsistence farming in tropical forests all over the world for a very long time. The north-eastern hill region of India, which includes the states of Assam, Meghalaya, Manipur, Nagaland,

Tripura, Arunachal Pradesh and Mizoram, is home to a large number of practitioners of this system. Additionally, Andhra Pradesh, Bihar, Madhya Pradesh, Odisha, and Karnataka states practise it to varying degrees. Referred to as '*jhum*' in the hilly regions of north-eastern India and '*podu*' in the states of Andhra Pradesh and Odisha, it is especially detrimental to forest areas.

The practice of shifting cultivation causes a significant acceleration of soil degradation since it leaves cropland uncultivated for durations varying from one season to multiple years. With the advancement of agroforestry's better fallow system, shifting cropping is now enhanced. Recovering depleted soil nutrients is the aim of an enhanced fallow system. Because of the severe lack of available land, crops are reintroduced once the soil has recovered for one or more seasons, with the fallow periods getting shorter.

Choice of species

Fallow areas are typically used to maintain or replenish soil fertility and minimise soil erosion, while certain species are planted particularly for commercial purposes. Plants having commercially viable products should also be taken into consideration when selecting species; 'soil improvers' should not be the only ones used. Improved fallows should contain plants that are beneficial to crops in the future, do not harm the soil physically or chemically, and do not compete with crops that will be grown on the same spot in the future.

A multitude of techniques can be employed to establish improved fallows at different phases of the fallow cycle. These techniques may include direct sowing of clean, harvested and tilled plots; selective cutting of bush, followed by tall seedling enrichment planting; introducing tall seedlings and cuttings into fallows of lower quality on degraded land; and planting tree seedlings in deeply spaced, closely spaced planting holes or furrows within blocks of cleared cropland. The precise methods vary depending on the land's past usage, its worth as fallow, its state as fallow vegetation, and how long the fallow is projected to last.

TAUNGYA SYSTEM

Under the taungya method, which is a modified version of shifting cultivation, the right to produce agricultural crops is granted to the local population in the early phases of the creation of forest plantations. The procedure involves clearing the ground, planting trees, raising agricultural crops for one to three years, or until the canopy cover is too dense, and then starting the process over in a new location. Crops may occasionally be planted a year ahead of trees in certain situations. Depending on the soil and climate, a wide range of crops and trees are planted.

Trees and associated Crops

Trees: *Acacia nilotica*, *Anacardium occidentale*, Bamboo spp., *Ceiba pentandra*, *Hevea brasiliensis*, *Tamarindus indica*, *Tectona grandis*, etc.

Crops: Millets, pulses, oilseeds, cotton, tapioca, banana, vegetables, *etc*.

There are three kinds of taungya systems. When the forest department employs many people on a daily wage basis to produce agricultural crops and plantations, the practice is known as departmental taungya. Here crops are mostly grown to minimise weed growth, generate extra revenue, and make the best use of available land. In leased taungya the highest-bidder who guarantees crop maintenance and grows crops for a predetermined number of years is granted a lease on the plantation property. In village taungya, the people who have made their home in a hamlet inside the forest are allowed to cultivate crops. Every household typically has usually less than two hectares of land on which to grow food and nurture trees for three to four years. Out of the three taungya methods, this one is the most effective.

Table 8.1: Crops grown in taungya system in India

State	Tree crop	Associated agricultural crops
Uttar Pradesh	*Shorea robusta, Tectona grandis, Acacia catechu, Dalbergia sissoo, Eucalyptus* spp, *Populus* spp.	Maize, paddy, sorghum, pigeon pea, soybean, wheat, barley, chick pea, rapeseed
Andhra Pradesh	*Anacardium occidentale, Tectona grandis, Bombax ceiba*, Bamboo, *Eucalyptus* spp.	Hill paddy, groundnut, sweet potato
Kerala	*Tectona grandis, Eucalyptus* spp., *Bombax ceiba*	Paddy, tapioca, ginger, turmeric
Assam	*Shorea robusta*	Paddy
Tamil Nadu	*Tectona grandis*, Bamboo, *Santalum album, Tamarindus indica, Acacia nilotica, Acacia mearnsii, Ceiba pentandra,* Cashew, Rubber	Millet, pulses, groundnut, cotton
West Bengal	*Tectona grandis, Shorea robusta, Schima wallichii, Cryptomeria japonica, Quercus* spp., *Michelia doltsopa*	Paddy, maize, millets, turmeric, ginger, lady's finger, pineapple, sunhemp
Andaman & Nicober Islands	*Pterocarpus dalbergioides*	Sugarcane, maize

Source: Tejwani, 1994

ALLEY CROPPING

The technique of growing annual crops in the 'alleys' or gaps between rows of multifunctional woody perennials is known as hedgerow intercropping or alley cropping. As an alternative to the conventional tree fallow method, it is more promising. The woody crops are routinely harvested, and the cropped alleys are mulched with leaves and twigs to reduce evaporation from the soil's surface, inhibit weed development, and replenish the soil with necessary nutrients and organic matter. The main elements of hedgerows in areas where crop productivity depends on nitrogen are nitrogen-fixing plants.

Alley cropping's main goals are to ameliorate microclimate, suppress weed growth, and improve soil fertility in order to maintain or raise crop yields. The hedgerows may provide farmers with a variety of tree products, such as food, fuelwood, fodder, building poles, medicine and more. Hedgerows and mulching with trimmed plant parts helps in controlling soil erosion on sloping terrain. Alley cropping often performs better in areas where there is a need to increase crop yield but issues with soil fertility.

The main uses of tree hedgerows are for the cycling of nutrients from deeper soil layers, the benefit of green manure and mulch to companion food crops, the suppression of weeds by mulch and shade, the creation of favourable conditions for beneficial soil organisms, the control of soil erosion, the provision of green animal fodder, the use of prunings for firewood and construction poles, and the supply of biologically fixed nitrogen to the system.

Alley cropping model

In agricultural areas, woody plants are planted as hedgerows to enhance the benefits of trees and minimise their negative impacts on crop management and yields. Trees unaccountably compete with agricultural crops for light, moisture, and nutrients in the soil. But fundamentally, croplands become more productive when the right sort of trees are chosen and planted at the proper spacing, together with effective crop management.

In an alley cropping system, the location and spacing of crop and hedgerow plants are determined by the kind of plant, soil, climate, slope, and amount of space needed for intercultural activities. Hedgerows should ideally be placed east-west, parallel to the contour of the slope, and perpendicular to the predominant winds. Maximum solar exposure for the "alley" (crop) and maximum erosion control are provided by this orientation. Depending on the slope, the location and layout of soil and water conservation buildings, hedgerow positions and spacing vary. Hedgerows on sloping terrain should always be positioned according to their contour. If they do not have the ideal east-west orientation, they may need to be thinned on a regular basis to avoid overshadowing nearby crops.

Alley width is the primary consideration for designing high-intensity hedge row system. The intensity of crop-tree root competition and the ratio of tree products/services to crop products are determined by alley width. Tree hedges develop and agricultural yields drastically decrease if alleys are too small. Trees' ability to improve soil is diminished if alleys are very broad. It is advised to have a breadth of 4 to 8 metres based on proven techniques.

Two lines of trees spaced 0.5 metres apart make up the suggested hedge for the production of biomass, feed, or mulch material. Trees inside the lines should be separated by 0.5 metres. Trees inside the lines of a timber-based alley cropping system should be spaced one to three metres apart. Closer spacing promotes the growth of fewer branches and more leaves. In humid areas, the wider spacing is often used, whereas in sub-humid or semi-arid areas, the closer spacing.

Choice of species

The choice of tree species for alley cropping is extremely important and to a large extent determines the success or failure of the system. The following attributes are to be considered when selecting a tree species for an ideal alley cropping.

Bushy form: Multi-stemmed woody perennials usually yield a greater biomass of a leaf: stem ratio than bigger, single-stemmed species.

Quick growth and biomass output: More production of short branches and leafy biomass indicates a build-up of excess litter/mulch, more feed, and/or more fuelwood.

Simple to establish: Growing trees from seed, whether directly in the field or in containers, is a straight forward process.

Easily re-establishing: The stems and leaves regrow repeatedly and quickly following the silvicultural practices such as pruning, coppicing, pollarding, or lopping.

Nitrogen fixing: Trees that fix nitrogen from the atmosphere can provide an additional advantage to the system.

Deep rooting: To avoid competing with crops and absorb nutrients and water beyond their reach, it should develop a deep taproot system with a small number of lateral root branches close to the surface.

Ease manageable: It should possess thin lateral roots that enable for easy pruning by ploughing along the hedgerow without causing significant harm to the crop plants. This makes it readily controllable.

Easily decomposable: Either all or part of the leaf litter should break down quickly, releasing nutrients to facilitate the cropping cycles within the system.

Multipurpose: Alley cropping should yield wood, food, medicine, fodder, and other things that are useful to growers and the surrounding community. It should be multipurpose.

Broadly adaptive and stress-tolerant: The chosen species must be able to flourish in the face of the site's unique constraints, including acidic or salty soils, severe winds, droughts, and flooding.

Free from diseases and pests: Trees should not serve as a home for fungi or insects that harm crops.

Gliricidia sepium, Senna siamea, Leucaena leucocephala, Sesbania grandiflora, Sesbania sesban are a few suitable alley tree crops.

Hedge management

Pruning hedges on a regular basis is required to keep agricultural crops from being shaded out by tree crops. The shade tolerance of a certain root crop at a given stage of growth will determine how much and how often trimming is necessary. As a general guideline, the hedge height should be equal to the separation between it and the closest row of crops. Frequency of pruning also depends on the intended yield mix. More green manure is added to the soil or more fodder may be harvested with more frequent pruning. On the other hand, there will be more fuelwood diameter twigs with less frequent trimming. In tropical regions where breakdown leaves and organic matter mineralization occur quickly, the timing of prunings is particularly crucial. The pruning timing should be decided withthe time required for litter decomposition that will release nutrients when the crops need it most.

HOMEGARDENS

Homegardens are always considered as a major additional source of food and nutrition security as well as a means of subsistence across the world. Homegardens may have started in prehistoric times when hunters and gatherers unintentionally or purposely scattered the seeds of extremely important food-producing trees near their campsites. The most traditional and long-lasting type of farming is the production of food on small plots next to populated areas. Homegardens have been a vital part of local food production systems and family farming for millennia.

Home gardening, in general, refers to the cultivation of a small plot of land that is closer to the house or accessible by foot. The homegarden is a small-scale food production system that provides food for humans and animals, as well as consumables that are either hard to come by, prohibitively expensive, or easily accessible through field agriculture, retail marketplaces, hunting, fishing, and gathering. Due to reasons of security, convenience, and particular maintenance, homegardens are typically found closer to the house. They utilise labour that is marginal to primary home economic activity and occupy land that is marginal to agricultural output. Homegardens are characterised by a lower capital input and straightforward operating technology, with a variety of environmentally appropriate and complimentary plants.

A homegarden is a clearly defined, multi-levelled, multipurpose space close to the family home. It functions as a small-scale, supplemental food production system that is maintained by the family members and includes a variety of plant and animal species that resemble the ecosystem of the natural world. The ability of homegardens to provide goods for subsistence and generate extra revenue has been the main draw for interest in them. They are renowned for their minimal input requirements, consistent or recurring harvests throughout the year, wide variety of crops, and steady yields. Modern homegardens are primarily used for the production of food, fodder, and lumber. A mixed cropping system that includes decorative plants, medicinal herbs, plantation crops, fruits, vegetables, and spices, together with animals that can be raised there is what is known as a homegarden.

This system is widespread in tropical areas with considerably high rainfall. Homegardens are examples of land use systems where multifunctional trees and shrubs are intentionally managed in close proximity to both annual and perennial crops as well as animals. A homegarden's exact dimensions vary from household to household, although on average, they are less than the household's total area of arable land. The majority of homesteads in India have between 0.20 and 0.50 hectares of land, which indicates that the tradition is sustained. Kerala is one state where homegardens are common.

The great diversity of species with various living forms in the homegarden is one of its most remarkable structural features. Numerous types of herbaceous plants, including trees, shrubs, and vegetables, are cultivated in dense, haphazard, or spatial configurations. The majority of homegardens in the tropics may be categorised as agrisilvipastoral systems, which include herbaceous crops, woody perennials, and animals, depending on the kind of components used. Nonetheless, some homegardens only include the first two elements of an agrisilvicultural system. Family members are actively managing every animal unit in the tree crop. As a result, homegardens are extremely practical, sustainable, and prolific.

Species diversity in homegardens

Plant density and species diversity are location-specific and depend on a variety of ecological and socioeconomic variables. Homegardens were classified as the most biologically diverse man-made agroecosystems, according to a categorization based on species diversity and production strategies. One recent study, for example, examined the species utilisation, structural compositions, and agrobiodiversity of 330 homegardens in Thrissur, Kerala. It found that out of these, 525 plant species, representing 320 genera and 102 families, were the most prevalent; 237 of these were trees (> 20 cm girth at breast height), representing 140 genera and 49 families, 100 shrubs (62 genera and 32 families), and 187 herbs (138 genera and 54 families) (Kunhamu, 2015). Usually, homegardens are developed on a small scale due to custom, taste, or availability of planting materials. They preserve species and types that are not profitable in field agriculture.

Structure and composition of homegardens

The most notable features of all homegardens are their layered layout and combination of friendly species. Most of the homegardens may not have any defined structure since trees, shrubs, and herbaceous plants grow in a complicated variety without any discernible pattern. The absence of any obvious planting pattern, typically there are no blocks, rows, or precise planting spacing between the components, also contributes to the structure's forest-like appearance. A closer examination of the vegetation reveals that the plants actually do appear to be arranged in various levels or patterns. Because homegardens have many canopy layers, they are also known as 'multi-tier' systems.

The layered configurations and combination of compatible species are the most important characteristics of all homegardens. The gardens are carefully structured systems where each component has a specific place and function. Every homegarden usually consists of an herbaceous layer near the ground, a tree layer at the upper level, and intermediate layers in between. The lower layer is usually divided into two, with the lowermost (less than 1 m height) dominated by different vegetable and medicinal plants, and the second layer (1-3 m height) being composed of food plants such as banana, papaya, yam and so on. The upper tree layer is also partitioned into two, consisting of emergent, fully grown timber and fruit trees occupying the uppermost layer of over 25 m height, and medium-sized trees of 10-20 m height occupying the next lower layer. The intermediate layer of 3-10 m height is dominated by various fruit trees.

Averrhoa carambola, *Anacardium occidentale*, *Psiduim guajava*, *Phyllanthus emblica*, *Citrus* spp., *Artocarpus heterophyllus*, *Mangifera indica*, *Swietenia macrophylla*, *Tamarindus indica*, *Tectona grandis*, and other tall trees make up the uppermost canopy layer. Spices, coffee, banana, papaya, pepper, and other ingredients predominate in the centre layer. Amaranthus, arrowroot, okra, brinjal, beans, cowpea, ginger, onion, pumpkin, pineapple, tapioca, turmeric, sweet potato, yam, etc. make up the lowest layer. The livestock components consist of goats, chickens, rabbits, and cows.

Significance of homegardens

Sustained food and nutritional security: Homegardens indirectly contribute significantly to food production by improving the soil's capacity to sustain agriculture and by directly producing edible goods like fruits. Seldom do the homegardens provide all of the family's fundamental essential food needs. They do, however, compliment other croplands. Homegardens are therefore a part of the household's bigger farming system. In fact, the species composition of the homegarden would be dominated by food crops such cassava, yam, vegetables, banana, moringa, etc. if it is the only land the household has access to.

Household energy requirement: For rural households, traditional homegardens are the main source of biofuels.

Supply of wood: Homegardens meet the small timber requirement by the households. In Kerala, a study reported that homegardens supplied more than 70% of the wood needed, and of the state's total bamboo requirement, homegardens accounted for 63% and forests for the remaining 37%.

Employment opportunity: Homegardens create employment opportunity throughout the year.The family labour is thus fully utilized. The use of labour for the farm is also very efficient due to proximity of the garden to the house, within the walking distances.

Increased economy: Homegardens serve as stores for non-timber goods such green leaf manure, gums, resins, fragrant and medicinal plants, and ornamentals. Furthermore, a variety of goods grown in homegardens are frequently essential in bridging diverse cultures within the village ecology. The people that weave bamboo use bamboo poles that are available from a household's gardens to weave mats, baskets, fish nets, and other goods. The weavers give some of the finished products to the people who supply the raw materials and will sell the remaining goods in the market.

Protective function: The protective function of homegardens is very high with soil conservation, water retention and favourable microclimate for humans and animals.

Carbon sequestration: The numerous plant species in homegardens, particularly the woody perennials and soil, have a great capacity for storing and sequestering carbon. By reducing the pressure on natural forests, they aid in the conservation of the carbon stock in already-existing forests.

Ecosystem services: Through a nutrient pumping mechanism, the close relationship between the annual and perennial components improves nutrient recycling and lowers susceptibility to soil erosion. The biomass of the tree components and the species composition have a significant impact on nutrient turnover. The multi-species, multi-strata composition of homegardens allows for an improvement in resource use efficiency, in contrast to monoculture or dual component systems.

Preservation of traditional knowledge: Through experience and observation, the locals are better informed about the flora and animals in their homegardens, their uses, their interactions, the seasonal fluctuations in the environment, how to manage and utilise plants sustainably, and other such intricacies of how ecosystems work. It is possible to maintain this customary wisdom for the benefit of next generations.

Pleasant living environment: Homegardens ensure a pleasant living environment, shade and provision of privacy. They can also provide aesthetic values and open patches in the homegarden often serve as family or village gathering places.

WINDBREAKS

Windbreaks are collections of linear plantings, such as strips made of one or more rows of trees or shrubs, that block wind flow, slow down air currents in the lower atmosphere, and create a sheltered area around them. The following advantages come from windbreaks that are properly planned and maintained.

- Decrease the wind's velocity
- Protect pastures and crops from sweltering, dry winds
- Enhance the microclimate for crop growth
- Lessen water evaporation from water surfaces, like irrigation ponds, canals, or streams
- Lessen water evaporation from soil and plants
- Protect people and animals from cold weather
- Lessen/prevent soil erosion
- Create habitat for wildlife
- Lessen water evaporation from farmlands
- Serve as delineation and fencing

Furthermore, wind barriers may yield a variety of beneficial items, such as fruit, fibre, fodder, and mulch, in addition to poles and fuelwood.

Designing windbreaks

Permeability: The most effective windbreaks filter and break wind power while still allowing some wind to flow through. A cushion of slowly flowing air is created on both the upwind and downwind sides by the air passage via the windbreaks. This keeps the majority of the wind from dropping for a while by deflecting it upwards. As a result, the shielded area's wind velocity may be lowered to 25–75% of its peak wind speed.

Dense wind barriers provide a small region of quiet air behind the trees in a short strip, although there may be significant turbulence further downwind. Dense windbreaks are preferred when a high degree of protection is required for limited areas, such those around homesteads and work areas, or for livestock that is particularly sensitive, like newborn lambs or calves. The right permeability may be achieved by carefully choosing the kinds of trees and shrubs.

Orientation: For maximum effects, windbreaks should be planted at right angles to the winds from which protection is required. However, if the wind direction changes often, shelter belts may be constructed in quadrangles.

Height: The sheltered area's dimensions are determined by the height of the wind barrier. The area that a windbreak shelters increases with its height. A windbreak on flat land will slow the wind over a distance that is roughly 20 times the height of the trees on the downwind side.

Width and length: When windbreaks extend more than 25 times the mature height of the trees, they are most efficient when they do so without any noticeable breaks. Although it should be around 1:10, the height to width ratio might vary depending on the local circumstances. In India, shelterbelts up to 50 metres wide are said to be optimum.

The number of rows: For most agricultural farmlands, windbreaks with three to five rows work better and are less impacted by gaps created by missing trees. Modest shrub species should be planted in the outer rows and tall growing tree species in the central rows.

Spacing of trees: The distance between trees varies depending on how important the windbreak is for protection or productivity. Farmers may choose shorter strips and a higher percentage of tiny trees and shrubs that provide fuelwood and feed in areas where windbreak productivity is highly valued. The distances between wind breakers and their spacing can be extended if the produce is wood.

Gaps: Gates and railway lines need gaps, yet gaps generate a funnelling effect that can significantly increase wind velocity in these regions. This may be avoided in multi-row windbreaks by angling the gap at an angle of around 45 degrees to the direction of the prevailing wind. As an alternative, to widen the opening and lessen the funnelling effect, a few trees or bushes may be planted on either side of the gate or pathway. Other options include planting a second short row to fill in the gaps or planting five or six trees as a wing at an angle to the main belt.

Choice of species: In order to establish a windbreak, trees and shrubs should be chosen based on their species' hardiness, attractive form and foliage, longevity, deep root system that prevents spread into neighbouring fields, drought resistance, low maintenance requirements, resistance to pests and diseases, suitability for the planting site, and compatibility with the windbreak design. A perfect windbreak would have two rows of shorter spreading species, like *Cassia* spp. or *Leucaena* spp., on either side of the central core, and a central core planted in double rows with fast growing species, like *Acacia auriculiformis*, *Eucalyptus* spp., *Casuarina* spp., *Dalbergia*, *Grevillea*, *Syzygium*, *Tamarix articulata*, *Parkinsonia aculeate*, etc.

As trees mature, their forms change, thus it is usually required to combine several species with varying growth rates, diameters, and shapes in multiple rows. It is best to use a quickly growing multipurpose species to get the desired effect as soon as possible. Furthermore, it is possible that some of the trees chosen would not survive as long as others. To increase the windbreak's usable life, a combination of trees with varying life spans and species should be planted. Combining several species also offers defence against pests and diseases that may quickly wipe out stands of a single species.

Features of windbreak design

Single row windbreak: A single row windbreak is designed only in cases where land is so expensive. Single row windbreak is used to protect orchards and field crops. Shrubs and dense evergreens, and densely branched deciduous trees, particularly with thin crowns, that maintain lower limbs and leaves should be planted in the single row windbreak. The primary drawback of a single row windbreak is that the efficacy of the entire windbreak is decreased if one tree is lost to pests or diseases.

Two row windbreak: In urban settings, fields and orchards, as well as some wildlife assets, are protected by two-row windbreaks. These are planted closely together, much like in a single row. Plants are planted in rows that are closely spaced apart, with a gap in one row being filled in by a tree in the other. Two rows of thick evergreens or one row of evergreens and one row of shrubs or deciduous trees are used as a traditional two-row windbreak, with regular spacing between rows.

Three row windbreak: With the row of food-bearing shrubs, the three-row windbreak protects larger fields, farmsteads, and feedlots, while also aiding in the conservation of wildlife. It is composed of two rows of deciduous trees, bushes, or pines with at least one row of thick evergreens.

Four row (or more) windbreak: In addition to being great for wildlife, the four row windbreaks are designed to protect farmsteads and feedlots. It makes use of a variety of shrubs, one row or more of thick evergreen trees, and deciduous trees. Within rows, it might switch up related species. Greater design freedom allows for the usage of vast tracts of land. Generally speaking, windbreaks with three to five rows work better in most agricultural contexts and are less impacted by gaps left by missing trees.

SHELTERBELT

The purpose of a shelterbelt is to provide general protection to cultivated areas against wind erosion and the desiccating effect of hot winds on the leeward side. It is a wide belt of trees, shrubs, and grasses planted in rows that cross the land at a right angle to the direction of the prevailing winds.

The triangle cross-section of a conventional shelterbelt may be created by placing towering trees in the middle, followed by lesser trees, tall shrubs, low spreading shrubs, and grasses on either side. In shelterbelts, a certain degree of penetrability is preferred since it causes the zone of effect to be significantly larger and the velocity curve to exhibit a smooth, gradually diminishing trend.

The width of a shelterbelt depends on the soil type, wind speed, and local climate. A shelterbelt should be positioned as closely at right angles to the direction of the predominant wind as feasible. When winds are coming from several directions, quadrangles should be used to increase the shelterbelt. The major objectives of the shelterbelts included inverting air currents, lowering the prevailing winds' velocity, shielding the leeward regions from wind erosion, reducing the drying impacts of hot wind and supplying food, fodder, timber, and other necessities.

Attributes of shelterbelts

Shape and composition: The form of shelterbelts is commonly triangular. Tall trees may often be raised in the core to achieve this.

Density and width: A certain amount of wind penetration is intended since the leeward side's protection rapidly diminishes if the barrier becomes a solid wall. In India, shelterbelts up to 50 metres wide are generally considered to be optimum.

Height and spacing: A height to width ratio of around 1:10 is appropriate. But depending on the local factors this can be changed.

Orientation: The direction and speed of the predominant winds determine how shelterbelts are oriented. In the event that the wind direction fluctuates often, shelterbelts may be raised in quadrangles.

Length: Length of a shelterbelt is a crucial factor. A shelterbelt should be at least 25 times longer than its height.

Species selection: The following species are suggested for a shelterbelt.

Grasses: *Panicum turgidum*, *Panicum antidotale*, *Saccharum spontaneum*, *Saccharum munja* and *Cenchrus barbatus*.

Shrubs: *Calotropis procera*, *Agave* spp., *Sesbania aculeata*, *Cassia auriculata*, *Dodonaea viscosa*, *Jatropha curcas*, *Leptadenia spartivm*, *Calligonum polygonoides*, and *Clerodendron phlomoides*.

Small trees: *Salvadora oleoides*, *Acacia jacquemontii*, *Acacia leucophloea*, *Balanites aegyptiaca*, and *Capparis aphylla*.

Trees: *Albizzia lebbek*, *Azadirachta indica*, *Dalbergia sissoo*, *Lannea coromendelica*, *Parkinsonia aculeata*, *Acacia arabica*, *Acacia Senegal* and *Acacia cyanophylla*, *Prosopis cineraria*, *Prosopis juliflora*, *Acacia tortilis*, *Tecoma undulata*, *Pongamia pinnata*, *Tamarix articulata*, *Eucalyptus* spp.

MULTIPLE CROPPING

Growing two or more crops on the same piece of land in one calendar year is known as multiple cropping. It is the intensification of cropping in time and space dimensions i.e. more number of crops within a year and more number of crops on the same piece of land at any given period. Multiple cropping makes effective use of inputs such as soil, water, fertiliser etc. Thus output per unit area increases. It can be practised in annual food crops, fodders, vegetables, fruit plants and perennial crops. It includes intercropping, mixed cropping, sequence cropping, crop rotation, relay cropping and paira cropping, etc. For successful multiple cropping, availability of irrigation facility and other input resources including labour should be ensured at right time.

Principles of multiple cropping

- The crops should be agro-climatically suitable and technically feasible.
- The most profitable crops capable to utilize growth resources and land in best manner should be chosen.
- The crop combinations and input rations should be compatible with farmer's skill, enterprise preference, capital, assured irrigation facility.

- The essential inputs like improved seeds, fertilizers, pesticides and labour should be available locally.
- Marketing, storage and transportation facilities of the produce are to be ensured before choosing the crops.

Advantages of multiple cropping

- Land is better utilized through multiple cropping.
- In multiple cropping, incidence of crop failure owing to biotic agents is minimised. The loss of any crop due to biotic or abiotic stress may be compensated to some extent by another crop.
- Overall production and yield per unit of area is increased.
- Cost of production is reduced in multiple cropping as compared to individual crops grown for the same level of production.
- Pest and disease infestation is checked.
- Different type of products can be produced at a time in multiple cropping.
- Multiple cropping provides a balance diet for a farm family
- Multiple cropping helps in maintaining soil fertility provided suitable crops such as legumes are included in the cropping system. It improves or maintains the soil health.
- Weed menace is reduced in multiple cropping. It narrows the space available for weeds to grow and sometimes may hamper their growth through exudation of allelochemicals.
- An important aspect of multiple cropping is the utilisation of nutrients more efficiently as the crops growing on the same piece of land simultaneously would have different nutritional requirements.
- Multiple cropping is also important from marketing point of view. As more than one product is available simultaneously if the selling price of one product is less in the market, the other will be there to compensate.
- It generates more employment opportunity by having crops round the year.

Limitations of multiple cropping

- Multiple cropping may reduce the farming efficiency of a farmer. When a farmer grows different type of crops, he cannot specialize in production technology of all the crops. He has to acquire the latest technologies for the crops grown which may not be possible at all the time.
- Sometimes the crop management decisions become difficult and often

the advice of the professionals may be required.

- The survival of pests becomes easier in a multiple cropping. Pests can easily shift from one crop to another crop. Sometimes the insect pests and diseases get more favourable environment to flourish, thus diminishing and deteriorating crop yields.
- Inter-cultural practices are sometimes difficult to be carried out in the crops grown together.
- Weed management may also be difficult.
- Initial investment may be high with procurement of different types of inputs including equipments and machineries for a number of crops at a time.

INTERCROPPING

Practice of intercropping is as old as agriculture. But scientific approach to it is relatively new. Intercropping refers to growing of two or more dissimilar crops simultaneously on the same piece of land. The recommended optimum plant population of the base crop is suitably combined with appropriate additional plant density of the associate crop. Various workers have classified and defined intercropping in many ways. However, the objective remains the same, i.e. the intensification of cropping both in time and space dimensions and to raise productivity per unit area by increasing the pressure of plant population. It has better utilization of growth resources than sole cropping. Intercropping can be practised as annual plants with annual plants intercrop, annual plants with perennial plants intercrop and perennial plants with perennial plants intercrop.

Types of intercropping

Intercropping may be of various types based on the method of sowing and objective of cropping.

Mixed intercropping: It is an intercropping where two or more crops are grown simultaneously intermingled in the same piece of land with no distinct row arrangement. This type of intercropping can be suitable for grass-legume intercropping in pastures.

Row intercropping: It is an intercropping where at least one crop is planted in regular rows, and crop or other crops may be grown simultaneously in row or randomly with the first crop.

Mixed row intercropping: It is growing two or more crops simultaneously in the same piece of land intermingled within a distinct row arrangement.

Strip intercropping: It is growing two or more crops simultaneously in different strips wide enough to permit independent cultivation, but narrow enough for the crop to interact with each other.

Relay intercropping: Growing two or more crops simultaneously during at least a part of the life cycle of each. A second crop is planted after the first crop has reached its reproductive stage but before it is ready for harvest. This practice is beneficial in terms of resolving time conflict for plantation of various crops. For example delayed sowing is an important reason for yield decline in field pea or chickpea that can be avoided through relay cropping of these crops in standing rice crop. Relay cropping can also fetch certain other benefits such as usage of residual nutrients and moisture from the previous crop and reduced planting costs.

Parallel intercropping: In this practice two crops are selected which have different growth habits and have a zero competition between each other and both of them express their full yield potential. Greengram or blackgram with maize and greengram or soybean with cotton are the examples of parallel intercropping. Annual agricultural crops may also be grown with a tree crop or a perennial fruit crop.

Companion cropping: In companion cropping the yield of one crop is not affected by other. In other words, the yield of both the crops is equal to their pure crops. The standard plant population of both crops is maintained. *Toria*, wheat, potato, etc. can be taken as companion crops with sugarcane.

Additive intercropping: Here the plant population of the main or base crop remains unchanged as a pure crop stand but one or more minor crops are grown simultaneously with it by utilizing the inter-row space. Pulses, rapeseed or potato can be taken as intercrops by utilizing the inter-space of two rows of a fruit crop like mango, coconut or guava during initial years of establishment planted with its normal spacing.

Substitutive intercropping: Here one or more rows of the main crop are replaced with same number of rows of a minor crop. Two rows of rice can be replaced with red gram or groundnut after each four or six rows.

Alley intercropping/alley cropping: Food crops are grown in alleys formed by hedge or shrubs or trees. *Gliricidia* + groundnut or any suitable annual crop is an example of alley cropping. It is a well-defined agroforestry system.

Trap crops: These crops are grown in the main cropped field in definite rows to trap insect pests.

Filler cropping: Growing of short duration crops in between the newly established perennial crops for few years to fill the space and to utilize the resources. Coconut + pineapple, mango + greengram, guava + turmeric are the examples of filler cropping. It is also an agroforestry system.

Principles of intercropping

- Before putting any intercrop with the main crop it is very essential to know the prerequisites of the companion crops such as soils and water requirement compatibility, competition for space, sunshine and air, compatibility for pests and diseases, duration and yielding potential and time of sowing and harvesting. The selected intercrop should be compatible with the main crop in its water, nutrients and soil requirement.
- The crops grown in association should have complementary effects rather competitive effects. The subsidiary crop should be of shorter duration and of faster growing habits to utilize the early slow growing period of main crop and they must be harvested when main crop starts growing.
- The proper spatial arrangement is required to maximize cooperation and minimize the competition between or among the crops selected for intercropping. An intercropping is successful only when a suitable companion crop is selected to grow with the main crop.
- Overcrowding should always be avoided by optimising the population of intercrops. A standard plant population of main crop should be maintained and plant population of subsidiary crops can be increased or decreased as per demand of the situation.
- The crops of different maturity dates should be preferred to reduce the competition for resources during peak demand period.
- In an intercropping system the component crops should have different root depths so that they can draw water and nutrients from different layers of soil. This will minimize competition for nutrients, water and root respiration
- The intercrop should not be an alternate host for pest and diseases of the main crop. Component crops of similar pests and disease pathogens and parasite infestations should not be chosen. Crops selected should be of different families to avoid pests and diseases.
- Component crops should have similar agronomic practices.
- Erect growing crops should be intercropped with cover crops like pulses. These check the weeds and reduce soil erosion. The losses of water due to evaporation are also reduced.
- The planting method and management should be simple, less time consuming, economical and profitable so that it may have wider adoptability.

- Fast growing crops should be taken with slow growing crops.
- Crops should have least allelopathic effect.

Advantages of intercropping

- Additional income is realized from the companion crop or crops. Thus, intercropping gives higher income per unit area than sole cropping.
- If the base crop is damaged due to any unfavourable climatic conditions or spread of insect pests and diseases, the companion crop may give some income. So it acts as an insurance against failure of crop in abnormal year.
- Legumes grown as companion crops always benefit the principal crop through N-fixation. Legumes also leave a huge quantity of nitrogen rich organic matter in soil in the form of roots and leaf litter.
- Quick growing companion crops always suppress the growth of weeds thriving in the inter-spaces of the main crop.
- Intercropping practices encourages better utilization of growth resources like nutrient, water, light and space.
- Incidence of insect pests and diseases is low.
- Employment opportunity is created throughout the year. It offers best utilization of labour, machine and power throughout year.
- Soil erosion due to runoff is checked as the soil is covered with vegetation most part of the year.
- Intercrops maintain soil fertility as the nutrient uptake is made from both layers.

Limitations of intercropping

- The fertilizer management is difficult in intercropping practice because the nutrient requirement of the crops is different.
- Harvesting of different crops become difficult as the seeding and harvesting time of the crops vary.
- If the selection of crops in an intercropping system is not compatible, the growth of one crop may be suppressed by another crop. There may be competitive effects among component crops.
- Incidence of insect pests and diseases is increased because sometimes one crop may serve as alternate host for pests and diseases.
- The grower has to acquire the latest technologies for the crops grown which may not be possible at all the time.

- Inter-cultural practices are sometimes difficult in an intercropping system since the time and requirements of agronomic practices for different crops in a system are different. Mechanization is also difficult.
- Application of irrigation water, nutrients, pesticides or herbicides is difficult in an intercropping system and for proper management of the system some skill is required.
- It is labour intensive.
- There may be allelopathic effect.

MIXED CROPPING

Growing of two or more crops simultaneously on the same piece of land, without any definite row arrangement is termed as mixed cropping. It is mainly practised in areas where climatic hazards such as flood, drought, frost etc. are frequent and common. Mixed cropping is also practiced with a view to achieve multiple requirements of food, feed and fibre. Mixed cropping is especially important for fodder crops where it can provide enormous quantities of feedstuff for supporting sustainable livestock production. In mixed cropping, the time of sowing of all the crops is almost the same. However, they may mature at a time (wheat+gram, wheat+barley or wheat+mustard) or they may mature at different times (redgram+jowar, sesame+greengram, wheat+mustard or bajra+groundnut).

Types of mixed cropping

Based on method of sowing and objective of cropping mixed cropping can be classified into the several groups.

Mixed cropping: The seeds of different crops are mixed together and then sown either in lines or broadcasted. The system is not scientific. Agricultural operations and harvesting of the crops become difficult.

Guard crops: The main crops is grown in the centre, surrounded by hardy or thorny crops such as safflower around pea or wheat, mesta around sugarcane, jowar around maize etc. with a view to provide protection to the main crop.

Augmenting crops: When sub crops are sown to supplement the yield of the main crop, the sub crops are termed as augmenting crops. Mustard is sown on the borders of wheat fields to harvest mustard leaves for greens during the initial stages.

Multistoried cropping or multi-tire cropping: Growing plants of different height in the same field at the same time is termed as multistoried cropping. It is mostly practiced in orchards and plantation crops for maximum use of solar energy even under high planting density. Examples of multi-tire cropping are

Eucalyptus + papaya + berseem, sugarcane + potato + onion, sugarcane + *toria* + potato, and coconut + pineapple + turmeric/ginger.

Principles of mixed cropping

- All crops in a mixed cropping system do not fail under adverse climate conditions.
- An epidemic attack of any insect, pest or disease kills only one crop without affecting the rest of the crops.
- The different types of crops are grown with an objective of fulfilling the daily need for cereals, pulses and oilseeds of a farm family.
- The sub crops should not affect the growth of the main crop.
- The sub crop should mature earlier or later than of the main crop.
- One of the crops should preferably be a legume.
- The crops in a mixed cropping system should have different growth habits and nutrient requirements.
- The crops should have different rooting depths and ramification.
- The crops should not be very exacting in climatic requirements.

Advantages of mixed cropping

- Mixed cropping ensures efficient utilization of land, labour, equipment and other resources.
- It utilizes available space and nutrients to maximum extent possible.
- It provides safe guard against hazards of diseases and pests.
- It also reduces the effects of adverse weather conditions for the farmer as planting and harvest is at different times of the season.
- Mixed cropping secures daily requirements like cereals, pulses, oilseeds, etc. to a farm family.
- It improves yield. According to some scientific basis there is 10-25% increase in the yield in mixed cropping than monocropping.
- It gives balanced cattle feed with both legume and grass fodders.
- Costs of input decrease as compared to individual crop growing cost.

Limitations of mixed cropping

- Some crops are very specific in the type of soil they need for maximum profits.
- Crops for mixed cropping if not chosen properly have chances of competition between the crops for nutrients.

- It also may reduce the fertility of the soil as more than one crop is grown at a time in the same piece of land.
- Besides these, all the limitations for intercropping are also applicable for mixed cropping.

SEQUENTIAL CROPPING

Sequential cropping refers to growing crops in sequence within a crop year, one crop being sown after the harvest of the other. Here crop intensification is only in the time dimension without any temporal intensification like intercropping. One has to manage only one crop at a time in the same field. Irrigation facility, availability of short duration, photo-insensitive and thermo-insensitivity high yielding varieties and other production inputs should be ensured for realizing the maximum benefits from the sequence cropping.

Types of sequential cropping

Double cropping: Cultivation of two crops in succession on a piece of land in a year. Growing rice-wheat, rice-groundnut or red gram-maize in a year in sequence are the examples of double cropping.

Triple cropping: Cultivation of three crops in succession on a piece of land in a year. Growing rice-maize-greengram, rice-potato-okra or red gram-maize-cowpea in a year in sequence are the examples of triple cropping.

Quadruple cropping: Cultivation of four crops in succession on a piece of land in a year. Growing rice-fodder maize-greengram-greens, rice-potato-greens-okra or greengram-fodder maize-fodder cowpea-okra in a year in sequence are the examples of quadruple cropping.

Ratoon cropping or ratooning: It refers to raising a crop with regrowth coming out of roots or stalks after the harvest of the crop. Here, more than one harvest is done from one sowing/planting. Thus, ratooning consists of allowing stubbles of the original crop after harvesting and to raise another crop. In sugarcane 3-4 ratoon crops can be taken, but here a cropping year is not the time limit as it is a 10-18 months duration crop. Seasonal vegetables like okra, chilli etc. can be taken as ratoon crops from the stubbles in the following season within a cropping year.

Advantages of sequential cropping

- It increases the overall productivity of the system.
- It also influences the soil conditions favourably.
- It ensures efficient utilization of available resources like nutrients, water and solar energy.

- Positive residual influence of previous crop with respect to plant nutrients on succeeding crops is realized.

Disadvantage of sequential cropping

- Pests and disease from a crop may be carried over to the succeeding crop.
- Weed population and species may sometimes be shifted from one crop to second crop.
- The growth of second crop may be adversely affected by allelopathic effect of first crop. Sunflower has allelopathic effect on succeeding crops like mustard and soybean by inhibiting germination of their seeds.

CROP ROTATION

Crop rotation refers to growing of different crops alternatively on the same piece of land in a definite sequence or process of growing different crops in succession on a piece of land in a specific period of time with an objective to get maximum profit from least investment without impairing the soil fertility. This may also be defined as the repetitive cultivation of an ordered succession of crops (or crops and fallow) on the same land and one cycle may take one or more years to complete.

Principles of crop rotation

- Crops best suited to the climatic conditions and soil type should be selected.
- The crops with tap roots should be followed by those with fibrous root system. This helps in proper and uniform use of nutrients from the soil.
- The leguminous crops should be grown after non-leguminous crops. Legumes fix atmospheric nitrogen in the soil and add more organic matter to the soil in the form of huge quantity of root and foliage.
- More exhaustive crops should be followed by less exhaustive crops. Exhaustive crops like sugarcane, maize, etc should be followed by low inputs demanding crops like oilseeds and pulses.
- The crop of the same family should not be grown in succession because they act like alternate hosts for pests and diseases.
- The crop rotation should provide maximum employment to the farm family, farm machineries and equipments available.
- Selection of the crops should be based on local demand, storage, transport and marketing facility.

- Grower's preference and his socioeconomic conditions and resource mobilizing power should also be considered.
- The selection of the crops should be based on farmer's knowledge and skill.
- The latest production technology of the crops should readily be available.
- On slopes which are prone to soil erosion, alternately erosion-promoting crops like millets or other row crops and erosion-resisting crops like legumes should be adopted.
- Under dryland farming or partially irrigated areas, the crops like pulses, sunflower which can tolerate drought spells to some extent should be selected. Similarly in low lying and flood prone areas the crops like rice and jute which can tolerate water stagnation should be selected.

Benefits of crop rotation

- Soil fertility is restored by fixing atmospheric nitrogen by integrating legumes in a rotation.
- Proper selection of crops in a rotation encourages soil microbial activity and improves physico-chemical properties of the soil.
- The succeeding crop utilizes the residual nutrients.
- The soil is also protected from erosion, salinity and acidity.
- It controls pests and diseases if dissimilar crops are taken in succession.
- It also restricts the weeds growth.
- Farmers get better price for their produce because of its higher demand in the locality or in the market.
- The family labour, power, equipment and machineries are well employed.
- Proper utilization of all the resources and inputs could be made.

Limitations of crop rotation

- Specialization in many crops is not possible always.
- Requirement of equipments and machineries varies from crop to crop.
- Allelopathic effect of preceding crop affects the succeeding crop.
- Similar types of crops serve as alternate hosts for pests and disease.

ENERGY PLANTATIONS

An energy plantation is a collection of carefully chosen tree and shrub species that are particularly grown for fuel and can be harvested in a short rotation.

These plantations contribute to the supply of wood for industrial and residential uses. In order to address the ever-increasing need for energy, energy plantations have been planned in a big way. The energy plantations offer nearly limitless renewable energy sources that are locally sourced. For every cycle, the total time requirement is limited to 3-8 years.

Several initiatives have been launched for afforestation and forest management, including social forestry, silviculture, and agro-horticultural activities on waste and barren areas, in an effort to generate a sizable quantity of biomass. High density energy plantations (HDEP) that are resistant to pollution, drought, and salt are being grown in waste and desolate areas as part of these projects. The method of planting trees closely together is known as high density energy plantations, or HDEP. Here, trees battle to survive, which causes them to grow quickly. It offers quick and substantial rewards together with several chances for long-term employment and income.

Over 133 million tonnes of firewood are being used in India every year, with the majority going towards cooking. One kg of food requires 1.2 kg of fuel to cook. If food has to be cooked before consumption, it is evident that India needs to generate more wood than food. An estimated 400 million tonnes of cow dung, are burned each year in India for domestic use (Sharma and Gupta, 2010). This quantity of animal manure might improve soil productivity if it is applied in the soil. The main justification for tree harvesting is fuelwood. Thus, to reduce the pressure on natural forest, for fuelwood and to preserve ecological sustainability by preserving trees in natural forests, fuelwood tree planting should be integrated into agriculture through agroforestry practices.

Characteristics of energy plantations

- Wood has a good level of heat content
- Wood is non-polluting and low in sulphur
- Planting trees in areas that are prone to erosion helps to prevent soil erosion
- It also helps to provide jobs in rural areas
- Ash from burned wood is a good fertiliser

Selection criteria of tree species for energy plantation

- Tree species should be fast growing with excellent photosynthetic efficiency, leading to higher biomass production.
- Tree species should possess a high potential for coppicing and pollarding.
- Conical or cylindrical-shaped tree species should be used for energy plantations.

- Tree species should possess wood with a high calorific value, dense wood, and the ability to burn without producing hazardous smoke or sparks.
- In dry regions, tree species should have the capacity to decrease transpiration loss.
- Tree species should be able to fix nitrogen, which can increase soil fertility.

Suitable species for energy plantation

Tropical dry region: *Acacia catechu, Acacia modesta, Acacia nilotica, Acacia senegal, Acacia tortilis, Anogeissus pendula, Albizia lebbek, Azadirachta indica, Cassia siamea, Cordia rothii, Dalbergia sissoo, Emblica officinalis, Eucalyptus camaldulensis, Erythrina superb, Gmelina arborea, Parkinsonia aculeate, Peltophorum ferrugineum, Pongamia pinnata, Prosopis cineraria, Prosopis juliflora, Tamarindus indica, Tamarix troupe, Tecomella undulate, Zizyphus maurtiana* etc.

Tropical humid region: *Adina cordifolia, Acacia auriculiformis, Acacia catechu, Acacia nilotica, Albizia procera, Azadirachta indica, Cassia siamea, Casuarina equisetifolia, Dalbergia sissoo, Dendrocalamus strictus, Ficus* spp., *Eucalyptus* spp., *Kydia calycina, Leucaena leucocephala, Madhuca indica, Melia azedarach, Morus alba, Salix tetrasperma, Syzygium cuminii, Tamarindus indica, Trewia nudiflora, Gliricidia sepium* and *Gmelina arborea.*

Sub-tropical region: *Acacia catechu, Acacia melanoxylon, Acacia nilotica, Aesculus indica, Ailanthus excels, Celtis australis, Grevillea robusta, Michelia champaca, Populus deltoids, Populus nigra, Robinia pseudoacacia, Salix alba* and *Toona ciliate.*

Temperate climate: *Acer* spp., *Aesculus indica, Alnus nepalensis, Alnus nitida, Celtis australis, Populus ciliate, Quercus semecarpifolia, Salix alba* and *Toona serrata.*

Charcoal

Charcoal is a lightweight black carbon residue produced by heating wood above 400° C in an oxygen-starved environment to remove all water and volatile constituents. The process, called pyrolysis, is exothermic, meaning it gives off heat once started. Charcoal may be made more efficiently out of the firewood. Charcoal finds diverse applications, including metallurgical fuel in iron and steel production, industrial fuel, cooking and heating fuel, reducing agent in chemical processes, and as a raw material in pyrotechnics.

Certain firewoods emit unpleasant and poisonous smoke when they burn. But cooking with charcoal produces no smoke. Charcoal has a combustion efficiency of around 28%. Therefore, it is preferable to convert firewood into charcoal rather than firewood itself for use as fuel. Charcoal is devoid of sulphur. One kg of charcoal can be substituted with 2.38 kg or more of firewood.

Suitable species for charcoal making are *Acacia nilotica, Adina cordifolia, Anogeissus latifolia, Casuarina equisetifolia, Pinus roxburghi, Quercus leucotrichophora, Quercus semecarpifolia, Tamarindus indica, Terminalia arjuna, Terminalia bellerica, Terminalia chebula* and *Terminalia catappa.*

The quantity of heat generated when one gram of fuel is burned entirely without the addition of air or oxygen is known as the calorific value. A single gram of carbon may generate around 30 KJ/g of heat when it is fully burnt. Consequently, fuel with a high calorific value is considered to be good fuel since carbon has a calorific value of 30 KJ/g.

Table 8.2: Species used in energy plantations with their respective calorific value and specific gravity

Sl. No.	Species	Sp. gravity	Calorific value K cal/kg
1.	*Acacia auriculiformis*	0.60-0.78	4800-4900
2.	*Acacia catechu*	1.00	5142-5244
3.	*Acacia dealbata*	0.70-0.85	3500-4000
4.	*Acacia leucophloea*	0.78	4899-4886
5.	*Acacia mearnsii*	0.70-0.85	3500-4000
6.	*Acacia nilotica*	0.67-0.68	4800-4950
7.	*Acacia senegal*	-	3200
8.	*Acacia tortilis*	-	4400
9.	*Adina cordifolia*	-	3855
10.	*Aegle marmelos*	0.91	4495
11.	*Albizia lebbek*	0.55-0.64	5163-5166
12.	*Albizia odoratissima*	0.73	5131-5266
13.	*Albizia procera*	0.68	4870-4865
14.	*Alnus nepalensis*	0.32-0.37	4600
15.	*Anogeissus latifolia*	0.94	4948
16.	*Anogeissus pendula*	0.94	4900
17.	*Anthocephalus cadamba*	0.94-0.53	4800
18.	*Artocarpus heterophyllus*	0.51	5318
19.	*Azadirachta indica*	0.75	-

20.	*Baringtonia acutangula*	0.58	5078
21.	*Bauhinia retusa*	0.72	5027
22.	*Bauhinia variegate*	-	4800
23.	*Butea monosperma*	0.54	4909
24.	*Bischofia javanica*	0.74	5162
25.	*Cajanus cajan*	-	4594
26.	*Cassia siamea*	0.60-0.80	-
27.	*Casuarina equisetifolia*	0.80-1.2	4950
28.	*Cedrela toona*	0.57	5113-5168
29.	*Chloroxylon swietenia*	-	4759
30.	*Dalbergia sissoo*	0.75-0.80	4908-5181
31.	*Diospyros melanoxylon*	0.79-0.87	4957-5030
32.	*Diospyros Montana*	0.70-0.80	5125
33.	*Dodonaea viscose*	1.20-1.28	5035-4939
34.	*Emblica offcinalis*	0.70-0.80	5200
35.	*Eucalyptus camaldulensis*	0.6	4800
36.	*Eucalyptus globules*	0.80-1.00	4800
37.	*Eucalyptus grandis*	0.40-0.70	4900
38.	*Eucalyptus tereticornis*	0.70	4800
39.	*Gmelina arborea*	0.42-0.64	4763-4800
40.	*Grevillea robusta*	0.57	4904-4914
41.	*Grewia spp.*	0.67	5292
42.	*Hardwickia binata*	1.08	4891-4952
43.	*Holoptelia integrifolia*	0.63	5228
44.	*Lannea coromandelica*	0.55	4933
45.	*Leucaena leucocephala*	0.55-0.70	4200-4600
46.	*Madhuca longifolia*	0.56	5043-5156
47.	*Mangifera indica*	0.58	4610
48.	*Melia azedarach*	0.56	5043-5176
49.	*Morus alba*	0.63	4371-4773
50.	*Michelia champaca*	0.45	5068
51.	*Ougeinia oojeinensis*	0.85	5178
52.	*Pithecellobium dulce*	0.64	5177-5600
53.	*Pongamia pinnata*	0.75	4600
54.	*Populus euphratica*	0.48	5008-5019
55.	*Prosopis chilensis*	0.80-0.92	5000-5500

56.	*Prosopis cineraria*	0.77-0.94	5000
57.	*Prosopis juliflora*	0.70	4800
58.	*Pterocarpus marsupium*	0.79	4904-5141
59.	*Pterygota alata*	0.25-0.62	5160
60.	*Quercus leucotrichophora*	0.74	4633
61.	*Schleichera oleosa*	0.91-1.08	4928-4950
62.	*Sesbania grandiflora*	0.55	4407
63.	*Shorea robusta*	0.68-0.82	5095-5433
64.	*Syzygium cuminii*	0.67-0.78	4834
65.	*Tamarindus indica*	0.91-1.28	4909-4969
66.	*Tamarix aphylla*	0.60-0.75	4835
67.	*Tectona grandis*	0.55-0.70	4989-5535
68.	*Terminalia alata*	0.71-0.94	5047-5373
69.	*Terminalia arjuna*	0.74-0.82	5030-5128
70.	*Terminalia chebula*	0.77	3967
71.	*Trema orientalis*	0.48	3095
72.	*Xylia xylocarpa*	0.92	4975-5044
73.	*Zizyphus mauritiana*	0.93	4900

Source: http://ecoursesonline.iasri.res.in

Petro-plants or Petro-crops

Petro-crops are those plants that produced petroleum substances as supplementary to petrol. These plants produce hydrocarbon compound or ethanol. Because these species may act as supplements to petroleum-based chemicals, they are referred to as 'petro-crops'. These plant species provide abundant energy and might be utilised as sustainable energy sources.

Based on exudates and extractives, plants may be classified as latex, vegetable oil and waxes, resins, essential oils and tannins or phenolic compounds bearing plants. Plant species yielding latex belong to family apocynaceae, asclepiadaceae, euphorbiaceae, moraceae and sapotaceae.

The potential use of vegetable oils as a source of hydrocarbons or as a liquid fuel is enormous. Some of these can be blended with diesel.Oil tree species that yield inedible seeds can be grown on wastelands, marginal areas, and impoverished soils. Several noteworthy species of plants that produce seed oil are *Antinodaphe hookeri, Aleurites triloba, Anacardium occidentale, Garcinia indica, Jatropha curcas, Azadirachta indica, Calophyllum inophyllum, Cocos nucifera, Croton tiglium, Madhuca indica, Madhuca longifolia, Mesua ferrea, Melia azedarach, Mimusops elengi, Pongamia pinnata, Hydnocarpus wightiana*, *Ricinus communis*, *Sapium sebiferum*, *Samecarpus anacardium*, *Shorea robusta*, *Simmondsia chinesis*, *Vateria indica*, *Pittosporum resiniferum*, *Salvadora oleoides*, *Schleichera oleosa*.

Resins include volatile oils (turpentine) and non-volatile resins (rosin), which are mostly extracted from pinaceae family members. Turnpines are highly flammable and may be utilised in a variety of fuel formula for vehicles. They are also the primary source of resins for synthetic rubber and other polymers.

Advantages of energy plantations

- Fuelwood releases less nitrogen dioxide and sulphur dioxide than fossil fuels.
- Energy plantation restores deteriorated areas.
- It creates employment opportunity in rural areas.
- The quantity of greenhouse gas absorbed by living and actively growing plant biomass is roughly equal to the amount released when they burn or decay. Therefore, they are referred to as 'carbon neutral' fuel sources.
- Growing energy crops releases carbon underground through the tree root system, creating a 'carbon sink'.
- It reduced energy costs per unit area due to the need for fewer inputs in comparison to agricultural crops.
- The nutrients taken by the energy plants from the soil can be recovered as ash and restore it to the planting areas by using the thermochemical process of biomass conversion.
- It is a reliable and sustainable energy source that both creates jobs and reforests marginal areas.
- Energy plantation provides aesthetic value.
- In some cases, it may be used as windbreak and shelterbelts.
- Energy plants produce fodder, and other non-timber forest products.
- Handling and disposal of by-products is safe.

SILVIPASTORAL SYSTEMS

A management technique known as 'silvipasture' involves integrating trees into a land unit with livestock, such as pigs, poultry, or ruminants, when doing so benefits the agricultural system directly on an ecological or economic level.

Benefits from Silvipastoral Systems

The two main types of advantages linked with silvipasture are agroecological (both on and off farm) and economic benefits. Both of these advantages can and frequently are provided by trees, but the relative importance of each will depend on the farm's management goals and the surrounding environment.

Economic benefits

The availability of a second tree crop might serve as a major incentive for adopting silvipasture in addition to generating revenue. Alternatively, it can be a major side benefit if trees have been incorporated to give other advantages, such as shelter. In order to maximise the return on investment, maximising the economic value of trees does require careful management. This usually takes longer to realise because, for example, timber trees must mature before being harvested. Another crucial element is the farming environment. Higher-quality fruit or timber can be produced faster in more protected lowland silvipastoral systems situated on better soils than in exposed upland silvipastoral systems.

Agroecological benefits

Numerous agroecological advantages are provided by trees. Although these advantages are sometimes more intangible and more difficult to quantify economically, they are frequently vital to the long-term viability of agricultural systems. The precise combination of benefits will vary depending on the farm's aims and circumstances.

Improvements in soil health: The long-term soil fertility of pastures is preserved in part by trees. Through the turnover of their roots and litter, trees replenish the nutrients that have been lost below the grass rooting zone. Trees increase the soil's ability to retain nutrients and water. Trees can improve infiltration and reduce animal compaction, or poaching. Trees may draw water and nutrients from deeper within the soil by investing in their mycorrhizal relationships during times of heightened stress, such as drought. The root systems of trees greatly lessen soil erosion-related soil loss. Additionally, trees support healthy soil microbes. The ratios of bacteria and fungus as well as the quantity of earthworms grow significantly in silvipastoral systems. This rise is a sign of better soil for most systems.

Reduction of effects of wind exposure: Livestock can profit greatly from silvipastoral systems when they are located in regions with considerable wind exposure. In exposed situations, livestock require a notably higher calorie intake to maintain their body condition. Animal welfare is increased and feed costs are reduced when shelter-seeking animals consume less energy to maintain their body temperature. The agricultural system's profitability may rise as a result of these advantages. With careful planning, silvipastoral systems may significantly lower the rates of cattle mortality during harsh weather or at birth.

Reduction of heat stress: Livestock production can be significantly impacted by overheating. Heat stress can impair milk production, breeding efficiency, and livestock's ability to acquire live weight by making them eat less. Heat stress can impair ewes' chances of getting pregnant and decrease rams' fertility. Heat stress causes chickens to consume less feed, lay fewer eggs, and have a compromised immune system. On silvipasture, where sun radiation may drop by up to 58%, skin temperature is 4°C lower than on open pasture. Other typical behavioural habits, such eating and sleeping are therefore better maintained. Livestock will often cluster in places with little shade, which raises the risk of illness, compacted soil, and plant loss. This impact can be mitigated by utilising silvipasture to provide more equal shade.

Reduced incidence of pests and diseases: In wetland pastures, trees dry up the soil, which makes it less conducive to the growth of bacteria that cause foot rot or the snails that are a component of the liver fluke cycle. Although trees provide a home for flies, which might increase the infestation in animals, well-planned silvipastoral systems also have higher dung beetle populations. When paired with additional predators, these expedite the removal of faeces, leading to fly levels that are 40% lower than on open grassland. Planting trees in chicken systems enhances the welfare of the birds and lessens their stress levels. Higher-quality eggs and more output follow from this. Additionally, since wild birds that cluster in more open settings pose the highest danger of avian influenza, it can lessen the chance that poultry will come into contact with these species.

Supplementing livestock diet: In general, micronutrient levels are greater in trees than in grasslands. In India, the use of tree fodders is gaining popularity, especially for treating nutritional deficiencies and for their anti-parasitic qualities linked to secondary chemicals (tannins) present in the leaves. Studies conducted in the Netherlands revealed that when willow coppice was added to a dairy system, the animals preferred to graze it. The consumption of sodium (Na), zinc (Zn), manganese (Mn), and iron (Fe) was between 2 and 9% of the daily needs, despite the fact that intake was typically low (0.6 and 0.4% of the needed dry-matter intake for dry and lactating cows, respectively). During drought times, trees may also be used as a source of feed. Harvested fodder can be kept for up to 24 months before being used.

Broader environmental benefits: There are other environmental advantages to trees that extend beyond farms. Increasing the area under silvipasture will sequester a sizable quantity of carbon. Benefits to biodiversity can also be obtained by silvipasture. It is linked to increased soil biodiversity and offers a variety of birds, animals, and other wildlife with semi-natural habitat and habitat networks.

Different Types of Silvipasture System

Trees in silvipasture can be categorised into three main groups based on their arrangement.

Table 8.3: Different types of silvipasture

Silvipasture system			Examples
Trees within livestock pastures	A	Boundary tree systems	Shelterbelts, riverside planting and hedgerows
	B	Regularly spaced tree systems	Grazed orchards, row systems, clump systems
Livestock within woodland systems	C	Woodland grazing	Pannage systems, silvipoultry, parklands

Linear tree systems

In silvipastoral systems, linear tree systems are typically employed when a buffer of some kind is needed, mainly to maintain the quality of the soil, water, and wind. The by-products of these systems include biomass, wood, fuel, and fruits, mostly from the shrub layer.

Regularly spaced tree systems

Under these systems, trees are planted in pastures in rows or in regular patterns, usually with the goal of producing or preserving a high-value wood product such as timber or fruit. Better soils and less exposure increase the chances of these systems successful. Sheep in orchards save mowing costs, improve nitrogen cycling, and lessen the strain of grazing on other agricultural areas. These designs are also particularly effective for tree fodder systems, when trees are purposefully included into rows to offer an additional food source.

Designing for Livestock Benefits

Reducing cold stress

Linear features such as hedgerows and shelterbelt systems help lessen wind stress in cattle. These advantages are still present in pastures with tree integration, especially in clump or row systems that have the potential to include a layer of nursing trees or shrubs. The trees will not only provide shade for the cattle but also create a microclimate that enables the fodder to green up earlier in the year and resist dry conditions more readily.

Reducing heat stress

The finest shade is provided by mature trees with expansive canopies. These are often restricted to field margins as part of the hedgerow network or as mature standards inside fields since they are thought to compete with grazing.

When animals have few alternatives for shade, this can lead to issues since it drives animals to gather around them. This may lead to unsanitary conditions, poaching, and other issues. Trees spaced regularly provide more uniform shade and prevent animals from crowding together.

Encouraging natural behaviour

Livestock behave better as a group and utilise well-designed silvipasture more equally than open grassland. This is partially because they can conceal themselves behind and beneath the trees, and partially because there is less competition for scarce resources like shade and shelter. As a result, social relationships among animal groups in silvipastoral systems get better.

Silvipasture for ruminants

Since shade and shelter are the main advantages, silvipasture in all its forms can be advantageous to ruminants. In highland farming systems or intensively grazed farms, shelter is probably very vital which is supported by the well designed silvipastoral systems. Although they provide more consistent shade, silvipastoral systems provide less shelter than shelterbelt systems made specifically for this purpose.

Silvipasture for poultry

Farmers interested in organic and free-range chicken systems, where the hens have access to an outside run, might consider silvopasture. Because trees give hens a place to live, they are more inclined to range, which benefits their wellbeing and enhances their health and productivity.

Silvipasture for pigs

Access to grazing areas in woodlands is beneficial for pigs. In the summer, they especially benefit from shade. Because they are omnivores, pigs may find a variety of foods in forests, such as berries, nuts, roots, and plants. However, their erratic behaviour necessitates careful supervision. If left unchecked, they can have disastrous effects on trees and the entire ground cover.

Maximising the Value of the Trees

Choosing the right tree

Choosing the proper tree species for the right location is essential since planting new trees is costly, time-consuming, and even may not be possible to replace the trees once established. For instance, trees in highland agricultural systems usually have to withstand significant wind exposure and/or intermittent water logging. Hardier tree species will thrive, particularly those that are native to the area and are therefore more likely to be suited to the environment. Instead of producing high-value tree crops, agroecological advantages are probably going to be the main motivators for early planting in these farming systems. Overstocking and poorly drained pastures can lead to poaching and raise the prevalence of lameness. By minimising water accumulation on pasture and intercepting runoff, trees can help prevent poaching and related problems.

Tree arrangements

Depending on the type of tree being planted and the livestock system, trees can be planted equally at broad spacing at densities ranging from 100 to 400 trees per hectare. For the first 12 years after introduction, the tree canopy will not overshadow the grassland for the majority of species at these densities. Quick-growing plants could shade early. Grasses which can withstand shade should be preferred to reduce the impacts of shadowing. The tree crop should selectively be thinned once the canopy restricts the solar radiation reaching on the ground. The thinned plants may be used for fencing or firewood. Mixed tree species can use the thinning cycle to produce a variety of goods.

With row systems, trees may be planted in one or more rows at a closer spacing from one another, and different tree mixes can be employed to provide the finished wood crop nursing benefits. To provide access, the rows themselves must be sufficiently broad. The canopies will gradually start to shade out the grass. Planting the rows in a north-south direction will help to reduce this. These systems are capable of producing a variety of goods, fruit, nuts, and even browsing. Trees which are easily coppiced can be planted on considerably shorter rotations to decrease the shadowing impact if the rows are being used for biomass or fodder.

When compared to planting individual trees, clumps may be more productive and have a smaller environmental effect. For clusters, the expense of tree preservation is less. As in conventional forestry, it is feasible to choose high-quality trees inside clumps by gradually thinning the trees until just a limited number of final crop trees remain in each clump. By choosing trees that generate a thick, evergreen or early flushing cover along the margins, clusters might have a higher shelter value.

Light competition and grass growth

The influence of trees on forage growth is often quite little during the early stages of their growth. However, as trees get older, they may require management like pruning and thinning to minimise the shadowing from the tree canopy in order to preserve the quality of the sward. Although the trees may not provide as much light to the forage canopy as they formerly did, they do have other advantages. It gets greener sooner in the spring because trees provide cover for grazing. Trees also provide a microclimate that shields the forage under them from heat stress. Thus, silvipasture improves the farming system's resilience.

By drawing nutrients from deeper in the soil and reintroducing them as leaf litter, trees may also help the forage. The amount of fertiliser required may be decreased or maybe eliminated if nitrogen-fixing plants are included in the tree combination.

Tree protection

Cattle and sheep are examples of grazing animals that harm tree stems, roots, and ground vegetation. Since it is their normal habit to graze and rub or trample, in the case of cattle, establishment is difficult without protection and regular supervision. For the first few years until the trees are big enough, fodder can be cut for hay or silage. It costs more to protect individual trees than it does to protect groups or rows of trees. Access of wild animals can be restricted through either permanent or temporary fence systems which helps trees to grow naturally.

Importance of management

Trees require just as much care as grass to thrive. In all circumstances, during the first three to five years of tree establishment, weeding will be necessary around the base of each tree. When necessary, tree protection must be routinely inspected, maintained, and replaced if necessary. For the best-quality wood trees, trees that are larger will require frequent trimming in the winter, when the tree is less active. Pruning trees with bad shape should be done carefully. Tree management may yield a variety of secondary products for use on farms, such as woodchip bedding for animals, fuel, and fodder.

9

Agroforestry Planning and Management

Agroforestry is a land use system that involves two or more plant species at least one of which must be a woody perennial. When perennial woody and herbaceous components are grown together on the same piece of land their performance would largely depend on their ability to share various growth resources in a given environmental situation. Due to difference in growth pattern and resource requirement of the components in agroforestry situation, a close interactive relation is obvious. Thus, careful management practices for the components are required to establish a successful agroforestry system. The characteristics of the trees and crops, and their interactions, can be modified with good management practices in order to take advantage of the positive characteristics and minimize the effects of the negative ones. Effective and efficient agroforestry management may be divided into two groups; tree management and agricultural crop management.

FACTORS AFFECTING THE SELECTION OF TREE SPECIES

Trees in an agroforestry system contribute significantly to the total output of the system and therefore, success of the system depends on the selection of the tree species. The following factors are to be considered to choose the best tree species for the agroforestry systems in a particular site.

Knowledge of the species

The grower should have a sound knowledge of the species he intended to include in the system. Previous successful experiences and existing preference for the species by other farmers provide a better assurance of success and adoption of the species. Trees should be adaptable to local climatic conditions.

Commercial value or local use of the species

The trees should produce timber of medium to high quality. The optimum would be to introduce timber trees with the highest quality and market prices, when the environmental and service factors and growth rate are appropriate. The marketing options and the local and external demand for the products are the most influential factors in the species selection. Some of the species provide secondary products, such as poles and firewood during pruning and thinning.

Rapid growth

Species should be fast growing so that benefits become available to the farm family as soon as possible. Rapidly growing tree species quickly generate a product and have a lower maintenance cost.

Self-pruning in open light conditions

Trees in agroforestry systems tend to have poor trunk form compared to trees growing in block plantations due to the wide tree spacing and lack of lateral competition. Trees in open conditions tend to have more branches and more persistent branches, higher harvesting costs, more waste at the sawmill and lower timber quality due to knots in the wood. However, in some timber species the lower branches are self-pruned, which means they die and fall when in open conditions. These species are preferable for agroforestry systems because they have lower production costs and higher timber value.

Availability of certified germplasm

The probability of success is increased if certified seeds or seedlings are used. It is better to use high quality genetic material that produces strong and healthy plants, and the use of genetic material of dubious quality should always be avoided. Choosing healthy and vigorous plants in the nursery or during thinning of natural regeneration is also important.

Lack of susceptibility to pests and diseases

Species susceptible to the pests and diseases of the area, or with other known problems, must be avoided. The trees can host pests and diseases that damage the agricultural crops.

Minimal negative effects for the associated crops

Even with crops that are generally grown under shade there are sometimes negative effects from competition for light, nutrients and water. The leaves of some trees can contain substances toxic to some crops. These types of problems are almost inevitable in permanent agroforestry systems. In addition to managing the competition it is important to make sure that the trees do not have alleolopathic effects on the associated crops.

Small and open crown

In order to minimize the competition for light with associated crops the tree species should have open and narrow crown. Harvesting of the trees is less expensive and there is less risk of damage to the crops since the majority of damage is caused by the fall of the crown rather than the fall of the trunk. It is preferable to have non-continuous foliage that provides patches of shade rather than a uniform canopy that creates a low quality light environment for the crops. It is an advantage if the species has small leaves in order to avoid the effect of unifying the raindrops into strong steady streams of water that fall off the leaves, causing damage to both crops and soil.

Deep rather than surface root system

A root system on the soil surface creates strong competition with crops and is more susceptible to damage during crop management. The root systems and root growth characteristics should ideally result in the exploration of soil layers that are different to those being tapped by the agricultural species.

TREE PROPAGATION METHODS

Trees that provide materials for propagation are called mother trees. Selection of mother trees is important because the young trees will be expected to inherit the favourable characteristics from the mother tree, such as fast growth, upright or spreading shape of the tree crown, good flowering and fruiting, and tolerance of diseases or pests. There may be sexual propagation by seed or vegetative propagation by using other parts of the tree. The genetic make-up of the seeds differs due to recombination of the genes, resulting in seedling variation. Thus, all the seedlings may look like the mother tree in some respects, but no two seedlings are the same. In case of vegetative propagation through cutting, layering, grafting or budding, the genetic make-up of the young tree is exactly the same as that of the mother tree. Consequently, all new trees grown vegetatively from one mother tree have the same set of gene pairs and therefore the same characteristics; they form a so-called clone. Differences between plants of a clone can only be caused by different growing conditions.

The number of trees can be increased through natural regeneration of existing trees or through man's propagation activities. Spontaneous growth of new plants, without the interference of man, is called natural regeneration. Seeds are the main source of new plants in natural vegetation. Some plants also spread through suckering. Regrowth from the stump after cutting down certain trees (e.g., many *Eucalyptus* species)is also a way of natural regeneration. Natural regeneration can be encouraged by improving conditions for growth of young trees and by protecting them against hazards such as fire and browsing animals.

Farmers may collect the seed and sow it where they want these trees or shrubs to grow. Instead of collecting and sowing seed, young seedlings or rooted suckers may also be gathered from where they are growing spontaneously for direct field planting. Field establishment of these wildings needs care and favourable growing conditions, because these plants were not raised to be transplanted and often most roots are lost when the plants are lifted.

If natural regeneration or direct field planting is not successful for the tree species, the trees must be multiplied in a nursery. A nursery is a plot of land where young trees can be raised under more or less controlled conditions like protection against animals and birds, an assured irrigation facility, shade against the hot sun, and shelter against strong wind, and improved soil conditions. It takes about from two months to a year to raise seedlings of different agroforestry species in the nursery, depending on how fast the plants grow and on the preferred size for main field planting.

Seed

Seed can be collected locally or obtained from distributors, seed centres or research institutes. Success of planting mainly depends on seed quality. The seed should be available at the right time in sufficient quantity. If collection is local, it is important to know where the best trees are located. Seed should come from an area that is as similar as possible in altitude, quantity of rainfall, type of soil, etc. to the area where the seedlings will be grown. Seed should be collected from healthy and vigorous mother trees with the desired characteristics. Seed should not be collected from isolated bushy or very young trees. Seeds should be collected only from ripe fruits. If fruits are collected too early, the seed might be immature. If collection is delayed too long, the seed may be attacked by insects and fungi. Seeds of some tree species lose their viability if they are stored even for a week after their extraction from the fruit. These seeds are called recalcitrant seeds. Thus, recalcitrant seeds should be left in the fruit till they are sown. Once the seeds have been extracted from the fruits, they are to be cleaned thoroughly to remove all bits of fruit flesh, pulp or husk. Seeds are sieved to remove undersized and broken seeds. In most cases, when seeds are immersed in water the good ones sink and bad ones will float. Then the good seeds are removed and dried in shade carefully.

If seeds cannot be planted immediately after collection and extraction, they should be stored carefully. Some seeds can be stored at normal temperature for many years if they are kept dry. However, recalcitrant seeds should be sown immediately after extraction from the fruit. Seeds are dried before storage to prevent infection by fungi and bacteria. But extreme heat must be avoided as this may destroy the seed viability. Seeds should not be exposed to direct sunlight. Seeds are dried in a shaded and well ventilated place.

Seed treatments to accelerate germination

Germination of seed of some tree species may take many months. Germination may also be delayed by a hard seed coat or by presence of certain substances in the seed. Such seeds may be treated to speed up germination and it is very important to ensure more simultaneous germination.

Soaking in water or acid: This is a simple method involving soaking the seeds for 2 days in water prior to sowing. This makes the seed coat soft and the substances that delay germination are leached out. In some cases, acid may be added to make the liquid more abrasive. This method works for seeds from many tree species.

Hot water treatment: In this case, seeds are immersed in boiled water immediately after taking it off the fire. The water with seeds is allowed to cool overnight and seeds are rinsed with clean water the next day. This makes the hard outer cover soft and the sprout can break through it more easily.

Scarification: To facilitate germination a shallow incision is made in the hard seed coat using a file or the tip of the seed is cut off. A simple way of scarifying legume seeds is to rub them over a rough surface, e.g., sandpaper. This scratches the seed coat, but the scratches should not go right through the coat.

Stratification: When trees from the temperate zones are to be grown in tropical highlands seeds of these trees may require winter chilling to break seed dormancy. Here, the seeds in a pot with moist sand are kept in the refrigerator for about two months.

Wildings

Wildings are planting material such as seedlings, cuttings, etc. collected from natural vegetation. Wildings should have at least 2 to 4 fully developed leaves. Larger wildings have stems of about the diameter of a pencil. When the soil is moistened with rain water, the wildings can be collected either with or without a root ball. The network of roots and the soil clinging to them when a plant is lifted from the soil is called its root ball. Wildings with root ball are dug out by making cut on soil at both sides of the young plant at an angle of 45 degree using a spade. The plant is then lifted on the spade, holding the stem with the other hand and removed with the root ball intact. It is difficult to dig up small wildings with a root ball, because the root system is too small to hold the soil. Bare-rooted wildings are uprooted by loosening the soil around the roots with a pointed stick. The taproot and large sized roots are often trimmed. Leaves except for the top ones can be removed to reduce transpiration. The wildings should be planted in as quickly as possible after their removal.

Cuttings

The stem, leaf or root of a tree may be cut for rooting. But stem cuttings are most common. The hardwood cuttings are generally taken during the dry or cool season, when shoot growth is minimal. In deciduous trees the cuttings are best taken when the trees are leafless. The leaves are stripped from the branch or stem cutting that is inserted in the soil. Usually, a few leaves are left at the tip of the cutting. The cuttings are likely to dry out if the leaf area is large. The diameter of cuttings ranges from pencil-thick to about 3 cm. The upper cut is oblique so rainwater runs off. The lower cut is usually made just below a node, because rooting generally occurs mainly at the node. There may be rotting which leads to failure of the cutting if a cut is not smooth and clean. Improper and unclean cutting may also infect the wound on the mother tree. Straight branches and twigs are preferred for cuttings; they grow upright after rooting, forming a tree with a proper trunk. Several pruning techniques such as cutting back, pollarding, and coppicing may be practised to stimulate trees to form additional shoots from which cuttings can be taken.

Cutting back: Cutting back branches or twigs stimulates buds to break and grow into new shoots. These shoots are of uniform in age, shape and size, and can be used as cuttings.

Pollarding: Pollarding is a drastic pruning where the tree trunk is cut at a height of about 2 m. A number of water shoots emerge below the cut. After about a year these shoots can be used as cuttings which are termed as live stakes. *Gliricidia sepium* or *Erythrina* species produce good live stakes. The cuttings from these species strike root easily when planted during the wet season.

Coppicing: Coppicing is cutting the tree trunk off close to the ground in order to stimulate the growth of new shoots on the stump. These shoots can then be cut off and used for planting.

Cuttings for propagation are also available from tree roots, e.g., *Casuarina* species. Some trees even produce root suckers spontaneously. These suckers if detached from the mother tree and planted in another place they develop their own roots and grow to a new tree. In some tree species, roots are cut using a sharp spade to encourage the emergence of root suckers. These soot sucker cuttings are planted in a nursery under shade. If a high humidity is maintained rooted suckers will be ready after a couple of months for planting.

Layering

Some tree species cannot be propagated through stem cuttings because cuttings from these trees do not form roots readily. So in some cases, trees may be induced to form roots on their shoots before these are detached from the mother tree to be planted. This method of propagation is called layering. These techniques are particularly suited to home gardeners. In simple layering, long flexible twigs of certain shrubs are bent down to ground level and part of the twig behind the tip is covered with soil. The twig may be wounded by cutting a notch in it or by removing a ring of bark. Rooting will take place near the wound on the side of the tip. But it is difficult to bend an upright branch down to the ground to be layered. The alternative is to bring the soil up to the branch called air layering. A ring of bark is removed from the twig or branch that is to be air-layered and the soft cambium layer is scratched off, so that the wound cannot heal. A ball of moist friable soil, shredded coconut fibre or other rooting medium is fixed around the ring by wrapping it with polythene sheet so that it does not dry out. Roots are formed just above the ring and after 2-6 months the rooted air layers can be cut. In homegardens fruit trees such as guava and litchi are often propagated by air layering.

NURSERY MANAGEMENT

In an agroforestry system agricultural crops are generally directly sown in the field but planting materials of tree crops are usually raised in nurseries and the prepared seedlings are then planted in the main field. However, before going ahead with nursery plans, a thorough economic analysis is required since a good nursery can be an expensive undertaking. For small and medium scale plantations, procurement of seedlings from an authenticated commercial seedling grower is the best and the cheapest source of quality planting material.

A successful nursery operation depends on many factors like selection of site, planning and supervision, timely operations, etc. The selection of nursery site is important to ensure efficient production of good quality seedlings and easy nursery management. The following factors should be considered while selecting a site for nursery.

Water: A nursery should be established near a regular source of water. The nursery requires between 10 and 20 litres of water per square metre bed per day, depending on the temperature and the size of the young plants.

Soil: A good source of sandy loam or loamy soil, preferably forest topsoil, should be as close as possible to the site to reduce transport costs. This is particularly important if a large number of potted seedlings are to be produced.

Topography: The site should be relatively flat with 1-2% slope. The gradual slope allows runoff and prevents water logging and erosion. In hilly areas, nurseries must be built on terraces.

Microclimate: The site should not be exposed to dry winds. The site should receive sunlight for the major part of the day.

Materials: There should be easy access to manure, chemical fertilizers, compost, polybags, pesticides, etc.

Location: To minimize transportation time and cost, the nursery should be located as close as possible to the planting area.

Access: The site should be close to a good road so that the nursery will be accessible throughout the year.

Labour: There should be enough local labour available near the nursery site.

Size: This is determined mainly by the number of plants to be produced, the time they will spend in the nursery and their density in the beds.

Fence: The nursery must be protected against intruders, particularly stray animals and birds.

Shade: A nursery must be shaded. Shade prevents overheating during the hot season and raises the humidity in the beds by impeding air movement.

Raising bare-rooted seedlings and wildings in the nursery

Growing bare-rooted seedlings is the simplest form of tree propagation in a nursery. The seeds are sown in a well-prepared bed and the seedlings are allowed to grow till they are ready for field planting. A more common method is to sow the seed closely in seedbeds and to prick out the young seedlings into plant beds at the right spacing for their further growth in the nursery. Pricking out is the term used for transplanting young delicate seedlings by lifting them gently, using a pointed bamboo stick. There are good reasons to separate the seedlings at germination stage (seedbed) and to allow further growth in the nursery (plantbed).

1. The seeds of many trees germinate slowly and unevenly. By pricking out the seedlings that have reached the right stage every few days, a uniform stand is achieved in the plantbed. If plants grow uniformly, they can be given the right treatment (e.g., topdressing with fertilizer, root pruning) at the right time.
2. The requirements for germination are not the same as those for seedling growth. A seedbed needs careful levelling to prevent washing down of seeds to a lower edge during watering. The seedbed requires a fine tilth so that the seeds can be sown at the proper depth. The soil in a seedbed does not need to be fertile because the seedlings are pricked out before the food reserves in the seed have been exhausted. A plantbed does not require such a perfect surface, but friable, rich soil is required to encourage growth and branching of the roots.

Seed-trays of 40 cm × 30 cm size or so and about 5 cm deep, filled with coarse sand may also be used as seedbeds. The soil in seedbeds must be compacted to ensure good contact between the top layer in which the seeds are sown and the deeper layers. Light watering after sowing helps to settle the soil around the seed. Seeds should not be sown too densely, as crowding may result in weak seedlings. If seeds are sown in seedbeds or trays and are to be pricked out later, it is better to broadcast or dibble in rows. Small seeds (e.g., *Eucalyptus* species, *Acacia mangium*) are mixed with fine sand in a sand-seed ratio of 2:1. This makes even sowing at the right density much easier. Broadcasting is limited only to small seeds, because large seeds cannot be placed at the right depth by broadcasting. Sowing in rows ensures an even distribution of seeds and facilitates weeding and pricking out.

If seeds are sown directly in their final position in the nursery without the practice of pricking out, they are sown in pits at the proper spacing in a regular pattern. The spacing depends on the tree species and the required seedling size for planting out. Seed in seed-trays is often covered with coarse sand, which is easier for the seedlings to break through. If germination takes several weeks it is better to mulch seedbeds or seed-trays. The mulch softens the impact of watering and keeps the topsoil moist. However, the mulch is removed as soon as the first signs of germination are observed. After emergence of the seedlings they are thinned leaving only a single seedling in each grid position.

Pricking out

Seedlings in seedbeds or seed-trays are to be transferred to the plantbeds. Pricking out is usually done when the seedlings have one or two normal leaves. If seeds are sown at the right spacing, the seedlings fill the bed or tray by the time they have reached the stage for pricking out. For most species pricking out is done within a few weeks from emergence of the sprout. Simultaneous germination makes it possible to prick out all the seedlings at the same time. But even with proper pre-germination seed treatments, seeds of many tree species do not germinate simultaneously. Thus, it is often necessary to prick out repeatedly at an interval of several days.

In most tree species seedlings are large enough for pricking out at about 10-20 days after germination, but conifers are often pricked out sooner, even 2-4 days after emergence. If the seedlings are too young, they will be very delicate and likely to be damaged. However, if seedlings are allowed in the seedbed for long, they become overcrowded and separating the roots becomes difficult. Pricking out must be completed within a few hours per day, avoiding work during the hot hours. Successive sowings are planned in such a manner that batches reach the pricking-out stage one after the other, not simultaneously.

Raising seedlings in pots in the nursery

Sowing or planting in pots is an alternative to sowing or planting in beds. This requires extra work in the nursery and greatly increases the cost of transport if the seedlings are to be planted in a remote site. The extra expense involved in using polythene pots is compensated by better quality of the planting material, resulting in a high percentage of survival and a uniform stand in the field. The main advantage is of course that the roots are left undisturbed in the potting soil when pot seedlings are planted out.

Soil for filling the pots should be light and loose textured for good aeration and drainage, so that root development is not impeded. A commonly used mixture is 3 parts topsoil, 1 part compost and 1 part sand. The pots are watered and left to stand for a few days till the pot soil has settled. The pots are placed close together in blocks of similar length and width as the plantbeds in the nursery. The pots should always be placed upright even though it is often easier to lean them against each other. Pots can be seeded directly, or they can be planted with seedlings pricked out from seedbeds or seed-trays. Wildings or cuttings can also be planted in pots. For most tree species pricking out seedlings into pots has mainly two advantages over direct seeding in the pots.

1. Pricking out stimulates branching of the root system, so that the pots are better filled with roots to hold the soil during field planting.
2. Each batch of pricked out seedlings are uniform in height and growth.

Care for plants in the nursery

Watering: Light watering should be provided to the seedbeds regularly until the seedlings emerge. Alternatively, the seedbeds may be mulched during this period to save water and to keep the topsoil moist. Mulch is removed as soon as the first seedlings emerge.

Shade: When seeds have been sown, seedlings have been pricked out or cuttings or wildings have been planted, shading is necessary to prevent drying out and scorching of the leaves. A 50 cm-high shade frame can be erected by using bamboo or wooden sticks. Trees should never be planted within a nursery, because full light is required for hardening seedlings before planting.

Weeding: Seedbeds must be kept weed free so that the seedling roots do not get tangled up with the roots of the weeds. The competition between weed and young seedlings for nutrients and water is also checked.

Windbreak: A windbreak consisting of at least two rows of shrubs and one row of fast-growing trees should be planted in areas having high wind velocity. These should be placed on the side of the nursery facing the wind. The trees should be far enough from the beds so that their crowns will not shade the beds.

Fertilizer application: Liquid fertilizer is ideal for nursery application. Chemical fertilizer is usually not recommended. The mixture of one part of cow dung and five parts of water is left for 10 days. Before spreading the liquid, it should be diluted by adding more water. Seedbeds require no fertilizer, but plantbeds should receive liquid manure once in two weeks. If the soil in the plantbed is deficient in plant nutrients chemical fertilizers may be applied in small quantity.

Damping off control: Damping off is a common and serious disease in seedbeds. Several moulds in the soil may infect germinating seeds and emerging seedlings. Sowing in clean soil, such as fresh river sand, is the best way to prevent damping off. High humidity, heavy wet soil, too much shade, crowding of the plants in the seedbed, and a high organic matter content of the soil can all encourage fungus growth in a seedbed or seed-tray. The risk of damping off is reduced if the soil is well drained and aerated.

Field planting

Hardening-off: Five to six weeks before field planting, watering to the nursery should gradually be cut down and eventually stopped. Reduction of the shade can start much earlier depending on the tree species. This is known as hardening-off and is done to accustom the young plants to hardy conditions in the main field. The seedlings in the nursery may be watered on the day before planting out to facilitate lifting of the plants.

Lifting the plants: Bare-rooted plants are usually dug up by carefully lifting the plants with a fork. The plants are shaken gently to remove excess soil from the roots and packed in sacks, polythene bags with holes, banana leaves, or crates for transportation to the planting site. Diseased and plants that are not true to type should be removed.

Planting out: The best period for planting is after the first shower of monsoon rains. The land should be weed free and the soil loosened to improve the uptake of water. Planting spacing of trees depends on species, environmental conditions and the purpose for which these are grown.

Fertilizing the seedlings: The young trees are provided with extra nutrients by mixing compost, manure or fertilizer in the topsoil before or during planting. Fertilizer should be used in small quantities and only in moist soil. Manure and compost release nutrients slowly over a longer period of time and they also help to retain soil moisture.

TREE MANAGEMENT

A silvicultural regime or management strategy is a set of prescribed methods of manipulating the growth of trees so as to satisfy the objectives of the grower. These strategies must be developed with clear objectives appropriate to the existing physical and biological conditions, resources and technologies available. The main objective of tree management in an agroforestry system is to maximize the value of tree production without reducing the agricultural production significantly. The trees are to be managed in such a way that the agricultural crops do not suffer from inadequate levels of solar radiation, plant nutrients and soil moisture due to competition from the tree component. Agricultural production is expected throughout the life of the trees in an agroforestry system.

Aftercare in the main field

In an agroforestry system planting materials or seedlings of tree crops are generally raised in nurseries and the prepared seedlings are then planted in the field. Seedlings may be procured from a known source if the requirement is small. Raising a good nursery may sometimes be expensive. Thus, commercial seedlings are the best and the cheapest source of planting materials. But these must be genuine and true-to-type. Once seedlings are planted in the main field the work is certainly not finished. Care of the young trees in the field will improve establishment and growth.

Protection practices

Fire: Fire is often a serious threat during dry seasons. Regular weeding and removing dead branches and dry leaves decreases the risk of fire. This, however, affects the layer of litter, and ultimately soil fertility declines. Firebreaks (strips of cleared land which stop a fire) can be constructed between or around tree stands.

Wind: Young plants can be protected against the wind by using stakes. If the tree is moving to and fro in the wind, young roots are ruptured and this may greatly delay establishment of the plantation. The stake is placed on the windward side, so that the tree is not rubbing against the stake when the wind blows.

Shade: Many fruit and tree crops benefit from shade during the first year after field planting. If the young trees are staked, a palm frond can be tied to the stake to shade the tree.

Animals: Animals browsing leaves or twigs cause much damage to the young trees. Small animals such as rats can be caught in traps. Other solutions include digging ditches or placing wire fences. Young trees can be protected from birds by fine netting. Domestic animals, such as sheep, goats and cattle, not only browse trees, they also trample small trees down. The livestock should be kept away, if necessary, by employing a guard.

Pests and diseases: Healthy plants are often able to outgrow pests and diseases. Using hardy species, healthy planting material and good maintenance techniques reduce the risk of damage. Chloropyriphos should be applied in the pits at the time of planting if white ants are a problem. Mulch containing insect repellent material such as *Azadirachta indica* or *Eucalyptus* leaves can sometimes prevent the seedlings. If insect pests or diseases are noticed immediate control measures are to be taken.

Maintenance practices

Watering: Seedlings are planted usually at the onset of monsoon. But the problem arises if monsoon ceases early or a dry spell continues immediately after the plantation. Watering should be done as and when required. The weeds around the trees should be cleaned before watering. A small basin around each tree or a trench running along a row of trees will help the water to flow towards the tree. Water should be given in the late afternoon. The quantity of water needed depends on the weather and the amount of rainfall.

Weeding: Climbing or winding weeds can quickly strangle young trees. A circle with a diameter of at least 50 cm around the tree should be kept free of weeds. This prevents competition between trees and weeds for water, light and nutrients. Apart from weeding, weed growth can also be suppressed by mulching using organic material or small stones around the tree.

Nutrition: Trees are usually fertilized twice a year, once just before the onset of monsoon and another immediately after the cessation of monsoon. Nitrogen, phosphorus and potassium are given 50 gram each per plant per year for three to four years after plantation. The fertilizer is spread in a circle around the tree and mixed with soil by hoeing superficially, taking care not to damage the roots. The nutrients applied to the agricultural crops provide the required nutrition to the tree crops once the seedlings are established. Mulching with organic material around the tree will also provide extra nutrients.

Replanting: Replacement with the seedlings of about the same size and same age is done if some of the young trees die. This is known as filling in or gapping up. Some seedlings in the nursery are retained for this purpose in seedling beds or large pots. In some agroforestry systems, such as shade trees for perennial crops like tea and coffee, or windbreaks, replanting is much more important than in pure plantations because uniform cover is fundamental to the function of the planting. In windbreaks it is particularly important to have 100% of the positions planted during the first year since open places in the line create turbulence and reduce the overall effectiveness of the windbreak. Replanting should take place within one or two months of the initial planting, especially in areas with a significant dry season, as it is difficult for trees planted in the following years to succeed due to competition with the more advanced original seedlings.

Planting spacing of trees

In case of pure timber plantation, seedlings are planted at high density with the intention of thinning the plantation later. But in agroforestry systems, trees are generally planted at their final density. In agroforestry systems it is therefore necessary to have high survival rates, rapid initial growth, and in the case of timber trees, good trunk form in all of the planted individuals. However, it may be recommended to plant more trees than required so that some degree of selection can eliminate unsuited, damaged, insect infested or diseased trees when they become evident. The genetic quality of the tree species influences the number of trees to be planted per unit area to guarantee a quality tree crop.

The initial spacing of the trees depends to a large extent on the objectives of the plantation, the growth habit of the species, site quality and management. If the farmer wants to produce firewood the initial spacing may be 2 m or less in small plantations. For saw-wood plantations the spacing should be 3 m, with later thinning. In boundary plantings initial spacing for timber trees should be 3-5 m, whereas shade trees for coffee or cacao should be at more than 8 m spacing. In agrisilvicultural systems, hedgerow spacing is maintained at 4-8 m and tree to tree spacing within the row should be 2-3 m. Windbreaks should be planted in a line perpendicular to the predominant winds, the lines separated by about 20 times the maximum height of the trees. The final tree population depends upon several factors which are to be considered before deciding final crop stock.

1. The relative value of tree crops and agricultural crops greatly influence the choice of stocking rate because of the competitive nature of the various components in an agroforestry system due to interactions among them. The more valuable the agricultural crop, the fewer will be the number of trees.
2. Performance of trees at various densities is an important consideration. A tree species performs differently when grown at different densities. At the same time trees of different species do not perform similarly when planted at the same density. High plant density leads to lower growth rate in stem diameter on individual trees. However, the total biomass production per unit area remains more or less the same over a range of stockings. The choice is either lots of a greater number of small diameter trees or lots of fewer larger trees. When timber production is the objective, larger individual trees are generally preferred since this gives higher proportion of usable timber, and reduces transport, logging and processing costs per unit timber.

3. The climatic factors like rainfall pattern, temperature, edaphic factors like soil fertility status, soil type and structure, water holding capacity, etc., latitude aspect of the site that affects availability of incoming solar radiation are also important factors to decide the tree population.
4. Solar radiation is the most critical limiting factor for production of agricultural crops in an agroforestry system. Generally, shade tolerant annual crops like turmeric, ginger, pineapple, etc. allow higher tree density than the less tolerant crops like cereals and pulses.
5. Management and maintenance operations like movement of equipments and machines in the field may be affected if the tree density is high. The routine operations like fertilization, hoeing, watering, sowing and harvesting of agricultural crops may also be affected. A careful planting arrangement can avoid these problems to some extent.
6. The personal preference is one of the most important considerations for selecting the optimal tree density in an agroforestry system. A grower may prefer to have a 'parkland' environment by having a small number of trees per unit area whereas another may prefer a 'forest' appearance for his land with higher tree density. The advantage of agroforestry is that there is very little 'waste' of land even with very low tree density because a smaller number of trees per unit area generally means higher agricultural production.

Thinning

Thinning is a useful silvicultural tool to manipulate the tree growth and form once the agroforestry system is established. It simply involves the removal of selected trees from the site. Thinning allows the grower to maintain the tree population within the limits acceptable from the point of view of the associated crops. It is also an opportunity to harvest some products from the trees, decide on the arrangement of the trees, control the shade they cast on the crops and facilitate animal management. Thinning allows the use of a greater share of the limited resources of the site like solar energy, soil moisture, plant nutrients and space for the remaining trees and associated crops. In addition to the density and arrangement of the trees, thinning takes into account the form and health of the trees, eliminating the unhealthy, crooked or bifurcated trees. If malformed or undesired trees are eliminated before they start to compete with the healthy and straight trees, the growth of retained trees will be unaffected and total production will increase. Early thinning of trees is also required to reduce the competition between the trees and agricultural crops in the agroforestry system, thus increasing the production of arable crops. The specific objectives of thinning of tree crops may be summarized as follows.

1. To have the desired tree density in the system
2. To improve the hygiene of the crop by removing dead, dying, insect infested and diseased trees
3. To ensure the best physical conditions for growth of both tree crops and agricultural crops
4. To obtain the desired type of tree crop from the system
5. To save marketable volume of wood that will be used as fuelwood, pole, staking material, etc. from anticipated loss
6. To improve the quality of wood
7. To increase the net yield and financial return from the system per unit of land
8. To meet the immediate requirements of the grower

Thinning should be done at a time when the resultant debris will cause the least damage to the underneath agricultural crops. Thinning operation should ideally be taken up immediately after harvesting or before planting of an annual crop or after heavy grazing to reduce the loss on agricultural production. Thinning of tree species whose foliage has some degree of fodder value, should coincide with periods of high fodder demands. The degree of thinning depends on the number of trees initially planted, the desired number of trees in the system and the major objectives of the agroforestry system. The resource requirements, characteristics and value of the associated agricultural crops also decide the degree of thinning of trees.

Pruning

Pruning is the removal of plant parts to improve the form and growth of trees. It is probably the most important silvicultural tool in agroforestry. Branches are removed with minimum damage to cambium or growing tissue so that the wound will close in the shortest period of time and with the least possibility of wound infection. The major objectives of pruning of timber trees in a block plantation are for training the plant, maintaining plant health, improving the quality of flowers, fruit, foliage and stems, and controlling growth. However, the principal objective of pruning in agroforestry systems is to reduce the shade on the underneath crops. Timber trees in agroforestry systems tend to branch more than those in block plantations due to the wider spacing and reduced lateral competition, and therefore require more frequent and intensive pruning. However, not more than 30% of the crown should be pruned in any case. The season for pruning depends on the seasons for tree growth and development and on the climatic conditions of the site. It is better not to prune during the dry season when the crops need protection from the summer sun.

In general, deciduous trees and shrubs should be pruned when they are dormant, preferably in early spring just before the start of growth. At this time, wound healing will begin almost at once and it will be the most rapid. Dormant pruning will have less detrimental effect on the growth of trees than pruning during active growth of trees. Another advantage of dormant pruning with deciduous trees is that it is easier to select branches which should be removed when the leaves fall. Pruning of dead branches can be done at any time of the year since no living tissue is affected. When removing diseased material, pruning tools should be disinfected between each cut. In order to make sharp, clean cuts, ensure that tools are in good condition.

Pruning of timber trees, especially in agroforestry systems, can make a significant difference in the quality of the timber and will provide additional products such as firewood, posts and fodder for the farmer. Pruning is really the best preventive maintenance a young plant can receive. Formative pruning is designed to produce a single, straight stem. It should be carried out when the branches and forks are light enough to be removed with secateurs or a pruning knife. Formative pruning is generally carried out while the trees are still young.

Trees grown at wide spacing tend to develop heavy side branches. This varies greatly with species and provenance. Side branches form knots in the timber, the bigger the knots and the more their number, the lower will be the value of the timber. Pruning of the side branches should be done without damaging the bark of the trunk. Careful pruning to remove side branches will, thereafter, produce knot-free timber at that level as the tree stem expands over time. Branch bases left projecting from the trunk produce dead knots in the wood, which reduce timber quality and may also provide entry points for disease and infection causing pathogens, insects or termites. Pruning of lateral branches has no effect on the straightness of the trunk of a tree. However, removal of too many branches at one time may retard the growth of the tree.

Ideally, branches should be removed when they are of a size which allows them to be cut cleanly with a pruning knife or a pair of secateurs. The need for pruning, the intensity and frequency, varies from species to species and also depends on the nature of the associated crops.

The most common types of tree pruning are crown thinning, crown raising and crown reduction. Selective removal of branches on young trees throughout the crown is called crown thinning. This promotes better form and health by increasing light penetration and air movement. Strong emphasis is given on removal of weak branches. This is generally not practised on mature trees. Crown raising is removal of lower branches on developing or mature trees. On the other hand, removal of larger branches at the top of the tree to reduce its height is called crown reduction. It is different from topping because branches are removed immediately above lateral branches, leaving no stubs. Crown reduction is the least desirable pruning practice. It should be done only when absolutely necessary. The new techniques of tree pruning are fixed-lift pruning, variable-lift pruning, selective pruning and tip pruning.

Fixed-lift pruning: It is the complete pruning of all branches below a prescribed point on the stem. The lift is specified as height from the ground and the choice of height is based on the shape of development attained by the stand.

Variable-lift pruning: It is the complete pruning of all branches below a prescribed variable point on the stem. This point may be specified either as a proportion of the height or as a diameter limit. The specification is often directed towards achieving the removal of a consistent fraction of green matter from the trees, regardless of size.

Selective pruning: It is the removal of some of the branches on the stem at various levels above the ground according to a particular prescription. It is used to correct incipient faults in the crown or to remove branches which may be difficult in subsequent pruning. It promotes a deep crown of fine branches and good apical dominance.

Tip pruning: It is the pruning of a branch at a point other than its junction with the stem. Tip pruning is used to retard the development of a branch.

Coppicing and pollarding

Coppicing and pollarding are techniques which can be used in managing certain types of trees. Coppicing involves cutting the tree species close to ground level to produce new shoots from the stump. It normally sends up a number of shoots instead of the original single stem. Coppicing is a useful practice when it is desirable to produce large quantities of foliage close to ground level, as is the case of many fodder species and for leaf extract products. In some species coppicing occurs naturally if the trees are damaged. Pollarding involves cutting back the crown of a tree in order to harvest wood and to produce regrowth beyond the reach of animals and/or to reduce the shade cast by the crown. It is commonly practised during the periods of extreme forage shortage to produce emergency fodder without killing the tree or greatly reducing its height. Vertical pollarding is the close pruning of branches along the stem. Regrowth after coppicing and pollarding is vigorous because the tree's root system has already been well established. *Acacia catechu*, *Albizia lebbek*, *Anogeissus latifolia*, etc. are considered as strong coppicers. *Aesculus indica*,

Chloroxylon swietinia, Hardwickia binata, etc. are good coppicers and *Adina cordifolia, Bambax ceiba,* etc. are bad coppicers. *Grewia optiva, Hardwickia binata, Morus alba, Salix* spp., etc. are suitable for pollarding.

Lopping

Removal of one-year shoots or fresh growth from entire crown of the tree/plant in order to get sufficient fodder for livestock is known as lopping. Lopping is extensively done in *Morus, Grewia, Bauhinia,* etc.

Bending

Restricting the development of bole to allow more food material to new leaf shoots is called bending. Bending is useful when it is desirable to produce large quantity of foliage close to ground level.

Bushing

Bushing is the horticulture operation commonly used to increase fruit production at a convenient height for harvesting.

AGRICULTURAL CROP MANAGEMENT

Agricultural production in an agroforestry system may come from various crops like annual agricultural food crops, agricultural cash crops, horticultural crops, pasture for grazing, forage crops for harvesting of fodder, etc. or any combination of these. The selection of compatible crops is as important as the choice of tree species. However, certain principles should be followed while choosing agricultural crops to be grown with commercial trees in agroforestry systems.

1. The crops should not produce dense shade before the trees are well established. For example, cocoa should not be planted before a timber species.
2. Climbing species such as yam, vanilla and pepper should not be planted while the trees are young.
3. Crops that compete severely for nutrients and water should not be planted during the tree establishment stage.
4. Crops such as banana, maize and sugarcane that exhaust soil nutrients can negatively affect the establishment of trees on the site if adequate fertilizers are not applied.
5. Root and tuber crops with extensive horizontal root systems should be planted at a sufficient distance from the tree roots to avoid competition and damage during harvest of root crops.

6. It is an advantage if the crop fixes nitrogen. Annual legumes should ideally be included in the crop rotation.
7. The crop should not have the capacity to behave like a weed.
8. The crop should not act as a host to pests and diseases that can affect the trees.
9. In permanent agroforestry systems the crop must be shade tolerant (like ginger, turmeric, arrowroot, pineapple, etc.) or shade demanding (like cocoa) when the tree canopy closes.
10. The crop should require minimum manual labour when it is intended to promote reforestation programmes. The manual labour required for the agricultural crop can limit the total area that can be reforested in agroforestry systems.
11. The crop should have good economic potential.
12. The crop should not cause physical damage to the trees. For example, felling bananas in a new tree plantation can cause damage.
13. Perennial grasses should not be planted before the tree establishment.
14. The crop should not have allelopathic effects on the trees.

Due to the presence of perennial trees, the nature of system continues to change throughout the rotation of the system. Thus, agricultural management practice is also varied during the development of the tree crops both to minimize damage to the trees and to maximize agricultural production. On the basis of tree growth and management, the development of an agroforestry system is divided into five stages such as preplanting phase, establishment phase, silvicultural management phase, maturing phase and harvesting phase.

Pre-planting phase

This is the phase where farmer prepares a site for the establishment of agroforestry system. This involves selection of site, crop species and varieties, collection of good quality seed, careful land preparation, and proper management of rodents, insects, white ants, etc. Farmers are encouraged to use locally adapted varieties to ensure good crop establishment and high yield. In this phase the site is prepared and fertility of the site is increased by growing one or two annual legume crops before planting the trees. This will improve soil structure, control weeds and prepare the land for plantation of trees.

Establishment phase

In this phase of management, the young trees are carefully tended until they are large enough to fend for themselves. Planting arrangement of trees, rooting patterns of trees and agricultural crops, duration of the crops and growth habits of both trees and crops decide how close crops can be grown to the trees. Arable crops should not be planted in close proximity to the base of trees in this phase because this increases competition between the trees and crops for nutrients and water, and will adversely affect the tree establishment and growth. To encourage the early establishment and root development of the trees, a metre-wide free zone should be left on either side of the tree row. The agricultural operations which retard the tree growth even for a short period should be avoided as far as possible.

Silvicultural management phase

Thinning and pruning are the two important silvicultural operations carried out in this phase of management. By the age of first thinning and pruning operation, the trees begin to compete with the agricultural crops on the site. Light becomes the most limiting factor to affect the understorey crops as the canopy closes. To allow light for these crops, the side branches of trees should be pruned. While thinning or pruning the trees, care must be taken not to cause any type of damage to the agricultural crops. Once the productivity is reduced due to shade effect, shade tolerant crops like ginger, turmeric, arrowroot or pineapple, etc. should be grown. Pasture crops may also be grown.

Maturing phase

Different management practices are implemented beneath the maturing trees in this phase of management, such as grazing, maintenance of moisture and nutrients and weed control, which improve and augment the aesthetic value of the stand.

Harvesting phase

In this phase of management tree crops are harvested. Timber harvesting from an agroforest should begin as soon as trees become large enough to fetch a suitable or profitable price. The optimum time for harvest depends on the value of the standing timber and its expected increase in value over time, availability of a suitable market then and in the near future, the owner's preferences. The shape, size and diameter of the timber and the cost of felling of tree and transporting it to the point of sale determine the value of the timber.

MANAGEMENT OF FRUIT TREES IN AGROFORESTRY SYSTEMS

Perennial fruit crops can be used for almost any objective that trees might be used for in agroforestry systems, depending on the needs of the farmer, the nature of the fruit species, and the other components of the system. Fruit trees may be used as hedges, individual trees or live fences. They may also be found in homegardens or as service trees among perennial crops. The fruits are important for domestic consumption and are also sold when a market can be found. However, certain criteria must be followed for the selection of fruit trees for agroforestry systems.

1. The crop should have a commercial potential at the local and regional level.
2. Its growth habit must be compatible with other crops or trees in the system, for example a crown that allows light to reach the associated crops.
3. Fruits should have the ability for prolonged storage and post-harvest processing and transport.
4. The individual fruit tree should have the potential for high production of fruits or biomass.
5. Fruit trees should not act as a host to pests and diseases that are a threat to the associated crops.
6. These should have deep rather than surface root system.
7. Fruits should have the desirable characteristics for industrial uses, for example pectin content for jelly making.
8. Fruit crops should possess the easier vegetative propagation methods.
9. Fruits should have good flavour and high quality.
10. There should have easy fruit harvesting methods.
11. Fruit crops should have short period of establishment.
12. There should be low incidence of pests and diseases.
13. Fruit trees should have multiple uses.

Optimizing the use of light, water and nutrients during the formation and development of the fruit increases the quantity and quality of the fruit produced. Therefore, treatments to improve the capture of solar energy and application of water and plant nutrients will maximize the number of fruits that reach maturity. Abortion of fruits and flowers occurs when the inflorescence does not receive sufficient resources.

It is important to prune and train fruit trees in agroforestry systems. This reduces competition for nutrients, water and light with the associated crops and competition among branches for the tree's resources. Pruning also promotes regular flowering and fruiting, and maintains a low crown that facilitates fruit harvest and the control of disease. Old, sick, damaged and diseased branches and sprouts should be pruned regularly.

10

Agroforestry for Different Agroecological Zones

Ecology of a place is influenced by a number of biotic and abiotic factors. An agroecological zone is broadly homogeneous in climatic and edaphic factors, but not necessarily contiguous, where a specific crop exhibits roughly the same biological expression. On the other hand, the characteristics of an agroecological zone influence the crops and cropping patterns of the zone. Thus, there are considerable variations in the types and management practices of agroforestry systems being followed in different agroecological zones.

AGROFORESTRY FOR HUMID AND SUBHUMID TROPICS

Humid and subhumid tropical regions have much variation in biological, social and economic parameters. The characteristics of humid tropics include high mean annual rainfall of more than 1500 mm, dry season of less than four months in a year and mean annual temperature of more than 22°C. Rainfall usually exceeds the evapotranspiration. However, demographic pressure has caused excessive deforestation and overgrazing which developed soil related problems like low soil fertility, low organic matter, high acidity and high rainfall erosivity. The agroforestry practices designed for this region should aim for improved fallows, soil fertility improvement and conservation, food production, etc. Thus, the common agroforestry systems in this zone are *taungya*, homegardens, alley cropping, plantation-crop combination, windbreaks, silvipastoral systems and various intercropping systems. Shifting cultivation is a common practice in tropics which should be discouraged as it causes environmental degradation and ecological imbalance, soil erosion and soil nutrient loss.

Trees selected for the agroforestry systems in the tropics must have capacity to produce food, fodder, wood and other agricultural and forest products throughout the year. Besides, trees must have the ability to contribute to soil and water conservation, and soil fertility improvement. Most importantly, these tree species should have low fertility requirement for their growth with fast growing habit.

Suitable agroforestry systems

Taungya: The *taungya* system is a temporal agroforestry system which combines production of forestry tree crops and agricultural crops on forest lands. The *taungya* system is found in areas where state forestry departments are establishing tree plantations. Food crop production takes place during the period between land clearing and the plantation establishment phase. But, the main objective of *taungya* is wood production, not food. Although through *taungya* system artificial regeneration of forest tree species is cheap and establishment of forest plantation is possible at a very low cost, the system involves in exploitation of human labour and exposure of land leading to soil erosion and loss of fertility.

Tropical homegardens: Homegarden is a type of stable traditional agroforestry system in the tropics, where subsistence food crops, tree crops, animals and sometimes cash crops are grown mainly around the homestead. Other terms used for homegarden are compound farming, village-forest garden and household garden. Homegardens tend to be located close to dwellings for security, convenience and special care. These traditional farming systems are developed by integrating ecologically adapted tree and crop species with low capital input and simple technology. Several features of the gardens show the imitation of the natural forest like diversity of cultivated plant species, multistorey vegetation structure and cycling and recycling of matter, resulting in maintenance of soil fertility which contribute to the stability and sustainability of the agroecosystem of homegardens. These are not static but change with new opportunities and socioeconomic conditions. Population pressure appears to have a profound effect on the stability of homegardens. Besides providing economic benefits like food, fodder, timber, fuelwood, etc. to the land owner, the tropical homegardens provide many community and environmental benefits. The gardens are carefully structured where each component has a specific place and function. All homegardens usually consist of an herbaceous layer near the ground, a tree layer at the upper level, and intermediate layers in between with shrubs, fruit trees and medium sized timber trees.

Alley cropping: In alley cropping, food crops, preferably legumes spaced 4-6 cm apart, are grown in alleys formed by hedgerows of trees and shrubs. The hedgerows are cut back at planting and periodically pruned during cropping to prevent shading and to reduce competition with the associated food crops. Root pruning is recommended during early stages of hedgerow establishment. The hedgerows are allowed to grow freely to cover the land when there are no crops. One major advantage of alley cropping is that cropping and fallow phases can take place concurrently on the same land. This allows the farmer to crop the land for an extended period without a fallow period. Although alley cropping was designed for small-scale farmers, it is adaptable to mechanized farming with appropriate machineries. Livestock may also be included in alley cropping by using a portion of the hedgerow foliage for animal feed. Hedgerows of fast-growingtrees and shrubs, particularly leguminous species like *Leucaena leucocephala* and *Gliricidia sepium*, do well and can provide green manure and mulch to crops grown between the hedgerows. This green manure/mulch contributes significantly to nutrient recycling, nitrogen supply, soil conservation, weed suppression and maintenance of soil productivity. In addition, trees and shrubs may provide fuel, staking materials and livestock feed. The choice of tree species for alley cropping is extremely important and to a large extent determines the success or failure of the system. The tree species should have a rapid growth rate, ability to withstand frequent cutting, good coppicing ability, nitrogen fixing capacity, and ability to withstand environmental stresses such as drought, waterlogging, and extremes of *p*H, etc. The suitable species for hedgerow planting are *Cassia siamea*, *Leucaena leucocephala*, *Gliricidia sepium*, *Calliandra calothyrsus*, *Sesbania sesban*, *Acacia mangium*, *Gmelina arborea*, *Albizia* spp., *Cajanus* spp., *Chamaecytisus* spp., *Desmodium* spp., *Erythrina* spp., *Flemingia* spp., *Inga* spp. and *Tephrosia* spp.

Shelterbelts: A shelterbelt is a wide strip of vegetation consisting of several rows of trees that slows wind speeds, thereby reducing wind erosion, evaporation and damage to crops by the wind. It is sometimes referred to as windbreak, although the latter often implies a single strip of trees and other vegetation. A shelterbelt presents a mechanical barrier to the impact of the wind and separates two zones - the windward and the leeward zones. The windward zone refers to the side from which the wind blows, while the leeward zone relates to the side where the wind passes. Properly placed field shelterbelts reduce soil erosion by wind, increase moisture for crop growth and reduce wind damage to crops.

AGROFORESTRY FOR ARID AND SEMIARID TROPICS

The arid and semiarid regions in India cover more than 317,000 square kilometers, mostly spread over the states of Rajasthan, Gujarat, Punjab, Haryana, Maharashtra, Karnataka, Telangana and Andhra Pradesh. These areas are characterized by hostile environmental conditions. The mean annual rainfall varies from 100 mm to 600 mm with over 80% of total annual rainfall received between June and September. The mean maximum temperature during summer is as high as 40° C and mean winter season temperature is less than 15°C. The potential evapotranspiration during summer is 7 to 9 mm/day, during monsoon 5.3 to 6.4 mm/day and in winter ranges from 1.8 to 2.9 mm/day. Thus, the evapotranspiration far exceeds precipitation throughout most part of the year. The infiltration rate is very high, ranging from 7 to 15 cm/day. Arid areas of north-west region of the country are characterized by high wind velocity during summer and monsoon seasons. The soils in this region are immature, structureless and very coarse in texture with low water holding capacity, low fertility status and low organic matter content.

Under such agroclimatic and edaphic conditions, returns from crop production are very often uncertain. Thus, people mainly depend on the livestock production although growing of some agricultural crops under tree crops is an age-old practice. However, ever growing human population and lack of arable land in the region aggravate the food scarcity. In areas of high population density and small land holdings, agroforestry plays an important role in many farmers' economic strategies. Improved agroforestry techniques help to mitigate the effects of deforestation, land depletion and rural poverty. Many farmers are concerned principally with meeting of household needs using tree products like fuelwood, small timbers, tree fodder for domestic animals and fruit. Using tree products for local consumption has the additional benefit of not having to rely on uncertain weather conditions for cash crops. This strategy minimizes risk and contributes to the overall diversification of family farms.

Selection of tree and crop species

Success of agroforestry systems in arid and semiarid tropics greatly depends on the correct selection of tree species. Since the water is a limiting factor for crop production in these areas, the trees should have drought tolerance mechanism like deep taproot system to draw water from deeper soil profile, leaf shedding in summer to conserve moisture, water binding mechanism and other xerophytic characteristics like wax coating, hairiness, sunken and covered stomata on leaves to minimize the loss of water through transpiration. Tree crops should also have tolerance to salinity and alkalinity, which are common features of these areas.

The trees suitable for arid and semiarid tropics are *Acacia nilotica*, *Acacia radiance*, *Acacia senegal*, *Acacia tortilis*, *Ailanthus excelsa*, *Albizia lebbeck*, *Azadirachta indica*, *Butea monosperma*, *Cassia auriculata*, *Cassia siamea*, *Dalbergia sissoo*, *Dichrostachys glomerata*, *Erianthus munja*, *Eucalyptus camandulensis*, *Eucalyptus* hybrid, *Hardwickia binata*, *Holoptelea integrifolia*, *Leucaena leucocephala*, *Parkinsonia aculeate*, *Prosopis chilensis*, *Prosopis cineraria*, *Prosopis juliflora*, *Prosopis tamarugo*, *Salvadora oleoides*, *Tamarix articulate* and *Tecomella undulate*.

Horticultural fruit trees recommended for agroforestry systems like hortisilvipastoral, agrihortisilvicultural or hortisilvicultural systems for arid areas are *Aegle marmelos*, *Annona squamosa*, *Carrisa carandas*, *Cordia myxa*, *Emblica officinalis*, *Feronia limonia*, *Grewia asiatica*, *Phoenix dactylifera*, *Psidium guajava*, *Punica granatum*, *Salvadora persica*, *Syzygium cumini*, *Tamarindus indica* and *Ziziphus mauritiana*,

The suitable grass species for silvipastoral or hortisilvipastoral systems are *Brachiaria mutica*, *Cenchrus ciliaris*, *Cenchrus setigerus*, *Chloris gayana*. *Dicanthium annulatum*, *Diplachure fusca*, *Lasiuruss indicus*, *Panicum antidotala*, *Panicum turgidum*, *Saccharum munja* and *Sporobolus* spp.

Agricultural crops for agrisilvicultural system in arid areas are pearl millet, cowpea, moth bean, sorghum, cluster bean, blackgram, greengram, chickpea, etc.

Suitable agroforestry systems

Several agroforestry systems are suitable for arid and semiarid regions. The most common agroforestry system is scattered trees in agricultural fields, constituting an agrisilvicultural system. Boundary plantations, alley cropping, silvipastoral systems, shelterbelts, windbreaks and block energy plantations are the other common agroforestry systems practised in arid and semiarid tropics. These systems could be suitably practiced based on situation, purpose and need.

Alley cropping: In arid region, animal husbandry is one of the major occupations of the people. The fodder needs for these animals may compete with the uses of hedgerow loppings for green manure and mulch which are the key components of the alley cropping technology to ensure improved soil organic matter and soil moisture, respectively. Alley cropping may have some drawbacks in areas where water is a limiting factor. Narrow hedgerows, spaced at a distance of 4 m each recommended for the humid and subhumid tropics are particularly competitive with crops in the arid and semiarid tropics. Despite these limitations, carefully placed and closely pruned rows of trees in agroforestry systems may have a role to improve system productivity. Such vegetative barriers could be located on field boundaries, providing feed for livestock and delineating cropped fields. Similarly, if situated at suitable contour intervals, they may serve to protect highly erosive soils. Alley cropping is considered to be successful in areas where the assured annual rainfall exceeds 750 mm on soils with good water holding capacity.

Silvipastoral systems: Usually many animal husbandry systems are in practice in the arid regions, but two extremes are well recognized. The first is where animals are a permanent feature in the farming system and are primarily stall-fed with supplemental pasturing. The other extreme is the nomadic system, where animals are moved from one site to another according to changing pasture conditions throughout the year. In both cases, livestock play an important role in sustaining field fertility through manure production. In the first case, farmyard manure is normally collected and spread on the fields. In both systems, livestock pressure must be balanced with rangeland productivity, if it is to be maintained. However, most grazing lands in the arid areas are public lands and are therefore freely accessible, a situation that may be leading to their overuse. This decline of communal pastures is aggravated by the growing tendency towards private ownership of grazing land; forcing poorer livestock owners to share less and less land. Adoption of agroforestry interventions such as silvipastoral or hortisilvipastoral systems with suitable fodder trees/shrubs and grasses having potentially excellent feeding value may be very useful to overcome this emerging crisis.

Windbreaks: Tree rows as windbreaks have long been used for protecting crops and soil against wind damage and wind erosion. Their effectiveness in increasing crop production has also been demonstrated in the drier parts of the arid and semiarid tropics. However, competition between the trees in a windbreak and the crops they are supposed to protect can also be a problem. Other than species selection, there are few management options for reducing underground competition. Canopy management may be effective in reducing competition for solar energy, provided it does not interfere with the aerodynamic efficiency of the windbreak.

Other modified systems: Where irrigation is available, or in river valleys where water tables are close to the soil surface, many traditional agroforestry systems can be practised. Such sites offer rare opportunities to maximize the output of tree products on a year-round basis and to take advantage of their many uses as vine supports, living fences, shelter, etc. Vegetative fences are widely used in the arid and semiarid tropics. Boundaries between fields and farms could be made productive by tree or shrub planting for production of leaf fodder and other uses. Trees can play important roles in stabilization of river and stream banks and waterways in vegetative contour strips for soil conservation and simply for shade. Tree plantings can be taken up by making use of farm areas that are not usually cropped, for example, bunds, field corners and rocky and unproductive field sections.

The best time to plant trees in the arid areas is the beginning of monsoon. Community based group nurseries should be encouraged for promotion of tree plantation in arid region. Training programmes and extension efforts are necessary to assist the farmers in establishing group nurseries. Application of compost or cattle manure 5 kg per pit helps in early establishment and quick growth of the tree seedlings. Soil moisture is a major limiting factor for successful establishment and early growth of young trees in these areas, particularly on sloping lands because the little water available from rainfall tends to run off down the slope. The mechanical measures like contour bunds, grading bunds, bench terracing, etc. and agronomic approaches like contour farming, strip cropping, etc. should be followed for conservation of soil moisture, reducing wind velocity and evapotranspiration.

AGROFORESTRY FOR TROPICAL WETLANDS

A wetland is an area of ground that is saturated with water either permanently or seasonally. A patch of land that develops pools of water after a heavy rain would not be considered a 'wetland' though the land is wet. Wetlands have unique characteristics; they are generally distinguished from other water bodies or landforms based on their water level and on the types of plants that thrive within them. Specifically, wetlands are characterized by the water table that stands at or near the land surface either permanently or seasonally. The water found in wetlands can be saltwater, freshwater or brackish. On the basis of water table fluctuation, wetlands can be classified into three major types; moist land, waterlogged lands and immersed lands.

Moist land: The water table fluctuates from the surface down to 30-40 cm in the ground. The ground may occasionally be flooded during heavy rains. Such lowlands are commonly located along river banks, lake shores and sea beaches. The vegetation commonly consists of moisture tolerant and hydrophillous trees, shrubs and herbs.

Waterlogged lands: The water table is close to the ground surface and often flooded during the rainy season. Soils are always water saturated except in a very dry climate and are rich in organic matter.

Immersed lands: Most of the ground surface is under water for a considerable period of time of the year. The depth of water may be as high as 100 cm during flood period.

Wetland plants and trees offer a number of benefits. Those found in coastal wetlands aid in the prevention of coastal erosion and the mitigation of its effects. They slow the flow of water and soak enough quantity to prevent nearby rivers and lakes from overflowing. In addition, these plants filter excess nutrients and pollutants from wetland waters. Wetland fish, aquatic wildlife and birds also find protection and breeding grounds within and among wetland trees and plants. Certain wetland plants are harvested for human consumption.

Measures for soil improvement

In wetlands too much water in the soil is the major obstacle to land use for either forestry or agricultural purposes. However, now several technologies are available to improve the site conditions. Wetlands may be converted for agriculture or tree production with proper water drainage technology. Agricultural drainage is the use of surface ditches, subsurface permeable pipes, or both, to remove standing or excess water from poorly drained lands.

Ditching: In moist land, ditching is commonly used to lower the water table by draining the excess water. Field ditches of shallow depth are opened connecting local depressions scattered over the field. All the ditches are converged to a common natural outlet. This method of drainage is most suited for uneven land with a number of local depressions. Tractors may also be used to turn over the subsoil for raising dykes on one or both sides of the ditches, which are then planted with trees. The depth and width of the ditches mainly depend on the water table and its fluctuation. The higher the water table with greater fluctuation, the deeper and wider are the ditches. The water table and its fluctuations are generally controlled below 50 cm from the ground surface.

Land smoothing: This method of drainage is suitable for flat lands with low depressions and for areas with a little slope. The elevated areas are cut off and the excess soil is spread over low areas so that the surface will be even with a uniform slope. All tillage operations are carried out parallel to slope. Field ditches are opened at the lower end to collect excess runoff from the smoothed slope.

Terracing: In waterlogged lands the water table is close to or even above the ground surface. The soil removed from the waterlogged and immersed lands are used to raise terraces on one side and to create ponds on the other. The size of the terrace and pond varies greatly in accordance with the water situation and agroforestry practice. In immersed land, terraces should be higher and ponds should be bigger and deeper. For planting trees with a deep root system the terraces should be much larger and higher. If fishery is considered as a part of the agroforestry system, bigger ponds with deeper depths may be required.

Sub-surface drainage: Subsurface drainage systems have a positive impact because they generally decrease the amount of surface runoff, thereby reducing the loss of substances generally transported by overland flow. Subsurface drainage facility promotes better root growth and plant health when soils have poor internal drainage. Unlike surface drainage systems, cropping area is not wasted due to subsurface drainage and there may not be any interference to farming operations. But in this system, high initial investment and regular maintenance are required.

Selection of tree and crop species

Ecological adaptation and economic benefits are two important factors governing the selection of tree species for a wetland agroforestry system. Despite developing terraces or opening ditches and adopting other surface or subsurface drainage systems, excess water is still a major problem for tree planting and crop cultivation in wetlands. Tree species for wetlands should have the following characteristics.

1. Tree species should have the ability to tolerate water logged soil for a longer period.
2. Tree species should be of relatively shallow root system.
3. Tree species should be deciduous with a narrow and thin crown so that solar radiation would not be a limiting factor for agricultural crops when canopy closes.
4. For agroforestry purposes fast growing multipurpose trees should be selected.

In the waterlogged areas, stagnation of water occurs for most part of the year because of imperfect drainage. Some areas in high rainfall zones where drainage is poor gradually become waterlogged. Waterlogged areas are also developed in the coastal areas mainly in the estuaries of rivers, in creeks and lagoons and lower reaches of tidal streams.

According to the ecological requirements and economic conditions the tree species commonly used for wetland agroforestry practices include *Acacia nilotica, Alnus cremastogyne, Alnus trabeculosa, Baringtonia* spp., *Bischofia javania, Casuarina equisetifolia, Eucalyptus robusta, Eucalyptus rudis, Lagerstroemia flosreginea, Morus alba, Paulownia tomentosa, Populus nigra, Salix babylonica, Salix xuchonensis, Syzygium cumini, Taxdoium distichum, Taxodium scandens, Terminalia arjuna*, etc.

Casuarina equisetifolia is the best recommended tree species for plantation along coastal wetland. Planting *Casuarina equisetifolia* in the outer area and growing perennial fruit crops like *Cocos nucifera* and *Anacardium occidentale* in the inner area is generally practised along the coast. Plantations are done in raised mounds where water table is very high. Other species found suitable are *Acacia auriculiformis, Acacia mangium, Acacia tortilis, Avicennia officinalis, Casuarina jhunjhuniana, Eucalyptus tereticornis, Nipa fruiticans, Pongamia glabra*, etc.

To obtain full utilization of environmental resources and to increase economic benefits, a suitable crop should be intercropped after the establishment of a tree plantation in accordance with site and climatic conditions. In wet season, rice and jute are the best suited crops. Wheat, barley, oat, soybean or toria are commonly grown as winter crops. Vegetables such as radish, carrot, cabbage, etc. can also be grown during this season. Watermelon, ladiesfinger and other vegetables may be grown in summer season.

When the canopy of tree crops closes causing considerable shade to the annual crops, shade tolerant crops like ginger, arrowroot, *Aloe vera*, *Curcuma amda*, edible mushrooms, etc. are selected for the system. But the other edaphic and climatic factors should not be ignored while choosing the crops.

Suitable agroforestry systems

Agroforestry provides an excellent opportunity for transforming long underutilized wetlands to highly productive lands to sustain life and the environment. However, appropriate choice of agroforestry systems with tree and crop species is necessary. Many of the regular agroforestry systems with some modifications may be well suited to different categories of wetlands. However, aquaforestry and agrisilviaquaculture are widely practised systems.

The aquaforestry system comprises of composite fish culture in the farm ponds and various trees and shrubs, leaves of which are preferred by fish (*Leucaena leucocephala*, *Morus alba*, *Gliricidia sepium*, *Moringa olifera*, etc.) are planted on the boundary and around fish ponds. Suitable arable crops may also be included in agrisilviaquacultural systems in addition to trees and fish culture.

In the low-lying coastal regions fish and prawn can be cultivated in saline water while growing coconut and other trees on bunds of ponds. Now fish culture in the mangroves is also advocated, which forms a rich source of nutrition to the aquatic life and breeding ground for juvenile fish, prawn and turtle.

A well-balanced system of animal husbandry including goatery, poultry, duck farming, turtle and fish farming in the small ponds makes a balanced system of high moisture, energy and nutrient use efficiency per unit area. In high rainfall areas, in low land paddy fields fish can easily be reared by planting trees on field bunds or boundary. Coconut plants can also be successfully planted on raised bunds of paddy field with an alley space of 5 m width depression which is utilized for fish culture. In shallow ponds and streams, aquatic plants such as lotus, water-chestnut can be planted as crops. For deeper ponds water-lily, water-lettuce, water-caltrop are cultivated as floating crops.

In silvipastoral and hortisilvipastoral systems grass species *Chloris* spp., *Cynodon dactylon*, *Panicum maximum*, *Sorobolus* spp., *Cenchrus* spp., *Spinifex squamosus*, *Sporobolus* spp., etc. are most suited.

AGROFORESTRY FOR HILL LANDS

Hill region in India is characterized by climatic and edaphic variations, depending on where it is located and its elevation. The Himalayas, extending from Kashmir in the west to Arunachal Pradesh in the east, is a great example for understanding the varying characters of hilly region. Hill farming is subject to a number of serious constraints.

1. Undulating topography, small fragmented and scattered land holdings, with very limited use of inputs
2. Though total annual precipitation is medium to high in most of the hilly regions, much of it goes waste through runoff
3. Due to the slopes, soils are prone to erosion, which is aggravated by heavy migratory grazing which has also led to soil degradation
4. Soils are shallow and stony, with low fertility status and low organic matter content and subject to periodic water stress
5. Despite sufficient water resources, irrigation facilities are meagre, and most agriculture depends on rainfall
6. Natural hazards like intense rainstorms, hailstorms, floods, epidemic diseases, insects and an erratic monsoon
7. In the high-hill temperate zone crops are grown only during the summer (April to September) as snow cover during winter does not permit sowing

Despite this, a variety of cereals, fruits, vegetables, flowers, and medicinal, aromatic and dye plants are grown in these areas. Hill farmers have also been growing a variety of native trees in their farmlands to maintain land productivity and to provide food and wood. However, in last few decades widespread deforestation and increasingly intensive use of land to sustain a growing population has increased soil erosion, lowered soil fertility and reduced agricultural productivity in the hills. This has raised concern over sustainability of the hill farming system. Agroforestry can be a promising solution to these problems and hence a key to the sustainability of the hill farming system.

The small farmers in hills depend mainly on the degraded lands for their living. In these areas agroforestry may be promoted by integrating more multipurpose trees in the already existing farming systems to improve their livelihood. Agroforestry in hills will increase agricultural productivity by nutrient recycling, reducing soil erosion, and improving soil fertility and enhancing farm income compared with conventional crop production. Agroforestry also has promising potentials for reducing deforestation while increasing food, fodder and wood production.

Agrisilvicultural systems are common in hills. A traditional homegarden with a number of tree species along with livestock, poultry or fish, mainly to satisfy the farmer's basic needs is also a common hill farming system. Agrihorticultural systems are practised in the mid-hills subhumid and high hills temperate wet zones where fruit trees and agricultural crops are grown on terraced beds.

Suitable agroforestry technologies for hills

Alternate contour rows of trees intercropped with shade tolerant crops: Trees are commonly grown in contour rows with close 1-2 m plant to plant spacing. The double row approach of trees in contour is more suitable. In the double row approach the row to row spacing is 1-3 m. Entire hillside can be brought into cultivation in this method. Within the tree rows, 6-10 m interspaces are kept for cultivation of seasonal crops. In the initial stage, different agricultural crops like paddy, maize, sesame, pulses and seasonal vegetables, etc. can be grown up to 3-5 years, but as the canopy of trees closes and the yield recovery of these crops reduces substantially, the shade tolerant crops like ginger, turmeric, arrowroot, pineapple, mango ginger (*Curcuma amda*), *Aloe vera*, etc. should be cultivated. Mulch practice in cultivation of shade tolerant crops is effective in soil conservation. Under zero tillage with mulch, the soil erosion decreases considerably.

Bench terracing method: The bench terraces of about 1.0 m wide and 3-5 m long are constructed along contour with 2-3 m interval throughout the selected hill side. The space between the bench terraces down the slope are of 3-6 m. The bench terraces and interspaces are cultivated with seasonal and perennial crops alternately. In this method, the soil erosion is less and water infiltration rate is high. However, high initial investment and regular maintenance are required in this system.

Differentiated slope agroforestry method: The major features involved with this modified agrihortisilvicultural system are the division of the entire hillside into upper, middle and lower part. The less fertile upper slope is used for cultivation of forest species, the moderately fertile middle slope is left for horticultural crops, and the lower fertile most portion of the hillside is retained under cultivation of seasonal crops.

Natural vegetative buffer strip: The natural vegetative buffer strip is a comparatively new practice of hill cultivation. In this method, the natural vegetation of local origin is intensively utilized for hedge purpose and trees and crops are commonly grown in the interspaces of vegetative strips. The main advantages of this technique lie on the easy growing of vegetative strip with less labour input and easy to manage.

Suitable tree species for hill agroforestry

The suitable tree species for different agroforestry systems in hills are *Albizia chinensis*, *Ailanthus excelsa*, *Alnus nitida*, *Bauhinia variegate*, *Bombax ceiba*, *Cedrus deodara*, *Celtris australis*, *Eucalyptus grandis*, *Ficus auriculata*, *Ficus semicordata*, *Grewia optiva*, *Mangifera indica*, *Melia azedarach*, *Pinus roxburghii*, *Prunus cerasoides*, *Prunus persica*, *Punica granatum*, *Pyrus pashia*, *Quercus himalaya*, *Quercus leucotricophora*, *Robinia pseudoacacia*, *Sapium insigne*, *Toona serrata*, *Rhus parviflora*, *Rubus ellipticus*, *Vitex negundo*, etc. The common horticultural crops are apple, pear, plum, apricot, cherry, peach, almond, etc.

11

Agroforestry for Watershed and Wasteland Management

A watershed is a geohydrological unit or all the land and water area bounded by a divide which contributes runoff to a common point. Watershed is considered to be synonymous with 'catchment basin' and 'drainage basin'. In watershed management approach, development is not only confined to agricultural lands but also covers the area, starting from the highest point of the land to the outlet of the natural stream. Watershed management becomes increasingly important as a way to improve livelihood of people while conserving and regenerating their natural resources.

Watershed management is the process of creating and implementing plans, programmes and projects to sustain and enhance watershed functions that affect the plant, animal and human communities within a watershed boundary. It implies the judicious use of all the resources,i.e., land, water and vegetation in an area to prevent soil erosion, improve water availability and increase food, fodder, fuel and timber on sustained basis.

PRINCIPLES OF WATERSHED MANAGEMENT

The main principles of watershed management based on resource conservation, resource generation and resource utilization are:

1. Utilizing the land based on its capability
2. Protecting fertile topsoil
3. Minimizing silting up of tanks, reservoirs and lower fertile lands
4. Protecting vegetative cover throughout the year
5. *In situ* conservation of rain water
6. Safe diversion of gullies and construction of check dams for increasing ground water recharge
7. Increasing cropping intensity through inter and sequence cropping
8. Alternate land use systems for efficient use of marginal lands
9. Water harvesting for supplemental irrigation

10. Maximizing farm income through agricultural related activities such as dairy, poultry, sheep and goat farming
11. Improving infrastructural facilities for storage, transport and agricultural marketing
12. Improving socioeconomic status of farmers

OBJECTIVES OF WATERSHED MANAGEMENT

The concept of National Watershed Development Project in Rainfed Areas (NWDPRA) is scientific utilization of natural and human resources to sustain production systems which can meet the socioeconomic need of the people and to maintain health, vigour and purity of production environment. The broad objectives of the NWDPRA are

1. Conservation, development and sustainable management of natural resources including their use
2. Enhancement of agricultural productivity and production in a sustainable manner
3. Restoration of ecological balance in the degraded and fragile rainfed ecosystems by greening these areas through appropriate mix of trees, shrubs and grasses
4. Reduction in regional disparity between irrigated and rainfed areas
5. Creation of sustained employment opportunities for the rural community including the landless

AGROFORESTRY- A SUSTAINABLE OPTION FOR WATERSHED MANAGEMENT

Adoption of agroforestry practices with the available technologies will fulfill all the objectives of the watershed management. This will help in conservation of soil and water from land surface and will always add to the productivity of resource poor small and marginal farmers. Though the agroforests are typically less diverse than native forest the significant number of plant and animal species in an agroforestry system can provide ecological resilience and contribute to the maintenance of beneficial ecological functions. Forest watershed functions include maintenance of high-quality water, regulation of water quantity and maintaining water-sediment balance in watersheds. These functions depend on the amount and type of forest and the various land use practices followed. Watershed functions are based on the three aspects of forests that together dominate the impact on the flow and quality of water.

Vegetation: Trees and understorey crops in an agroforestry system hardly allow any bare land. Trees are better in maintaining transpiration rates throughout the year than most other plants and their annual water consumption often exceeds that of other vegetation. Tree canopies intercept more rainfall than other vegetation and this is returned to the atmosphere by direct evaporation.

Soil conditions: Forest soils have a high surface infiltration rate and substantial macroporosity (due to soil biological activity and tree root turnover) facilitate deep infiltration and also sub-surface lateral flow of water.

Landscape: A landscape with a rough, uneven surface, including depressions and swamps provide temporary water storage and sediment filter functions and very few pathways for rapid surface runoff.

Maintaining a protective forest in the form of an agroforestry system as an integral part of the landscape provides a healthy watershed for reliable water supply, protection from soil erosion and a more secured supply of forest products for cash and home consumption. Watershed management through agroforestry practices offers a number of benefits to the marginal and small farmers living in a watershed area.

1. More income if the agroforestry is commercially successful.
2. Annual crops around tree crops protect the soil from erosion and help maintaining soil fertility.
3. Diversity of agroforestry provides both cash crops and food for home consumption.
4. Less mechanization and a greater role for draft animals and their manure reduces the need for chemical fertilizer.
5. Less chemical fertilizers and pesticides reduce input costs.
6. The diversity of agroforestry is more resilient to weather and market fluctuations.

CONTRIBUTION OF TREES TOWARDS IMPROVEMENT OF SOIL WATER REGIME

Trees contribute a lot towards maintenance and improvement of soil fertility (Nair *et al.*, 1984, Lefroy *et al.*, 1999). Since tree is the primary component in an agroforestry system it maintains or improves soil fertility through the following processes.

Processes which increase additions to the soil

1. Maintenance of soil organic matter through carbon fixation in photosynthesis and its transfer via decay of litter and roots
2. Nitrogen fixation by many leguminous and some non-leguminous trees
3. Uptake of nutrients released by rock weathering by tree roots

4. Increased water infiltration through better soil physical properties and effects of roots
5. Taking up of water from deeper soil layers by tree root systems

Processes which reduce losses from the soil

1. Protection from erosion and thereby losses of organic matter and nutrients
2. Reduction in the rate of organic matter decomposition, by shading and mulching
3. Reduction of water loss through evapotranspiration, by shade and litter
4. Increased water storage capacity, through better soil physical conditions

Processes which affect soil physical conditions

1. Maintenance of soil physical properties through organic matter and effects of roots
2. Penetration of compact layers by roots
3. Modification of extremes of soil temperature, by shade and litter

Processes which affect soil chemical conditions

1. Reduction of acidity or rate of acidification, through bases in litter
2. Reduction of salinity and sodicity by trees in association with other management measures
3. Reduction of soil toxicities caused by pollution

Soil biological processes and effects

1. Production of leaf litter of high quality, containing a balanced nutrient supply and its transfer to the soil by litter decomposition
2. Increased availability of phosphorus through mycorrhizal associations
3. Increased nodulation on roots of nitrogen-fixing trees in close proximity to roots of non-nitrogen-fixing plants, with possible direct transfer of nutrients between root systems

However, the greatest advantage of trees is in controlling soil erosion by checking the rate of soil runoff. The reduction in runoff is caused by canopy interception and direct transpiration is only to a small degree but the greater part of it results from higher infiltration capacity under trees. It helps in improving recharge of ground water, besides perennial flow of stream and better moisture availability through microclimatic interventions.

BIOLOGICAL MEASURES FOR WATERSHED MANAGEMENT

Choice of suitable cropping system and appropriate agronomic management practices like selection of suitable crops and varieties, sowing time, seed rate, planting techniques, use of fertilizers, etc. are the major biological considerations to tackle the problems of watershed areas.

Alley cropping

It is the practice of growing arable crops between hedgerows of shrubs and trees, which are periodically pruned to prevent shading of intercrops. When there are no crops, the hedgerows are allowed to grow freely or cut to meet the fodder needs. Alley cropping retains all the basic restorative attributes of the bush fallow such as nutrient recycling, fertility regeneration and weeds suppression so that all processes occur concurrently on the same land, allowing the farmer to crop the land for an extended period of time.

An important benefit of alley cropping is the addition of large amounts of organic materials from the prunings as mulch or green manure which can have favourable effects on soil physical and chemical properties, on microbiological activity and hence on soil productivity. There is also a reduction in the use of chemical fertilizers resulting in a decrease in environmental pollution with maintenance of soil health. The tree rows on sloping land act as a physical barrier to soil and water movement, resulting in significant reduction in erosion losses. The presence of prunings applied as mulch in the alleys also controls soil erosion. Alley cropping provides additional products such as forage, firewood or stakes when a multipurpose tree is used as the hedgerow.

The choice of tree species for alley cropping is extremely important and to a large extent determines the success or failure of the system. The tree species selected for hedgerow planting should have rapid growth rate, ability to withstand frequent cutting, good coppicing ability, nitrogen fixing capacity, deep root system with a different root distribution pattern from the crop and ability to withstand environmental stresses such as drought, waterlogging, and extremes of *p*H. The suitable species for hedgerow planting are *Cassia siamea*, *Leucaena leucocephala*, *Glyricidia sepium*, *Calliandra calothyrsus*, *Sesbania sesban*, *Acacia mangium*, *Gmelina arborea*, *Sesbania grandiflora*, *Albizia* spp., *Cajanus* spp., *Chamaecytisus* spp., *Desmodium* spp., *Erythrina* spp., *Flemingia* spp., *Inga* spp. and *Tephrosia* spp.

Boundary planting

This is a simple but effective practice for watershed management particularly for small farmers. It consists of planting trees all along the boundaries between the fields and farm or along the margins of roads and canals. It is also called four-sided forestry with the objective of gaining production from trees, while having no adverse effect on adjacent crops and possibly having a beneficial effect through fertilization by nitrogen fixing tree species or their leaf litter with protection from wind or aiding soil conservation, i.e., watershed protection. Tree species like *Acacia mangium*, *Acacia auriculiformis*, *Acacia nilotica*, *Sesbania grandiflora*, *Azadirachta indica*, *Albizia lebbeck*, etc. are most suitable for field bund or boundary plantation. Farmer's choice, land suitability, marketability of the produce, etc. are to be considered while selecting the tree and crop species for plantation.

Intercropping with multipurpose trees

In order to overcome the limitation of land availability and to attain self-sufficiency, the farmers can resort to intercropping for increasing the efficiency of land use, both under limited soil moisture as well as under adequacy of resources. Inclusion of annual pulse crop in inter-row spaces of widely spaced multipurpose trees offers additional productive benefits as well as protective benefits.

Homegardens

This is the most popular system of agroforestry being adopted by small and marginal farmers of the watershed areas. It is easier to implement and is the most rewarding component of agroforestry. Homegarden is the intimate association of multipurpose trees and shrubs with seasonal and permanent arable crops and livestock within the compounds of individual houses. Mostly the whole system, i.e., crop-tree-animal unit is managed by family labour. In a homegarden, production is very diverse and continuous with less risk and there is a regular flow of products such as food, fuel, fodder, fruit, spices and timber. The protective function of homegardens is also very high with soil conservation, water retention and favourable microclimate for humans and animals.

Agrihortisilvicultural system

In the agrihortisilvicultural system, suitable arable crops are grown in inter-spaces of perennial fruit or multipurpose tree rows. In between two trees one fruit crop like mango, guava or pomegranate may be planted. When yield of annual crops is substantially reduced due to canopy closure, the system is converted to a hortisilvicultural system. With appropriate management practices this can give a sustainable production system.

Silvipastoral and hortipastoral systems

In these systems, suitable fodder grasses like guinea grass (*Panicum maximum*), dinanath grass (*Pennisetum pedicellatum*), napier grass (*Pennisetum *purpureum*), thin napier (*Panicum polystachion*), anjan grass (*Cenchrus ciliaris*) and fodder legumes like stylo (*Stylosanthes hamata* and *Stylosanthes scabra*), cowpea (*Vigna sinensis*) are grown in inter-spaces of perennial fruit or multipurpose tree rows. These systems are very effective to control soil erosion in watersheds as the grasses give a permanent soil cover.

Random mix

Under this system, woody and herbaceous components are randomly mixed but the plants occupy their respective ecological niches, thereby avoiding competition.

CONSERVATION MEASURES FOR WATERSHED MANAGEMENT

Various types of soil and water conservation measures have been recommended depending on the soil type, topography, climate and size of the land holding.

Contour cultivation: Contour operations are done across the slope by cultivation of crops and trees on contour. The contour furrows created would form a multitude of mini barriers across the flow path of runoff. Contour cultivation remains the most effective on moderate slopes of 2-7%, besides *in situ* conservation of moisture and water harvesting.

Contour bunding: This practice comprises of constructing narrow based bunds on contour to impound runoff water behind them, so that impounded water is absorbed gradually into the soil profile. The bunds should be constructed from the top of the catchment and proceeded downwards.

Graded bunding: Graded bunds are constructed in relatively high rainfall areas to remove excess water out of the field to avoid water stagnation, especially in deep black soils. These bunds are outlets for safe removal of water. The channels of graded bunds are wide and shallow.

Bench terracing: Bench terracing involves the conversion of original sloppy land into level step like fields constructed by half cutting and half filling, which reduces the degree of the slope. In hilly areas this approach of bench terracing for agroforestry models is gradually getting popular.

Contour furrow: Contour furrow farming is also called as strip farming. The cropping is usually intermittent on strips or in rows with catchment area left fallow. The principle is to collect runoff from catchment area to improve soil moisture on the cropped area, particularly on sloppy land with undulating topography.

Agroforestry for Wasteland Development

India shares 17.4% of the world population, while its land is only 2.4% of the total geographical area of the world. Naturally, the pressure on the land is often beyond its carrying capacity. Therefore, the productive lands, especially the farmlands in India are in the constant process of various degrees of degradation and are fast turning into wastelands. At present, approximately 63.85 million hectare area of the land is lying as wastelands in India. Out of these lands, about 50% lands are such non-forest lands, which can be made fertile again if treated properly. In the last 50 years India's lush green village forests and woodlots have been deforested to the maximum. To restore this ecological imbalance by developing the degraded non-forest wastelands, Government of India had created the Department of Wasteland Development during July, 1992 under the Ministry of Rural Development, which has been subsequently reorganized and renamed Department of Land Resources, with a broader mandate.

National Wasteland Development Board was established in 1985 mainly to tackle the problem of degradation of lands, restoration of ecology and to meet the growing demands of fuelwood and fodder at the national level. The Board was reconstituted in August, 1992 and was made responsible for development of wastelands mainly in non-forest areas in totality by involving local people at every stage of development. It aims at creating a scenario where the Government acts as a facilitator and the people at the grass root level become the real executioner of the programme. The degradation of environment in the fragile Indian subtropical ecosystem is basically attributed to the following factors.

1. Increasing biotic pressure
2. Absence of adequate investments and appropriate management practices
3. High rate of population growth and high incidence of poverty in rural areas
4. Over-exploitation of natural resources
5. The break-down of traditional institutions for managing common village resources
6. Faulty land use practices

The consequences of this environmental degradation are soil erosion and land degradation, depletion of natural resources, lower productivity, ground water depletion, shortage of drinking water, reduction in species diversity and increase in the extent of wastelands. Development of wastelands mainly in non-forest areas aimed at checking land degradation, putting such wastelands of the country to sustainable use and increasing biomass availability especially that of fuelwood, fodder, fruits, fiber and small timber. Government of India is taking up this colossal task through its Integrated Wasteland Development Project (IWDP) by revitalizing and reviving village level institutions and enlisting people's participation. The basic objective of this scheme is development of wastelands through integrated approach. These plans are prepared after taking into consideration the land capability, site condition and local needs of the people. The major activities taken up under the scheme include

1. *In situ* soil and moisture conservation measures like terracing, bunding, trenching, developing vegetative barriers and drainage line treatment
2. Planting and sowing of multipurpose trees, shrubs, grasses and legumes
3. Encouraging natural regeneration
4. Promotion of agroforestry and horticulture
5. Wood substitution and fuelwood conservation measures
6. Drainage line treatment by vegetative and engineering structures
7. Afforestation

Wastelands are broadly classified into culturable and unculturable wastelands. The lands that cannot be developed for vegetative cover are called unculturable wastelands, which include barren rocky areas, snow covered glacier areas and steep slopy areas. The lands those are capable of vegetative cover but not being used due to different constraints of varying degrees are called culturable wastelands. These lands can be put to some productive use after proper treatment and reclamation measures. NWDB classified the culturable wastelands into the following categories.

1. Eroded gullied and ravinous land
2. Undulating upland
3. Waterlogged land
4. Salt affected saline and alkali soils
5. Shifting cultivation areas
6. Degraded forest areas
7. Degraded non-forest plantation area
8. Sand dunes, sand bar and coastal sand
9. Mine spoils and industrial wastelands
10. Strip lands
11. Pasture and grazing lands

The characteristics of the wastelands and their development vary with the categories. Agroforestry is being viewed as a restoration agent, rehabilitation process, bioremediation technology and mechanism to high input agriculture on fragile lands. This system involves the conscious and deliberate use of land for the concurrent production of agricultural crops including tree crops and forest crops.

MANAGEMENT OF WASTELANDS

Gullied and ravinous land

A narrow deep gash formed due to concentrated runoff is called gully and it is considered as the severest form of water erosion. Very deep and wide gullies with steep sides are known as ravines. The ravines are classified as shallow, medium or deep based on gully depth, bed width and area. Ravinous lands are mainly confined to the states of Uttar Pradesh, Madhya Pradesh, Rajasthan and Gujarat. Clearance of forest for agriculture, faulty soil works for agriculture, uncontrolled grazing, low annual rainfall and extreme dryness of soil due to easy and fast drainage and high soil porosity are the major causes of ravine formation.

The ravinous lands can be managed through integrated approach of soil conservation and afforestation. Maintaining dense multistoreyed vegetation is considered highly effective. This reduces slash erosion, reduces runoff, increases infiltration and improves soil structure through addition of organic matter. Shallow and medium gullies can be put under silvipastoral system with suitable tree and grass species. The deep and narrow gullies are best controlled by putting them to permanent vegetation. Gully heads and sides are stabilized by planting grasses. Reducing the slopes of the gullies and diverting the surface flows are essential for successful afforestation programme in the gullied and ravinous lands. The gullies have to be plugged with suitable mechanical measures to check soil erosion and conserve soil moisture.

The suitable tree and shrub species for these wastelands are *Eucalyptus* spp., *Dalbergia sissoo*, *Gmelina arborea*, *Emblica officinalis*, *Acacia auriculiformis*, *Acacia nilotica*, *Acacia catechu*, *Albizia lebbeck*, *Albizia procera*, *Leucaena leucocephala*, *Azadirachta indica*, *Dendrocalmus strictus*, *Prosopis juliflora*, *Prosopis chilensis*, *Pongamia glabra*, *Tectona grandis* and *Ailanthus excelsa*. Among the grasses and legumes *Cenchrus ciliaris*, *Cenchrus setigerus*, *Chrysopogon fulvus*, *Dicanthium annulatum*, *Saccharum spontaneum*, *Pennisetum pedicellatum*, *Cynodon dactylon*, *Heteropogon contortus*, *Stylosanthes* spp. are found suitable for such lands.

Shifting cultivation areas

Shifting cultivation is a cyclic land use practice of felling trees and burning of forest areas for growing agricultural crops leading to land degradation. Shifting cultivation areas are characterized with high rainfall, steep slopes and shallow soils. This system is prevalent in several parts of north-eastern hill regions of Arunachal Pradesh, Meghalaya, Mizoram, Manipur, Nagaland and Tripura. It is also practised in some parts of Andhra Pradesh, Assam, Odisha, Jharkhand and Madhya Pradesh. Burning of vegetation in shifting cultivation alters plant nutrients from organic to inorganic forms. Some nutrients are lost by volatization and a sizable quantity of nutrients is also lost through blown off ash due to strong wind velocity during dry period. Wrong tillage practices lead to runoff and soil erosion. Thus, shifting cultivation causes deforestation and denudation of hilly slopes. Integrated approach of conservation measures and improved farming including suitable agroforestry systems should be adopted for management of these degraded lands. Agrisilviculture, silvipasture, hortisilviculture and hortisilvipasture are the best suited agroforestry systems in shifting cultivation.

The main objective of introducing agroforestry systems in shifting cultivation fallows is to recover depleted soil nutrients. The best species for the fallow system should include good nitrogen fixation in soil. Plants included should be compatible with future crops, free of any negative physical or chemical effects on the soil and not in competition with the crops to be planted later on the same site. This includes direct seeding of clean tilled, harvested plots, selective cutting of bush, followed by enrichment planting with tall seedlings. *Acacia catechu*, *Dalbergia sissoo*, *Anacardium occidentale*, *Bombax ceiba*, *Santalum album*, *Acacia nilotica*, *Schima wallichii*, *Michelia doltsopa*, *Pterocarpus dalbergioides*, *Albizia lebbeck*, etc. are suitable for planting in the degraded soil.

The silvipastoral system comprising *Alnus nepalensis*, short duration fruit crops like pineapple and forage crops like *Panicum maximum* and *Setaria sphacelata* coupled with *Stylosanthes guyanensis* could be sustainable agroforestry system for hilly areas with 30-60% slope. Lower areas of hilly slope are terraced and stabilized with mechanical measures and planted with fodder legumes and grasses like *Stylosanthes hamata*, *Stylosanthes guyanensis*, *Setaria sphacelata*, *Clitoria ternatia*, *Panicum maximum* and *Panicum polystachion* on risers. Suitable field crops like rice, maize, millets, sesame, groundnut, blackgram, greengram, horsegram, and root crops like arrowroot, ginger and turmeric are grown on the terraces according to the requirement of the farmers.

Sand dune

In India, about 22.7% of total geographical area comes under sandy areas, mainly confined to Rajasthan, Gujarat, Punjab and Haryana. The sand dunes and sandy undulating plains are responsible for the creation of sandy wastelands. The new wastelands are being created due to faulty tillage and agricultural operations, cutting and lopping of trees and overgrazing of natural rangelands in arid and semiarid areas. These wastelands are characterized by loose structureless sands, poor fertility and water holding capacity and presence of alternate layers of calcium carbonate. The region is subjected to severe wind erosion, high temperature, scanty and erratic rainfall, frequent drought and high rate of evaporation.

These wastelands can be rehabilitated by suitable agroforestry practices. The plant species chosen should have deeply buried rhizomes or prostrate growth with stem anchored by abundant adventitious fibrous roots for rehabilitation of sand dunes. *Prosopis cineraria*, *Prosopis chilensis*, *Acacia tortilis*, *Acacia radiance*, *Acacia nilotica*, *Zyzyphus mauritiana*, *Erianthus munja*, *Dalbergia sissoo*, *Albizia lebbeck*, *Azadirachta indica*, *Leucaena leucocephala*, *Tamarix* spp. and *Parkinsonia aculeate* are the important tree species for these areas. Grass species like *Cenchrus ciliaris*, *Cenchrus setigerus*, *Lasiurus sindicus*, *Panicum turgidum*, *Panicum antidotala* etc. are suitable for silvipastoral system.

Coastal wastelands

Coastal sandy area in the form of narrow strips along the eastern and western coasts occupy about 8.4 million hectare of land in the country. The soil of this area is characterized with alkaline in reaction, high sodium salt content, poor in nutrients and low in water holding capacity. The water table is generally high. The occurrence of cyclones is a common phenomenon in the coastal area. Thus, afforestation of coastal areas in one kilometer width is recommended to moderate the effects of cyclones. *Casuarina equisetifolia* is the best recommended tree species for plantation along coastal sandy areas. Planting *Casuarina equisetifolia* in the outer area and growing perennial fruit crops like coconut (*Cocos nucifera*) and cashew nut (*Anacardium occidentale*) in the inner area is generally practised along the coast. Plantations are done in raised mounds where water table is high. Other species found suitable are *Pongamia glabra*, *Nipa fruiticans*, *Avicennia officinalis*, *Casuarina jhunjhuniana*, *Acacia auriculiformis*, *Acacia tortilis*, *Acacia mangium*, *Eucalyptus tereticornis*, etc. are suitable for plantations in coastal wastelands. In silvipastoral and hortisilvipastoral systems grass species *Panicum maximum*, *Cynodon dactylon*, *Sorobolus* spp., *Cenchrus* spp., *Chloris* spp., *Spinifex squamosus*, *Sporobolus* spp., etc. are most suited.

Mining and industrial wastelands

A large area in the states of Madhya Pradesh, Chhattisgarh, Bihar, Jharkhand, Odisha, Karnataka and other states are subjected to degradation due to open-cast mining of coal, bauxite, limestone, slate, etc. Surface mining requires operation of heavy machines to excavate the earth which leads to complete destruction of vegetation and the entire soil profile is disturbed. After mining these areas are left with boulders, lateritic heaps, stones and pits of various dimensions which give a very desolate look. Highly undulating topography with irregular slopes causes heavy water and soil erosion. Reclaiming these wastelands is a challenging task due to disturbed ecosystem with poor physical, chemical and biological soil properties. The species found suitable with high survival rate are *Eleagnusum berilata, Coriaria nepalensis, Indigofera pulchella, Acacia auriculiformis, Dalbergia sissoo, Eucalyptus camaldulensis, Grevillea robusta, Grevillea pteridifolia, Albizia lebbeck, Prosopis chilensis* and *Cassia siamea*. The area should be levelled to some extent with the help of dozer before undertaking plantation. Pits of one cubic feet are opened at the spacing of 2 m × 2 m. The pits are filled with the forest soils for mycorrhizal inoculation in case of plantation of tree legumes. Application of compost or farm yard manure 5 kg and plant nutrients N, P_2O_5 and K_2O at the rate of 50 gram each per pit helps in early establishment and quick growth of the seedlings. Silvipastoral system with suitable grass species is found highly successful in mined out wastelands.

High altitude and steep slopy areas

Indiscriminate grazing by cattle especially sheep and goat is the major cause of soil erosion in the hilly areas. In high altitude regions, the soil suffers from excessive runoff during monsoon and moisture stress during dry periods. Hedgerow planting is required under hilly terrain. Nitrogen fixing tree and shrubs suitable for agroforestry systems may be planted in closer spacing. To stabilize the structure and to make productive use of the land, in steeply sloping areas the risers or terraces are densely planted with multipurpose trees for fruit, fodder and fuelwood. The tree species are *Grevillea robusta, Acacia catechu, Pinus roxburghii, Acacia modesta, Prosopis juliflora, Populus alba, Populus nigra, Populus candicans, Populus ciliata, Alnus nepalensis, Leucaena leucocephala*, etc. are suitable. Fruit trees planted are ber, papaya, dwarf mango, guava, citrus, etc. The space between trees and within the rows may be utilized for growing perennial grasses.

Waterlogged and marshy land

In the waterlogged areas, stagnation of water occurs for most part of the year because of imperfect drainage. The central Board of Irrigation defined waterlogged land as an area where the water table rises to an extent that the soil pores in the root zone of a crop becomes saturated resulting in restriction of the normal circulation of air, decline in the level of oxygen and increase in the level of carbon dioxide. Large areas in the Punjab, Haryana and western Uttar Pradesh have become waterlogged due to canal irrigation.

Some areas in high rainfall zones where drainage is poor gradually become waterlogged. Waterlogged areas are also developed in the coastal areas mainly in the estuaries of rivers, in creeks and lagoons and lower reaches of tidal streams. Marshy lands are permanently or periodically inundated by water and characterized by vegetation which includes grasses and aquatic weeds. Proper drainage is important to reclaim these areas. Tree species suitable for waterlogged areas are *Eucalyptus robusta*, *Sygygium cumini*, *Salix babylonica*, *Populus nigra*, *Terminalia arjuna*, *Alnus cremastogyne*, *Alnus trabeculosa*, *Morus alba*, *Taxodium scandens* and *Acacia nilotica*. In marshy places, tree species best adopted are *Baringtonia* spp., *Bischofia javania*, *Taxdoium distichum*, *Eucalyptus robusta*, *Eucalyptus rudis*, *Casuarina equisetifolia* and *Lagerstroemia flosreginea*.

Arid and semiarid lands

The arid and semiarid lands are spread over the states of Rajasthan, Gujarat, Punjab, Haryana, Maharashtra, Karnataka and Andhra Pradesh. These lands are characterized by hostile environmental conditions,viz., productivity potential of the land is low, soils are immature, structureless, coarse in texture and poor in nutrient status with low water holding capacity. Correct selection of trees is of prime significance in this region for the development of agroforestry. The trees should have drought tolerance mechanism like deep root system, leaf shedding in summer to conserve moisture, water binding mechanism and other xerophytic characteristics. Tree crops should also have tolerance to salinity and alkalinity which are common features of these areas. The trees best suitable for these wastelands are *Tamarix articulata*, *Prosopis juliflora*, *Prosopis tamarugo*, *Salvadora oleoides*, *Eucalyptus* spp., *Dalbergia sissoo*, *Acacia nilotica*, *Albizia lebbeck* and *Azadirachta indica*. Horticultural fruit trees for arid areas are *Zizphus mauritiana*, *Punica granatum*, *Psidium guajava*, *Phoenix dactylifera*, *Aegle marmelos*, *Annona squamosa*. Agricultural crops for agrisilvicultural system in arid areas are pearl millet, cowpea, sorghum, cluster bean, blackgram, greengram, etc. The suitable grass species for silvipastoral or hortisilvipastoral systems are *Diplachure fusca*, *Sporobolus* spp., *Brachiaria mutica* and *Chloris gyana*.

Undulating land with or without scrub

These lands occupy high areas of topography but exclude mountainous terrain. The soils in these regions are highly prone to degradation and erosion. It is mainly associated with plateau and laterite formations found in the states of Maharashtra, Andhra Pradesh and Rajasthan. These areas are characterized with high rainfall and undulating terrain. The causes for the formation of such lands include deforestation and its rehabilitation is possible only through agroforestry. The tree species suitable for these wastelands are *Eucalyptus* spp., *Dalbergia sissoo*, *Gmelina arborea*, *Acacia auriculiformis*, *Emblica officinalis*, *Albizia procera*, *Anogeissus pendula* and *Dendrocalamus strictus*. *Agave sisalana* that has high efficiency to protect soil from erosion is also recommended with perennial forage grasses as intercrops.

Degraded pasture

This type of wasteland are found mainly on rocky topography, saline alluvial plains and eroded uplands and are characterized by stunted plant growth, irregular slopes, sandy and loamy soils. Overgrazing, deforestation, runoff and wind erosion are responsible for natural pasture and rangeland degradation. The recommended remedial measures for this type of wastelands are establishment of soil and water conservation measures like windbreaks, shelterbelts, mulching, contour bunding, etc. to upgrade soil productivity followed by tree cover with species like *Acacia senegal*, *Acacia tortilis*, *Acacia nilotica*, *Butea* spp., *Indigofera cordifolia*, *Heteropogon* spp., etc.

Degraded notified forest land

These lands are characterized by eroded and undulating topography with very steep slopes and are highly prone to water and wind erosion. The growth of trees and shrubs is very poor. To restore the productivity of land, nitrogen fixing tree species like *Acacia nilotica*, *Terminalia* spp., *Albizia lebbeck*, *Albizia saman* are found highly effective.

Barren rock hill ridges

The rock exposure is often barren and devoid of soil cover and vegetation. There is low canopy coverage at the top of hill due to slow growth as there is very little or no soil in the hills. The soils of these areas are very shallow, coarse, poor, eroded and degraded. The species suitable for barren rock hill ridges are *Dendrocalamus strictus*, *Eucalyptus tereticornis*, *Melia azedarach*, *Albizia lebbeck*, *Acacia catechu*, *Ailanthus excelsa*, *Hardwickia binata*, *Cassia siamea*, *Prosopis chilensis*, *Dalbergia sissoo*, etc. Tree species such as *Robinia pseudoacacia*, *Populus ciliate* and *Pinus roxburghii* are suitable for the areas of high elevation. The suitable grass species are *Cenchrus ciliaris*, *Cenchrus setigerus*, *Dicanthium annulatum*, *Panicum antidotala*, *Pennisetum pedicellatum*, etc.

Snow covered areas

These are created mainly by glacial and fluvial glacial activities in mountainous area. These lands are covered by extensive snow. The recommended species for rehabilitation are *Pinus gerardiana*, *Daplux acuminate*, *Desmodium* spp., *Butea minor*, etc.

12

Agroforestry Diagnosis and Design

In agroforestry systems the various components like tree, crop and pasture exist in different proportions and orientations. It is difficult to find out which agroforestry system is the best suited for a given land situation. Similarly, it is to be decided which technologies are required for refinement and improvement of the existing agroforestry practices. But without sufficient knowledge of the existing system in a particular land situation, it is very difficult to set the research priorities for modification and development of this system.

Diagnosis and Design (D & D) is a systematic and objective methodology developed by International Centre for Research in Agroforestry (ICRAF) to initiate, monitor and evaluate agroforestry programmes. D & D is based on the philosophy that knowledge of the existing situation (diagnosis) is essential to plan and evaluate (design) meaningful and effective programmes in agroforestry research for development. The methodology plays a strategic role in all the phases of the agroforestry research process. D & D in agroforestry is unique and it has been specially developed for the following purposes (Raintree, 1987).

1. To describe and analyse existing land use systems and their constraints
2. To design appropriate agroforestry technologies to address those constraints
3. To design appropriate research work such as trials and surveying

The basic unit of D & D analysis is the land use system (LUS). The LUS can be defined and analyzed at the level of a country, ecozone, farming system or any other unit. The structure and function of any LUS are determined by climatic, physical, biological, technological, economic, social and political factors. D & D focuses on the interactive effects of these factors on the LUS and searches for opportunities for improvement and development in the LUS.

KEY FEATURES OF D & D

Flexibility: D & D is flexible in its procedure which can be adopted to fit the need and resources of different users.

Speed: D & D has been designed with the option of a 'rapid appraisal application at the planning stage of a project with in-depth follow-up during project implementation.

Repetition: D & D is an open-ended learning process. There is always a scope for improvement of the initial designs. Thus, the D & D is not completed until further improvements are no longer required.

Diagnosis and Design is done at two levels- macro and micro level. Macro D & D is a large-scale analysis of an ecozone within a country. It is important for deciding agroforestry research agenda and extension policy of a country at the national level. But micro D & D focuses on the land use system within a larger ecozone that has special priority for agroforestry intervention. Micro D & D involves a detailed analysis of production systems of households in the LUS which in turn guides for research that will address the constraints of the prioritized LUS.

AGROFORESTRY SYSTEMS RESEARCH PROCESS

The basic objective of agroforestry research is to develop technologies to solve farmers' problems in land use systems in specific ecozones. An agroforestry technology should be defined with reference to at least its main components like multipurpose tree species, spatial arrangement, management and performance levels. ICRAF has developed a research process that uses a systems perspective and an interdisciplinary approach. The process is called 'Agroforestry Systems Research'. There are six main steps in the process.

1. Macro D & D (national and ecozone level)
2. Micro D & D (land use analysis at the production systems level)
3. Technology design
4. Component experimentation
5. Technology testing
6. Technology dissemination and adoption

MACRO D & D

There are four main objectives of macro D& D.

1. To identify broad issues and problems constraining all the land use systems in a given ecozone
2. To identify and prioritize areas for potential agroforestry interventions
3. To identify research priorities and formulate research programs
4. To identify needs, opportunities and mechanisms for inter-institutional collaboration for technology development

To meet these objectives, macro D & D uses rapid appraisal techniques. It relies heavily on secondary data which are verified and complemented by quick field surveys. There are seven steps in macro D & D exercise.

1. Identification of study ecozone
2. Delineation of land use systems within the ecozone
3. Description of land use systems
4. Analysis of land use system constraints and potentials
5. Analysis of potential agroforestry technologies
6. Definition of agroforestry research needs
7. Inter-institutional coordination

Identification of study ecozones

The first step in macro D & D is the selection of an ecozone for study. The zone covered in a macro D & D exercise is large, containing significant variations in land characteristics with respect to current uses and constraints. The choice of the study zone should reflect its biological and socioeconomic importance at the national level based on the following factors.

1. Zone's contribution to food production and/or income
2. Total population it supports and/or the area it covers
3. Urgency of its constraints
4. Extent of its unexploited potential for production
5. Level of its development with respect to other areas

Delineation of land use systems

A land use system within an ecozone level is defined as a "population sub-group in which the features and constraints of the farming systems are sufficiently homogeneous to yield similar results if a given agroforestry technology is introduced into those farming systems". The main guideline for distinguishing land use systems is that each system should display unique constraints and potentials differentiating it from other systems in the ecozone of interest. Thus, an LUS consists of a distinctive combination of trees, crops, livestock, soils and other production systems occupying a given unit of land where specific outputs are desired and obtained by a given management unit. Normally the smallest unit of decision-making is the household, but any unit that makes management decisions collectively and shares intimately in the input-output flows of a system is also considered to be an LUS unit.

Description of land use systems

All delineated LUSs are described by specifying the characteristics that are known to affect their current management and performance and would be expected to affect the introduction of potential agroforestry technologies. These characteristics include title of the system and its location with administrative and political divisions; ecological characteristics like topography, rainfall pattern, soil factors and vegetation; socioeconomic and land use characteristics; resources and supporting services available; and development activities and policies adopted.

Analysis of land use system constraints and potentials

Each system has to be evaluated for factors that prevent its households from obtaining optimal outputs from the available resources. This step requires analysis of farmers' needs and priorities to see how well these are being met by current performance of the LUSs. The performance gap is evaluated by comparing the present levels of outputs with the biophysical and socioeconomic potential of the resources. The range of yields obtained in different LUSs should be compared with yields obtained in on-station or on-farm research. Constraints analysis is based on problems facing households and emphasis is put on constraints which agroforestry can address. To diagnose constraints properly, the relationships between manifested symptoms and causal factors must be understood clearly. The team must be able to interpret the relationships between these factors and the objectives of the household. Furthermore, it must be able to determine what opportunities exist to address the constraints. Constraint analysis is done concurrently with LUS characterization. For example, if one observes steep slopes in cropland, one can conclude that soil erosion is a likely hazard if nothing is being done to prevent it.

Analysis of potential agroforestry technologies

Potential interventions are identified and assessed for their relevance to the constraints and their likelihood of increasing or sustaining productivity of the LUSs. First, all possible interventions are identified, whether they are from the areas of agronomy, forestry or agroforestry. For example, low soil fertility could be addressed by various technologies such as fertilizer, livestock manure, green manure from trees or shrubs, crop rotations. Next, each alternative is evaluated for its technical potential and suitability to farmers' resources and capabilities, infrastructure and support services. A judgement is then made on what interventions seem to have the highest potential. Agroforestry interventions are proposed only when they have a comparative advantage.

'*Ex-ante*' evaluation of a technology is part of technology assessment. It is carried out to determine a technology's potential for adoption. *Ex-ante* evaluation means the evaluation of the likely impact of a proposed technology before the technology has been introduced. It is based on appropriate assumptions using relevant data from other sources. This requires knowledge of technology management and performance under the specific conditions of the LUS.

Definition of agroforestry research needs

If a proposed technology is well known and some farmers are familiar with its management and requirements, then a recommendation for extension programme can be formulated. On the other hand, if very little is known about the technology, then the D & D team is to propose research activities. This research will address critical information gaps for designing viable and adoptable technologies. The team should next carry out a comparative analysis of the research needs for each agroforestry technology for each LUS within an ecozone. This analysis will be the basis for the design of appropriate research programme. Thus, the main output of the macro D& D exercise is the definition of a research agenda to develop relevant technologies for the ecozone of interest.

Inter-institutional coordination

A macro D & D exercise should initiate an inventory and review of past and present agroforestry research or development programmes. It will also suggest specific problem areas for complementary research in different institutions and better use of their scientific and physical resources. Multi-institutional participation in strategic phases of the research definitely facilitates the integration of individual efforts and the development of coordinated programme.

MICRO D & D

The major difference between macro D & D and micro D & D is that the former has a broad scope (i.e., an ecozone) whereas the latter focuses on detailed analysis of one prioritized LUS. There are three main objectives of micro D & D.

1. To describe and analyse an LUS in order to identify its constraints
2. To design and evaluate agroforestry technologies to address the constraints
3. To design and evaluate appropriate research programme aiming to develop these technologies

The basic principles for achieving these objectives are similar to that of macro D & D. The choice of the LUS for micro D & D depends mainly on three criteria.

1. Political and economic importance of the system
2. Technical potentials for improvement of the LUS
3. Scientific expertise and other resources in the national collaborating institutions for carrying out research in the LUS

Analysis of land use system and constraints

This phase of micro D & D aims at prioritizing the needs of the household, identifying production constraints and assessing potentials for system development. The basic framework used for this analysis is a farming system, where the decision-making unit is the household. The household usually manages a combination of crop, livestock and tree production systems, along with other non-agricultural and off-farm activities to satisfy its basic felt needs of food, fodder, cash, fuelwood, timber and security. Besides endogenous factors, the farming system is influenced by exogenous factors of a political, social, economic or technological nature. Understanding the interactions within the farming system and the effects of environment is essential for defining the aims of the micro D & D. Thus, the research team has to quantify the resources, management and yield of each component of the farming systems, including characteristics and priorities of the household. The constraints analysis aspect of micro D & D is well suited for planning the initial stages of research. However, later stages of research may require a reassessment or a more precise measurement of some of these constraints.

Design and evaluation of agroforestry technologies

The design and evaluation objective of micro D & D focuses mainly on technology specification and *ex-ante* evaluation of technology. Technology specification for any type of production system refers to a 'package' of husbandry practices and inputs which is specified in terms of the farming systems or households; its components and resource requirements; the management and implementation regimes to be followed by the farmers; and the estimation of real benefits and costs to the farmers under favourable and unfavourable conditions. In other words, technology specification should provide sufficient detail to permit technical feasibility analysis, socioeconomic analysis and assessment by farmers.

Technology specification demands a lot from the D & D team. It raises a large number of specific questions requiring knowledge of the farming systems and scientific expertise. If the questions raised cannot be resolved to the satisfaction of the team, then specific priorities for research have to be established. Because agroforestry is a relatively new science, there is a dearth of technical information on most components. For example, information on the biophysical productivity of multipurpose trees (MPTs) under different arrangements and management regimes is known only for a few species in selected environments. Successful technology design requires adequate research experience. The D & D team may design several technologies to address constraints of the farming system, in terms of the management requirements and performance levels.

Once a technology has been specified in its main components, it becomes possible to carry out an *ex-ante* evaluation based on data from relevant situations. *Ex-ante* evaluation of a technology is simply the analysis of its probable impacts and implications. This analysis looks at benefits and conflicts or problems likely to arise at the levels of farming system, community or village and region or catchment area. It should assess the production potential and technical feasibility of the technology. Economic viability, sustainability, farmers' acceptability and adoption potential of the technology are the four essential types of analysis required in *ex-ante* evaluation. The larger and more complex the technology, the more demanding is the *ex-ante* analysis.

Design and evaluation of research programme

There are four types of scientific research.

1. Basic research which is designed to generate new knowledge or understanding
2. Strategic research to solve specific research problems
3. Applied research to create new technology
4. Adaptive research to adjust technology to the specific needs of a particular set of biophysical or socioeconomic conditions

These are part of a continuum in the technology development process. Productive research requires an integrated and complementary research strategy consisting of on-station research and on-farm research. On-station research consists mainly of basic and applied research and it must be able to offer technical components, information and support to the on-farm research activities. On-farm research complements but does not substitute for on-station research. On-farm research provides feedback for setting on-station research priorities and adapting technologies or components coming out of on-station research. In the case of agroforestry, where basic and applied research is not well developed and where farmers have more experience than scientists with management of technologies, on-farm research may have a stronger role to play in the research strategy including applied research.

Research design criteria

A technology comprises a number of components. Experiments are designed to develop the technical components and to understand relationships among them. Agroforestry technology must be specified at least in its principal components like MPT species, spatial arrangement, management regimes and performance levels. Different types of trials are conducted to achieve these specifications. Within the D & D scheme of ICRAF, the three general categories of agroforestry trials are multipurpose tree species screening trials, MPTs technology screening trials and MPTs management trials. While all D & D teams have a mandate to design a programme of research, the scope of their proposals may vary. To achieve this, it may be necessary for the D & D teams to carry out multi-visit surveys and interactions with land users and to review scientific secondary information.

METHODOLOGICAL CONSIDERATIONS IN D & D

Research team

The nature of agroforestry systems suggests the need for setting up of multi-disciplinary and multi-institutional teams to achieve the objectives of D & D. For macro D & D, the research team comprises 5 to 10 scientists from biophysical and socioeconomic fields including agronomy, soil science, horticulture, plant protection, climatology, animal science, forestry, agricultural economics, rural sociology or anthropology. These scientists should have experience in research and extension. For micro D & D and follow-up studies, the expertise and composition of the team are based on the prioritized LUS, its identified constraints and the specific objectives of the study. Team leadership is critical for successful D & D implementation. Interdisciplinary interactions and specific contributions made by each discipline are used to formulate the overall research strategy. An objective D & D strategy is the key to bring various disciplines to work together to address the problems of the farmer in an integrated manner. Care is taken to provide an effective interface between the D & D exercise and the planning and implementation of technology development research.

Research domains and recommendation domains

Defining the target land use system or farming system is probably the most crucial step in the 'agroforestry systems research' process. Unless researchers have developed the technology and know how to manage it, what requirements it has and exactly what it can do, they do not have a solid basis to define an appropriate recommendation domain. Thus, in the preliminary stages of technology design, it is better to speak of a 'research domain' until there is sufficient understanding of the technology to determine precisely where it can fit. Accordingly, the research team should always be concerned about whether and how technology development is modifying the original research domain.

Precise definition of the target system is also important for the reason that every farming system or household is somewhat different. Some customizing or fine-tuning of technology is required for each case during the adoption process. Thus, the definition of research domains should strike a wise balance not too general so as to be useless as a guide to research, nor too specific so as to apply only to a small number of farms or households.

The research team can assess the effect of its work on the research domain by answering the following questions.

1. Which LUSs are most affected by the problem under investigation?
2. Which systems can benefit and to what extent can they benefit from the technologies being developed?
3. What are the major conditioning and determinant factors, endogenous or exogenous to the farming system, for technology management and performance?
4. What are the expected benefits and impacts of technology adoption?

The concept of research domain is a tool to facilitate and expedite the task of focusing on these key questions. If the team can answer them adequately, the concept has served its purpose.

Data collection methods

The D & D methodology employs several data collection methods appropriate to its specific objectives. Each method has its own strengths, weaknesses, degree of reliability of collected data and resource requirements. For example, the informal survey is fairly effective for identifying constraints, designing technology, promoting interdisciplinary interactions and contributing to research planning, but the reliability of the data is not up to the standard of other methods. D & D exercises should generate a minimum of raw data, a maximum of useful information and in a timely matter. For this reason, the preliminary D & D work will take a rapid appraisal approach using secondary information surveys. D & D work at later stages in the research and development process will use methods that make a maximum contribution within the limits of available human and physical resources.

Analytical methods and the role of farmers

The major decisions in the D & D analysis are derived using interactive and holistic methods with the principal actors being the D & D team and the farmers or households. This interaction should be based on solid information, consultation with development agents and policy makers, and a commitment to arrive at logical conclusions in the process. To ensure effective participation in discussions, all participants must show mutual respect and accept that each can make a valuable contribution. However, this does not mean that participants should accept everything said. There is a need to challenge, seek clarification and discover the root causes of disagreements or conflicts. In this respect, the farmers should be treated as equal participants.

Logistical and operational aspect

A macro D & D is usually completed in about three months. This includes two to three days to plan the study and orientation of the team, two to three weeks for review and synthesis of secondary information, two to three weeks to conduct field work and four to six weeks to analyse the information and prepare the report. The review work and report preparation do not require participation of the whole team; only two members of the team can do these with occasional assistance from the others.

For micro D & D, the total time requirement and logistic support are approximately similar. However, a formal survey of 5-100 farmers with a semi-structured questionnaire constitutes an important part of the study. In addition, the team may need to allocate relatively more time to the review of relevant agroforestry researcher and extension work to strengthen its analysis.

DESIGN AND ANALYSIS OF AGROFORESTRY EXPERIMENTS

Agroforestry systems are more complex than the agricultural systems. The long-term nature of trials involving woody species and the varying objectives of agroforestry trials demand that adequate caution be exercised in their design. Several characteristics of the trees like slow growth, long term effects on their surroundings, long life, age of trees, the area over which the influence of trees extend, etc. complicate the issue of designing experiments for these systems. Agroforestry experiments are often established on marginal sites which include sloping lands, and sites with infertile and degraded soils. It is thus difficult to find homogeneous sites of such problem areas, especially sloping lands where plots along contour lines are long and linear. Germplasm of many agroforestry tree species are usually collected from different origins which lack uniformity. Therefore, experimental materials in an agroforestry experiment may not be of uniform quality.

There are several features of agroforestry systems that require some modification of the methods usually used in agricultural experiments. Important issues related to agroforestry experiments include

1. Restrictions on land availability and topography
2. Larger variability in crop performance
3. Presence of tree crop interfaces
4. Effects of tree component that change over time and take several years to become apparent
5. General increase in soil heterogeneity with increasing land area, and modification of soil characteristics by imposed hedgerow trees or shrubs
6. Effect of types of land preparation on changes in soil fertility gradients
7. Need to reduce the number of factors and their levels to the basic minimum
8. Possibility of using farmers' plots (in on-farm trials) as replicates rather than as a complete experiment
9. Presence of neighbouring plot effects and the rows of discard
10. Presence of border effect

DESIGN REQUIREMENTS

Although design features change according to the aim of an experiment and the environment in which it is conducted there are some important design requirements which are common to all the agroforestry experimental studies.

Clear objectives

The kind of lay out used in agroforestry experiments depends on the precise research objectives and the resources available including the extent and nature of the site. Thus, the problem identified must be clearly stated. Then the objectives must be clearly defined and a hypothesis is developed. Since the agroforestry experiments are complex, multi-disciplinary and lengthy in time requirement, their objectives need to be specified very clearly so that they may be achieved consistently and with the desired precision by choosing the correct sampling procedures.

Control of variability

The experiments are then suitably designed to test the hypothesis. Designs of any long-term trial should aim to maintain control of both spatial and temporal variability and to minimize the input resources. When genetic, environmental and management factors are considered separately, the patterns in variability caused by the trees, plots, experimental blocks, farms and regional locations differ in relation to each other. Environmental changes in time due to tree component need to be determined. In an experiment designed to test different species, it is essential to test similar tree species in separate experiments as various agroforestry tree species are of different sizes, shapes and growth habits.

Data collection

A sampling strategy is determined both in space and time to ensure unbiased data. In agroforestry experiments, various plot sizes are used. Thus, an understanding of the likely changes in plot variability in time would help determine the appropriate initial plot sizes. In agroforestry trials data of tree component is usually measured annually. But, depending upon facilities and chosen variables, data may be recorded once before sowing and again after harvesting of intercrop. The choice of data and sampling procedure must be the same at each measurement date. Collected data are then analysed and interpreted properly which leads to drawal of conclusions about the hypotheses with more accuracy.

PRINCIPLES OF FIELD EXPERIMENTATION

Monitoring of the effects of certain inputs on a subject matter is known as experimentation. The allocation of treatments to an experimental unit or plot is referred to as an experimental design. A plot is the smallest unit of land receiving a treatment. The treatment is the material being forced on the subject and whose effect is to be studied. Experimental error is a measure of the difference between two units treated alike.

Replication

Experiments of the same nature, when presented under similar conditions, should yield similar results. The simplest way to achieve this is through the 'repetition',i.e., 'replication', of the same treatment on several plots or experimental units. By doing this, the researcher can average the responses of the same treatment in different plots, thereby obtaining a better sense of the typical response as opposed to a response from just one plot. Moreover, the use of several plots ensures minimization of the effect of uncontrolled variability in the field, which is an important parameter needed to quantify the reliability of the findings through statistical analysis.

Randomization

This refers to the allocation of treatments to plots in such a way that, within a specific experimental design, units are not discriminated for or against any treatment. Each unit is supposed to have the same chance of receiving a particular treatment. Randomization is a necessity as no two units or plots are exactly the same. Statistically, the randomization procedure allows elimination of researcher bias in the assignment of treatments to plots and ensures the computation of valid sampling errors.

Local control

Thisis the procedure by which the variability within the experimental materials and plots is reduced to make sure that the experimental units are as homogeneous as possible. One way to accomplish this is by blocking. A block is a relatively large area or several identical units receiving all or most of the treatments. This involves grouping experimental plots into relatively homogeneous units known as blocks, and then assigning all or most of the treatments of the experiment in each block within the overall experiment.

TYPES OF AGROFORESTRY FIELD EXPERIMENTS

The agroforestry field experiments may broadly be classified under following four heads.

Multipurpose tree screening and selection trials

Screening and evaluation of germplasms of multipurpose tree species (often of several species, but sometimes of provenances of one or two species) is the most common element of experimentation in all agroforestry field trials with the objective to identify the most promising species or provenance based on their early performance and also on specific outputs and benefits. The species or provenances of a species included in the experiment are determined by local researchers depending on various location-specific attributes. The experimental designs used are mostly randomized complete block designs or systematic designs. The trees have multiple uses and the management for one product may affect the output of other products. Thus, a given multipurpose tree species will have to be evaluated separately for each of its specific outputs and benefits. Another difficulty is the lack of standard procedures for evaluating the trees. Traditional forestry research procedures may not always be appropriate for multipurpose tree species because, the objectives of growing MPTs and traditional forestry species are different.

Component and system management trials

The objective of these trials is to improve specific agroforestry technologies like selection of hedgerow species, the method of planting them, hedgerow spacing, management of hedgerows (time, frequency and height of pruning), hedgerow-to-crop row distance, method of application of the pruned biomass and fertilizer application in conjunction with mulch, etc. Agroforestry experiments dealing with the nutritive value of tree/shrub fodder under silvipastoral systems, planting configurations and management aspects of other agroforestry practices are also common research topics. Another major set of management trials is the use of agroforestry for soil conservation.

Component interaction studies

The objectives of these studies include understanding and quantifying the interrelationships between components of agroforestry systems, mostly tree-crop interactions to investigate the sharing of resources below or above ground. These also include agroforestry soil related research.

Prototype evaluation trials

These trials are undertaken with the objective of evaluating specific packages of agroforestry technologies under realistic field conditions. They represent a transition between research and extension and are mostly undertaken either wholly or partly in farmers' fields.

PLOT SIZE AND ARRANGEMENT

The arrangement of tree and crop components in relation to one another within plots deserves important consideration, especially in interaction studies. Therefore, use of an appropriate design is a very important aspect of agroforestry experimentation. A general recommendation or solution cannot be given for all the situations since a specific problem requires a particular design. The plot size is also an important consideration in agroforestry experiments. The size of the plot depends on many factors like the objectives of the experiment, types of measurement to be made, the expected duration of the experiment, ultimate size of the trees, the requirement for extra space to avoid interference between plots, etc. Soil erosion studies require relatively large plots.

The type and nature of measurements to be made and the requirement of a guard area to reduce the influence of trees or treatments in one plot on the adjacent plots also influence the plot size and space requirement. Site variability is another major factor that will be a determinant of the size and arrangement of plots. Sites available for agroforestry experiments often exhibit wide variations even within a small area. These include variations in soil, topography, microclimate, and previous land-use. Blocking is one way to control the effect of such site variations, where the primary consideration is that each block is as uniform as possible in terms of these common variability factors. Site variability can also be controlled by making plot sizes smaller but this may present other problems such as the excessive use of experimental space for borders.

Arrangement of plots in an experiment depends on the type of design and topography. The arrangement of trees and crops in relation to one another within the plot is another important consideration in agroforestry experiments, especially in interaction studies. The choice of arrangement to be adopted depends upon the objective of the experiment as well as the available space.

EXPERIMENTAL DESIGNS

Experimental design refers to the way the researcher allocates treatments to plots. Type and number of treatments, degree of precision desired and size of uncontrollable variations are the three most important factors in selecting an experimental design. Scientific experiments are generally classified into two broad categories such as single-factor experiments and multifactor experiments. In a single-factor experiment, only one factor varies while others are kept constant. In these experiments, the treatments consist solely of different levels of the single variable factor. In multi-factor experiments (also called as factorial experiments), two or more factors vary simultaneously. The experimental designs commonly used for both types of experiments are classified as Complete Block Designs and Incomplete Block Designs. In a complete block design, each block contains all the treatments while in an incomplete block design not all treatments may be present in a block. The complete block designs are suited for small number of treatments while incomplete block designs are used when the number of treatments is large.

Randomized complete block design is the simplest and most commonly-used design in field experiments. However, there are many situations in agroforestry where this design may not be appropriate. The number of treatments may be too large in an agroforestry experiment, so that it is difficult to locate a block with sufficient uniformity to accommodate a complete set of all treatments to be tested. In such cases, incomplete block designs (e.g., lattice designs and confounded designs) may be used, where the number of plots in a block is less than the total number of treatments. The analysis of data for incomplete block designs is more complex than randomized complete block design. Among incomplete block designs, lattice designs are commonly used in species and variety testing.

The treatments in factorial experiments consist of two or more levels of the two or more factors of production. The advantages of the factorial arrangement are many. One major advantage is the reduction in the number of experiments, and a second, the possibility of studying the interactions among the various factors. A significant interaction implies that changes in one factor may be dependent on the level of the other factor. If this happens, interpretation of the results has to be done cautiously to avoid inaccurate general statements on the individual factors.

Another commonly-used arrangement of plots in field experiments is the split-plot experiment. These are factorial experiments in which the levels of one factor are assigned at random to large plots. The large plots are then divided into small plots known as 'sub-plots' or 'split plots', and the levels of the second factor are assigned at random to small plots within the large plots. This arrangement is often useful when certain treatments are combined, some of which require larger plots than others for practical convenience. Usually, the treatment on which maximum information is desired is placed in the split-plot or in the smallest plot. The split-plot is not a design as such but rather refers to the manner in which treatments are allocated to the plots. A split-plot arrangement in a randomized complete block design usually has two error terms; one for testing the treatments in large plots and the other for the sub-plot treatments and interactions. A common situation where the split-plot experiments are very useful is when there is a soil gradient such as slope or fertility difference in one direction. In such cases, the whole plots could be aligned across such gradients so that each whole plot will have an entire section of the gradient, within which the sub-plots will be allocated at random. Split-plot experiments are particularly useful if the treatments are susceptible to 'neighbour effects' as in irrigation experiments or in experiments involving tree species of varying growth habits. These split-plot experiments are commonly used in many agroforestry experiments. A split plot design can be further extended to accommodate a third factor through division of each sub-plot into sub-sub-plots. This is then called a 'split-split-plot' arrangement. Related to the split-plot experiments are the strip-plot experiments. These are specifically suited for two-factor experiments when both factors require large plot sizes, and when the interaction between the two factors is desired to be measured more precisely than the effects of either of the two factors.

Although factorial experiments provide opportunities to examine interactions among various factors, they are difficult to conduct when the number of factors and their levels are many. To overcome this difficulty, fractional factorial or confounding designs can be used. In a fractional factorial design, only a fraction of the complete set of factorial treatment combinations is included. Here the main focus is on selecting and testing only those treatment combinations which are more important. The fractional factorial design is used in exploratory trials, where the main objective is to examine the interaction between factors. In a confounding design all the treatment combinations of the factors and levels under study are tested with blocks containing less than the full replication of the treatment combinations.

Systematic designs

Many factors of interest in agroforestry research do not allow complete randomization, such as slope, orientation, rainfall or distance from tree or hedge. In these cases, non-random systematic designs are used. The concept of systematic design is that over a series of plots the level of the quantitative factor varies systematically but slowly (e.g., 10% change) through the sequence of plots. Each particular treatment level is then surrounded by treatment levels differing only slightly, and therefore guard areas are unnecessary. The harvested area on the systematic design is much more than in any other randomized design conducted on the same piece of land. For example, in alley cropping trials, the spacing between hedgerows (alley width) can increase gradually across a site rather than having plots of varying alley widths located at random. Another example is an alley cropping trial where distance of crop rows from hedgerows is the treatment variable so that the crop row nearest to the hedgerow will be compared against other rows that are progressively further apart from the hedgerow. The direction of the systematic change of the treatment level within each plot can and should be randomly selected.

One major difficulty with these systematic designs is that the statistical analysis of the data becomes somewhat complex. Randomization in the systematic design can be introduced at a higher level. This is done by repeating the sets of systematically arranged treatments at different locations within the experimental site. These designs despite their limitations have been proposed for agroforestry research because they minimize the requirement for field and experimental resources. Beehive design, Y design, augmented design, etc. are some of the commonly used systematic designs.

Huxley (1985) suggested that tree-crop interface (TCI) and orientation effects in an agroforestry system can be studied in a simple geometrical design described as 'Y design'with three strips (or arms) each containing the tree and crop components juxtaposed to each other. The component interaction studies depend on the precise research objectives and the resources available, including the extent and nature of the site. Orientation may have a significant role at TCI because of the shading effect of tall trees on the understorey crop component. Thus, orientation as a factor may also be considered. This design enables a study of the TCI in a relatively small space and reduces interfaces between the different arms, especially with tall trees. This design takes care of the orientation effects. Beehive designs suggested by Martin (1973) are used for competition experiments. In these designs there are two kinds of plant species that are intermingled on a triangular format in such a way that there is a wide range of competition.

ON-FARM RESEARCH IN AGROFORESTRY

On-farm trials are essential means of technology transfer and can be viewed as a means of providing feed-back from the farm to the research station. It is a very powerful and appropriate strategy for agroforestry research. The technology developed at research stations is tested under farm conditions to obtain information about its performance and acceptability by the farmers. Exploratory and survey-type investigations are usually the first step in a typical on-farm research (OFR) sequence. This enables the researcher to gather information on farmers' perceptions of existing land-use practices and thus identify the key elements that determine the social acceptability of any new technology. This information is then integrated into the technology design process. The designed technology is then tested under farm conditions to obtain information about its performance and acceptability by the farmers.

The methodology for conducting an on-farm investigation depends on its objectives and local conditions. However, there are some general guidelines for on-farm research in agroforestry.

1. A clear understanding is required on how, where and why farmers grow trees and what trees they grow before initiating the study.
2. Researchers should work together with farmers in exploratory or diagnostic trials with respect to new species and their management systems.
3. On-farm descriptive research may be a necessary substitute for the library research which usually precedes conventional agricultural experimentation.
4. The researcher should be acquainted with the experimental materials, such as new species, before they are used for on-farm trials. Such information can be obtained from simple on-station exploratory evaluations of the materials.
5. Selection of treatments is an extremely important factor. The heterogeneity of farmers' fields, the long-term nature of agroforestry trials, the need to minimize the number of treatments, the requirement of large plot-sizes for agroforestry experiments, etc., all necessitate the careful choice of relevant treatments.
6. Choice of experimental designs is an equally important task. The general rule for all on-farm experiments is to employ simple designs such as randomized block designs.
7. The selection of sampling units and sampling procedures in on-farm experimentation is also an important consideration. Random selection of farms or observational units spread over several farms and use of relatively large sampling units are two basic principles.

On-farm research, whether managed by the researcher or the farmer, requires simplicity in design. Sometimes this simplicity requirement may be due to limited resources such as land or treatments. The statistical implications also encourage simplicity. If the treatments are not kept to a basic minimum of two or three, the whole experiment, with or without replications, cannot be carried out on one famer's land. This would require the use of several farmers' plots, either as replicates or single plots. This could result in increased variability and might make it impossible to compare some treatments effectively.

Pinney (1991) has made use of augmented design for on-farm trials or prototype evaluation trials. An augmented design contains both treatments that are replicated in every block and those that are not replicated. It allows the farmer some flexibility to decide about what treatments are tested on his farm. The technology developed at research station forms the set of core treatments and the farmer-defined treatments are called the augmented treatments. The number of plots per farm depends upon the region, population density and farming system. The more the number of plots available per farm, the more is the scope for within-farm replication or for more treatment augmentation.

Modified stability analysis is a useful tool for analyses, especially of preliminary results of on-farm agroforestry experiments. This would be useful for both multi-farm OFR and making recommendations for more than a single inference space. It is a procedure by which the data from a wide range of environments can be evaluated using both researcher and farmer-oriented criteria. Modified stability analysis is based on statistical method to assess genotype x environment interactions. These physical or biophysical environments can be expressed through an index based on the yield of different components at each location. Given the long gestation periods of agroforestry experiments on the one hand, and the urgency of delivering recommendations on the other, modified stability analyses are potentially applicable to agroforestry, especially when recommendations may have to be developed based on preliminary on-farm analyses.

13

Agroforestry Systems Research and Development

Trees have been used in cropping systems since the beginning of agriculture. Throughout the world, at one period or another in its history, it has been the practice to cultivate tree species and agricultural crops in intimate combination. In the tropics, human beings underwent a transition from hunting/gathering to the use of domesticated plants and livestock. As a part of the process, they cut down trees, cleared the debris by burning and sowed crops in the ash-enriched soil. It was the 'slash-and-burn' agriculture, a primary forerunner of the present-day agroforestry and a practice that might have originated in the Neolithic period, around 7000 BC.

There are innumerable examples of traditional land-use practices involving combined production of trees and agricultural species on the same piece of land in many parts of the world. Trees were an integral part of these farming systems and they were deliberately retained on farmlands to support agriculture. These practices are now known as agroforestry. However, the ultimate objective of these practices was not tree production but food production.

HISTORY OF AGROFORESTRY RESEARCH

By the end of the nineteenth century, establishing forest or agricultural plantations had become an important objective for practising agroforestry. But the research on tree plantation was carried out by foresters. The foresters conducting the research never envisioned the system as being capable of making a significant contribution to agricultural development, or its potential as a land-management system.

In 1914, Russell Smith, an economic geographer at Columbia University in New York advocated the use of permanent tree-protected areas to maximize production on arable land. But the post-World War II economic and industrial boom lessened interest in agroforestry. However, a decade later new technologies and innovative approaches to agriculture and land-use systems coupled with increased interest in conservation and concern for degrading natural resources, once again brought attention to agroforestry.

In the 1960s and early 1970s there was increasing concern for the forested lands of the tropics. It was clearly recognized that they were under severe pressure. Some thought that commercial exploitation was the problem; others believed that fuelwood needs were the culprit, while still others thought that shifting cultivation was the root cause. In 1975, the International Development Research Centre (IDRC) located in Ottawa, Canada engaged Mr. John Bene, a retired forest industrialist to study the problem. Bene formed a small advisory committee in Canada, and recruited experts in the various continents to prepare studies pertinent to their area with the following major objectives.

- Identify significant gaps in world forestry research and training
- Assess the interdependence of forestry and agriculture in low-income tropical countries and propose research leading to optimization of land use
- Formulate forestry research programmes which promise to yield results of considerable economic and social impact on developing countries
- Recommend institutional arrangements to carry out such research effectively and expeditiously
- Prepare a plan of action to obtain international donor support

Based on these studies the IDRC published a report titled 'Trees, Food and People: Land management in the tropics' in 1977. This IDRC Project Report for the first time recommended the creation of an internationally financed Council for research on agroforestry, to administer a comprehensive programme leading to better land use in the tropics. The report suggested that the objectives of such a Council should be

- The encouragement and support of research in agroforestry
- The acquisition and dissemination of information on agroforestry systems
- The promotion of better land use in the developing countries of the tropics

With the recommendations of IDRC, an internationally financed organization, then known as the International Council for Research in Agroforestry (ICRAF) was established in 1977 and started its work in The Hague. First priority of the Council was to encourage the combined production system which would integrate forestry, agriculture, and/or animal in order to optimize the land use. The organization moved to its present headquarters in Nairobi, Kenya in 1978. In 1992, it was renamed as International Centre for Research in Agroforestry (ICRAF) when it joined the Consultative Group on International Agricultural Research (CGIAR). In 2002 the Centre's name was once again changed to

World Agroforestry Centre (WAC) in order to reflect its global reach, as well as its more balanced research and development agenda. However, the Centre's legal name - International Centre for Research in Agroforestry remains unchanged.

The World Agroforestry Centre has since been involved in the promotion of agroforestry research programmes with a heavy emphasis on Africa. Agroforestry research has accelerated rapidly since the early 1980s and has resulted in a greater understanding of the science. Much of this understanding has come from observations of existing practices and systems, although an increasingly important knowledge base is being established through designed agroforestry experiments.

WAC is the only institution that does globally significant agroforestry research in and for all of the developing tropics.The Centre conducts research in agroforestry, in partnership with national agricultural research systems with a view to developing more sustainable and productive land use. The focus of its research is countries/regions in the developing world, particular in the tropics of Central and South America, Southeast Asia, South Asia, West Africa, Eastern Africa and parts of central Africa. Knowledge produced by WAC enables governments, development agencies and farmers to utilize the power of trees to make farming and livelihoods more environmentally, socially and economically sustainable. WAC's work also addresses many of the issues being tackled by the Sustainable Development Goals, specifically those that aim to eradicate hunger, reduce poverty, provide affordable and clean energy, protect life on land, and combat climate change.

Building upon the success of the scientific studies carried out by WAC at its centres, agricultural scientists all over world began investigating the feasibility of intercropping in plantation and other tree crop stands as well as studying the role of trees and shrubs in maintaining soil productivity and controlling soil erosion. Livestock management experts also began to recognize the importance of indigenous tree and shrub browse in mixed farming and pastoral production systems. Many of these studies and efforts provided important knowledge about the advantages of combined production systems involving crops, trees and animals.

Agroforestry is now taught as a part of forestry and agriculture degree courses in many universities in both the developing and industrialized world. Agroforestry research is conducted by universities, government agencies, individual agricultural producers and other organizations. Indeed, agroforestry is fast becoming recognized as a system which is capable of yielding both wood and food and at the same time capable of conserving and rehabilitating ecosystems.

HISTORY OF AGROFORESTRY RESEARCH IN INDIA

Shifting cultivation in India is prehistoric and partly a response to agroecological conditions in the region. Horticulture as co-existent with agriculture is found to have been prevalent in India from early historic period. The plant husbandry or plant domestication started under arid climatic zone of India from about pre-Neolithic period. The role of many common trees such as khejri (*Prosopis cineraria*), aswattha (*Ficus religiosa*) and palasa (*Butea monosperma*) in Indian folk-life has been mentioned in ancient literature of Rig Veda, Atharva Veda and other ancient scriptures.

Maharishi Valmiki in the *Ramayana* had nicely illustrated the arrangement of trees ashok, neem and peepal in outer ring; jackfruit, mango, banana and guava in middle ring; and flowers such as *champa, chameli, juhi*, etc. in inner ring, keeping *sarovar* (water-tank) in the centre in Ashok Vatika (garden). This clearly indicates that growing forest trees, fruit trees and flower plants together in the garden was the development of an agroforestry practice. The agroforestry practices were improved during the *Mahabharata* period compared with those in the *Ramayana* period. It is evident from *Ramayan*a and *Mahabharata* that agroforestry practices such as agrisilviculture and agrihorticulture were in operation in their primitive state.

The most important agroforestry practice is known from the Kangeyam tract of Tamil Nadu, where *Acacia leucocephala* + *Cenchrus setigerus* in silvipasture system was practised. Similarly, in ravines of Yamuna and Chambal, trees, shrubs and bamboos with grasses were planted for rearing the milk-producing Jamunapari breed of goats and sheep. Scattered trees with *khejri* or *mehndi* in association with bajra, jowarand chillies were grown in the semiarid area of Tamil Nadu. Tree plantation continued as demarcation and control against wind erosion throughout the country. Plantation of *khejri* trees for various uses on farm was a common practice in Rajasthan. In the coastal areas, *Casuarina equisetifolia* and other trees were grown in association with crops on farm lands for cash and to generate small timber. In paddy-sowing areas, *Pongamia glabra* and *Sesbania grandiflora* were grown; lopped annually and their leaves applied to fields as green manure. Multistorey homesteads or homegardens were in existence in Kerala, Karnataka and Tamil Nadu as an important agroforestry practice. This practice is still followed in these states to fulfil the needs of fodder, fuelwood, fruits, vegetables, medicines, etc.

The farmers and land owners in different parts of the country used to integrate a variety of woody perennials in their crop and livestock production fields depending upon the agroclimatic conditions and local needs. Most of these practices are, however, very location specific and information on these are mostly anecdotal. Therefore, their benefits have remained vastly underexploited to other potential sites. It has now been well-recognized that agroforestry can address some of the major land-use problems of rainfed and irrigated farming systems in India, and that a great deal can be accomplished by improving indigenous systems through systematic research.

In India, research work on agroforestry was initiated during the late 1960s and 1970s by the Indian Grass Land and Fodder Research Institute, Jhansi, Central Soil and Water Conservation Research and Training Institute, Dehradun, Central Arid Zone Research Institute, Jodhpur and ICAR Research Complex for the North-eastern Hill Region. However, the organized agroforestry research in India began in May, 1979 when a seminar on agroforestry was organized at Imphal, Manipur by the Indian Council of Agricultural Research (ICAR), New Delhi to accumulate and compile the data on the 'Research and Development' of agroforestry in India.

The agroforestry research through the All India Coordinated Research Project on Agroforestry (AICRPAF) was conceptualized with the following six projects.

1. Diagnostic survey and appraisal of existing farming system and agroforestry practices including farmers' preference.
2. Collection and evaluation of promising tree species, cultivars of fuel, fodder and small timber for agroforestry interactions.
3. Studies on management practices of agroforestry systems.
4. Analyse economical relation of agroforestry systems.
5. Explore the role of agroforestry in environment protection.
6. Studies on post-harvest technology, fishery, apiculture, lac, etc. in relation to agroforestry systems.

The AICRP on Agroforestry was then initiated in 1983 at 20 centres of which 12 were in the State Agricultural Universities (SAU) and 8 at the ICAR institutes. At present there are 37 centres of AICRP on Agroforestry located in 27 State Agricultural Universities, 9 in ICAR institutes and one in Indian Council of Forestry Research and Education (ICFRE) institute representing all agro-climates of the country. National Research Centre for Agroforestry (NRCAF) was established at Jhansi on 8th May, 1988, which was later on upgraded as Central Agroforestry Research Institute (ICAR-CAFRI) on 1st December, 2014. Initially the coordinating unit of the AICRPAF was located at the ICAR headquarters, New Delhi. Later on, it was shifted to Jhansi on 1st April, 1997, under the administrative control of the Director, NRCAF, Jhansi with the additional charge of Project Coordinator.

The agroforestry research in the country includes diagnostic survey and appraisal of existing agroforestry practices prevalent in different agroecological zones of the country. One of the major thrusts of the agroforestry research is collection and evaluation of promising tree species/cultivars of fuel, fodder and small timber. A large germplasm of important agroforestry trees has been collected and evaluated in arboretum established by the AICRP on Agroforestry centres. On the basis of identification and evaluation of promising trees and 'Diagnosis and Design' survey of the existing systems, agroforestry interventions were initiated in different agroclimatic regions. The AICRPAF has completed 37 years and made substantial contribution in agroforestry research. The Central Agroforestry Research Institute, Jhansi has also been working on carbon sequestration potential of various agroforestry systems since 2000.

In addition to ICAR, Indian Council of Forestry Research and Education also conducts agroforestry research through its research institutes and advanced research centres in different parts of the country. Recognizing agroforestry as a viable venture, many business corporations, limited companies such as ITC, WIMCO, West Coast Paper Mills Ltd., Hindustan Paper Mills Ltd., and other institutions initiated agroforestry research with emphasis on production of improved planting material of the fast growing species with an objective to meet the demand of wood based industries (Dhyani, 2018).

FOREST POLICIES IN INDIA

The British were the first to officially recognize forest as the natural wealth of India and initiated the process of forming a forest policy during the second half of the 19th century. Thus, India became one of the few countries having a forest policy since 1894. The main thrust areas of this policy were to ensure maintenance of adequate forest cover for general wellbeing of the country, meeting needs of local people and after meeting local needs maximum revenue collection. However, main scheme of the British rulers was to exploit the natural wealth of the nation as much as possible, since timber trade was a highly lucrative trade during those times. After independence, two forest policies, National Forest Policy 1952 and National Forest Policy 1988 were launched. The Indian Forest Policy, 1952 was a simple extension of colonial forest policy. However, it became conscious about the need to increase the forest cover to one-third of the total land area. The National Forest Policy, 1952 was replaced by the National Forest Policy of 1988. This policy was considered to be better than the earlier one, as it laid emphasis on the conservation of the existing forests through the reforestation and soil conservation. Besides these two policies on forestry, the National Commission on Agriculture, 1976 also recommended some revised policies on forestry sector.

NATIONAL FOREST POLICY 1894

The first regular policy statement related to forestry was issued on 19th October 1894 by Government of India. This policy was primarily based on Dr.Voelcker's recommendations given in a report on 'Improvement of Indian Agriculture, 1893'. The major objectives of the policy were

1. Promotion of the general well-being of the country as being the sole purpose of management of forest
2. Maintenance of adequate forest for the preservation of climatic and physical conditions of the country and to fulfil the needs of the people subject to conditions
   - Permanent cultivation should get priority over forests
   - Satisfaction of the needs of local people at non-competitive rates, if not free, should over-ride all considerations of revenue
   - After fulfilment of the above conditions realisation of maximum revenue should be the guiding principle

Thus, this policy gave much importance to commercial exploitation of forest products, state custodianship and permanent cultivation. Through this policy the British administration encouraged the *zamindars* to convert the open forests into agricultural land for enhancing the revenue earning of the state. Forests were treated as a source of revenue to the state and not to meet the needs of the people.

In this policy, the forests were divided into four classes. The first class generally situated in hill slopes and essential to protect the cultivated plains from landslides and they played a conservation role for the benefit of cultivated plains and assured revenue to the state. The second class of forests consisted of valuable timber trees like *Shorea robusta* and *Tectona grandis*. Artificial regeneration of *Tectona grandis* was developed for commercial interest. The third class of forests included minor forests which yielded low quality timber, fuelwood and fodder and were meeting the demands of local people. The fourth class covered the pastures and grazing lands, the local people were allowed to use them with restrictions (Balooni and Singh, 2007).

NATIONAL FOREST POLICY 1952

The National Forest Policy was enunciated on 12th May, 1952 with the following main objectives.

- A system of balanced and complementary land use needed to be evolved so that it produces the most and deteriorates the least.

- Denudation in mountainous regions needed to be checked, on which the perennial water supply of the river system depended whose basins constitute the fertile core of the country
- Erosion needed to be checked in treeless bankers of rivers to prevent and arrest ravine formation.
- The need for establishing tree-lands, wherever possible, for the amelioration of physical and climatic conditions promoting the general wellbeing of the people.
- The need for ensuring progressively increasing supplies of grazing, small wood for agricultural implements, and in particular of firewood to release the cattle dung for manure to step up food production.
- The need for sustained supply of timber and other forest produce required for defence, communications and industry.

National Forest Policy 1952 recommended that on the basis of functions, all forest lands could be classified into protection forests, national forests, village forests or tree-lands.

Protection forests: 'Protection forests' denote forests found, or required, on hill slopes, river banks, sea-shores, or other erodible localities. In such sites the need for forest cover is dictated by purely protective physical considerations, such as prevention of erosion, conservation of moisture, and control of floods. The primary object of management of such forests should be to utilise in full their protective influence on the soil, the water regime, and the physical and climatic factors of the locality. The scientific management of such 'protection forests,' wherever possible, should include the production and exploitation of timber within the limits of safety.

National Forests: 'National forests' are those which have to be maintained and managed for valuable and quality timber to meet the needs of defence, communications, industries and other general purposes of public importance. They have to be managed chiefly in the interest of the nation as a whole, and their organization and development is one of the most important functions of the States. Their management on scientific and commercial lines is essential for maintaining a sustained supply of wood. Cultivation should not be permitted to encroach upon these valuable timber bearing tracts. The solution of the food problem of an ever-increasing population must be sought primarily in intensive cultivation and not in weakening the very basis of national existence by encroaching upon such forests.

Village Forests: 'Village forests', popularly termed 'fuel forests' are those which are intended to serve the needs of the surrounding villages in respect of small timber for housing and agricultural implements, firewood, leaves for manure and fodder, fencing thorns, grazing and edible forest products. The supply for such requirements should be made available at non-competitive rates, provided they are utilised by the villagers themselves and not traded in. The management of such village forests should aim at meeting the present as well as the future needs of the local population. The management of village forests should be entrusted to Panchayats. The cooperation of Panchayats should be enlisted in the protection and creation of village forests.

Tree-lands: Although 'tree-lands' are not part of regular forests in a country like India where their increase, management, and development are vital to the needs of the people, they cannot well be left out of any well-considered policy. Tree-lands are essential for the amelioration of the physical conditions of the country. Therefore, the area under tree lands could be immediately extended with the cooperation of the public and of local bodies. The cooperation of the public should also be sought in planting trees along canal banks, village roads and railway lines.

Protection of wildlife

The National Forest Policy emphasizes the protection to the animal kingdom and particularly to rare species such as the lion and the one-horned rhinoceros, which are fast disappearing. It is necessary, therefore, that bird and animal life should be controlled by special laws and rare fauna preserved by setting up sanctuaries and large-scale national parks. For this purpose, a Central Board for Wild Life was constituted by the Government of India.

Problem of grazing

The controversial question of grazing in State forests calls for a clear definition of policy. Grazing in the forests, particularly unlimited or uncontrolled grazing is incompatible with scientific forestry. There are indeed circumstances in many regions where a moderate amount of grazing does little direct harm, and may even do a great deal of indirect good in reducing the risk of fire and in suspending regression at a desirable stage. But efficient forest management requires that grazing should be regulated as regards the time and place. The formulation of the grazing policy should be based on the following principles.

- Continuous grazing on the same area by larger herds leads to deterioration of the grass complex. Wherever, it is permitted and is in great demand, efforts should be made to introduce rotational grazing, the benefits of which should be explained and demonstrated to the villagers.
- Cheap forest grazing has a demoralising effect on cattle and inadequate forest grazing reduces quality of the herds. Free and indiscriminate

forest grazing is, therefore, a serious threat to cattle breeding. It must be combated.

- Grazing should not be looked upon primarily as a source of revenue. There should be a reasonable fee for the privilege of grazing. Controlled and regulated grazing should be encouraged.
- Grazing must not be allowed in regeneration areas and young plantations.
- Grazing incidence should be kept at a minimum in 'protection forests'.
- Sheep and goat grazing should not be allowed in forests. The damage to young plants caused by the browsing of these animals is often irreparable, and their admission into the forest is incompatible with the aims and objects of forest management. Special fodder reserves under strict rotational control may be created for these animals.

NATIONAL COMMISSION ON AGRICULTURE 1976

The National Commission on Agriculture was assigned to study and recommend the basis for a revised national policy on the recommendation of the Central Board of Forestry. It recommended that the national forest policy should be based on important needs of the country, which included managing forest resources to provide maximum goods and services, preventing erosion and denudation, maximizing the forest productivity with a view to meet the growing demands and to augment employment potential with the ultimate aim at the enhancement of national prosperity. The commission identified providing small timber and fuelwood requirement and grass and grazing for livestock for the rural population as important needs. However, this was subject to a qualification. In this regard, the Commission observed that the free supply of forest produce to rural population and their rights and privileges should be reserved since these have brought about destruction to forests. The commission further stated that the rural people have not contributed much towards the maintenance and regeneration of the forests and having overexploited the resources they could not expect somebody else would take the trouble of providing them with the produce free of charge.

The National Commission of Agriculture 1976 proposed that forests could be classified into protection forests, production forests, and social forests.

Protection forests: The practice of forestry with a view to conserving flora, fauna, soil and water, reducing floods and droughts, amelioration of climatic conditions, etc. is called protection forestry. The main objective is to protect the site due to instability of terrain, nature of soil, geological formations, etc. Such areas where manipulation of the forest cover is not desirable may be classed as protection forests. The forests located on higher hill slopes, national parks and sanctuaries, preservation plots, biosphere or nature reserves and wilderness areas may be included under protection forests

Production forests: The practice of forestry with objectives of producing maximum quantity of timber, fuelwood and other forest produce is called production forestry. The production forests can further be classified into commercial forests, industrial forests, mixed quality forests, valuable forests and inaccessible forests.

Social forests: Social forestry refers to the management and protection of forests and afforestation on barren lands with the purpose of helping in the environmental, social and rural development. The basic objective of social forestry is to raising plantations by the common man so as to meet the growing demand for timber, fuelwood, fodder, etc., thereby reducing the pressure on the traditional forests. Various components of social forestry are community forestry, farm forestry, extension forestry, agroforestry and recreational forestry.

NATIONAL FOREST POLICY 1988

The National Forest Policy of 1952 was inadequate to reduce forest depletion. Forests being a state subject, there was no serious effort made for the preservation and conservation of the forests. Commercial outlook always dominated and the industrial demands were met without ensuring natural regeneration or compensatory afforestation and essential environmental safeguards. Maximization of the short-term benefits was the priority and the concepts of sustainability, protection and conservation of the forests was almost forgotten. It was, therefore, imperative to review and revise the National Forest Policy 1952, with a view to evolve a new strategy for forest protection in the future. Hence, the government came up with a new forest policy. The revised policy was called National Forest Policy 1988 and was placed and approved in the Parliament on 7th December, 1988. This led to a programme named Joint Forest Management, which proposed that specific villages in association with the forest department will manage specific forest blocks. In particular, the protection of the forests would be the responsibility of the people.

Objectives of National Forest Policy 1988

- The principal aim of forest policy must be to ensure environmental stability and maintenance of ecological balance including atmospheric equilibrium which is vital for sustenance of all
- The forest/tree cover in the country through massive afforestation and social forestry life forms human, animal and plant. The derivation of direct economic benefit must be subordinated to this principal aim.

- Conserving the natural heritage of the country by preserving the remaining natural forests with the vast variety of flora and fauna, which represent the remarkable biological diversity and genetic resources of the country.
- Creating a massive people's movement with the involvement of women for achieving these objectives and to minimise pressure on existing forests.
- Checking soil erosion and denudation in the catchment areas of rivers, lakes, reservoirs, in the interest of soil and water conservation, for mitigating floods and droughts and for the retardation of siltation of reservoirs.
- Meeting the requirements of fuelwood, fodder, minor forest produce and small timber of the rural and tribal population.
- Increasing the productivity of forests to meet essential national needs.
- Encouraging efficient utilisation of forest produce and maximising substitution of wood.
- Checking the extension of sand-dunes in the desert of Rajasthan and along the coastal tracts.
- Increasing substantially programmes, especially on all denuded, degraded and unproductive lands.

Strategy of National Forest Policy 1988

The strains and stresses experienced during the National Forest Policy of 1952 urged the planners to adopt the following strategies for the achievement of the objectives.

1. **Strategy of afforestation:** A massive need-based and time-bound programme of afforestation and tree planting, with particular emphasis on fuelwood and fodder development, on all degraded and denuded lands in the country, whether forest or non-forest land, is a national imperative.
2. **Strategy of urban forestry:** It is necessary to encourage the planting of trees alongside of roads, railway lines, rivers and streams and canals, and on other unutilised lands. Green belts should be raised in urban/industrial areas as well as in arid tracts. Such programme will help to check erosion and desertification as well as improve the microclimate.
3. **Common lands:** Village and community lands not required for other productive uses should be taken for the development of tree crops and fodder resources. Technical assistance and other inputs necessary for

initiating such programmes should be provided by the forest department through Panchayats. The revenues should be shared with local communities in order to provide an incentive to them. Certain ownership rights over trees could be considered for individuals, particularly from the weaker sections such as landless labourer, small and marginal farmers, schedule castes, tribals and women, subject to appropriate regulations.

4. **Effective land laws:** Land laws should be so modified wherever necessary so as to facilitate and motivate individuals and institutions to undertake tree-farming and grow fodder plants, grasses, and legumes on their own land. Wherever possible, degraded land should be made available for this purpose on lease basis. Such leasing of the land should be subject to the land grant rules and land ceiling laws. Appropriate regulation should govern the felling of trees on private holding.
5. **Wildlife conservation:** Forest management should take special care of the needs of wildlife conservation, and forest management plans should include prescriptions for this purpose.
6. **Shifting cultivation:** Alternative avenues of income, suitably harmonised with the right land use practices, should be devised to discourage shifting cultivation. Efforts should be made to contain such cultivation within the area already affected, by propagating improved agricultural practices. Area already damaged by such cultivation should be rehabilitated through social forestry and energy plantations.
7. **Prevention from fires and encroachments:** Encroachment on forest lands has to be arrested and effective action taken to prevent its continuance. Special precautions should be taken during the fire season. Modern management methods will be applied to deal with forest fires.
8. **Forest management:** Existing forests and forest lands should be fully protected with improved productivity. Forest cover should be increased on hill slopes and in catchment areas of rivers and reservoirs. The network of national parks, sanctuaries, biosphere reserves and other protected areas should be strengthened and extended adequately. Provision of sufficient fodder, fuel and pasture in areas near to forest is necessary to reduce pressure on forests. Afforestation should be intensified with special emphasis on augmenting fuelwood production to meet the requirement of the rural people.
9. **Tribal people and forests:** All the agencies responsible for forest management including the forest development corporations should try to involve the tribal people closely in the protection, regeneration and development of forests as well as to provide gainful employment to

people living in and around the forest. The rights and concessions from forests should primarily be for the *bona fide* use of the communities living within and around forest areas, specially the tribals. The rights and concessions enjoyed by the tribals and other poors living within and near forests should be fully protected. Their domestic requirements of fuelwood, fodder, minor forest products and construction timber should be made available through conveniently located depots at reasonable prices.

10. **Forest based industries:** Forest-based industries should raise the raw materials for their requirements, preferably by establishment of direct relationship between the factory and the individuals who can grow the raw materials by supporting the individuals with inputs including credit, technical advice and harvesting and transport services. Forest based industries should provide employment to local people on priority and also involve them fully in raising trees and raw-material. Natural forests should not be made available to industries for undertaking plantation and other activities since these serve as a gene pool resource and help to maintain ecological balance. Farmers should be encouraged to grow tree species required for industries on marginal/degraded lands available with them. The practice of supply of forest produce to industry at concessional prices should be discontinued. Industry should be encouraged to use alternative raw materials. Import of wood and wood products should be liberalised.
11. **Financial support for forestry:** There should be investment of financial and other resources in forestry sector on a substantial scale. Forests should not be looked upon as a source of revenue. Forests are a renewable natural resource. They are a national asset to be protected and enhanced for the well-being of the people and the Nation.
12. **Forestry research:** Scientific forestry research should be strengthened. Research priorities should be fixed on increasing the productivity of wood and other forest produce per unit area per unit time, re-vegetation of barren/marginal/waste/mined lands and watershed areas, effective conservation and management of forest eco-system, social forestry for rural tribal development, development of substitutes to replace wood and wood products, wildlife and management of national parks and sanctuaries.

Appraisal of the National Forest Policy 1988

This policy is comprehensive and superior to the forest policy of 1952 since it includes forest ecology, emphasises on the conservation and management of forests, and role of women and tribal people in the protection of forests. The positive aspects of this policy include

- It considers the importance of conservation of the forests.
- It emphasizes that 1/3rd of the plains and 2/3rd of the hilly regions must be covered with forests.
- It reiterates the need to carry out afforestation, social and farm forestry on a large-scale.
- It recognizes the customary privileges that belong to the people living in the forests and has enacted clauses to protect the rights and concessions of these people.
- It also tries to prevent the problem of encroachment of the forest lands by effective monitoring.
- Some efforts are taken in the policy to counteract any form of political intervention.
- It does not take up a pro-industrial stand, but focuses more on the conservation of the forests.

The policy also overlooked some of the important aspects which are summarised below.

- There is no scope for people's participation in the policy and no effort has been made to ensure the co-operation of the people, neither urban nor rural.
- The policy has ambitious programmes, but there is a lack of strong machinery to co-ordinate the activities of the department.
- The forest policy has not been region specific.
- There is no provision of awareness campaigns for various programmes on forest ecology, wildlife, conservation and protection.
- The policy does not fix any target, nor does it prescribe any well-defined methods for the success of the programmes.

NATIONAL AGROFORESTRY POLICY 2014

Need for agroforestry policy in India

Major policy initiatives, including the National Forest Policy 1988, the National Agriculture Policy 2000, Planning Commission Task Force on Greening India 2001, National Bamboo Mission 2002, National Policy on Farmers, 2007 and Green India Mission 2010, emphasize the role of agroforestry for efficient nutrient cycling, organic matter addition for sustainable agriculture and for improving vegetation cover. However, agroforestry has not gained the desired importance as a resource development tool due to the following factors.

- Absence of a dedicated and focused national policy and a suitable institutional mechanism
- Lack of an integrated farming systems approach
- Restrictive regulatory regime
- Inadequate attempts at liberalization of restrictive regulations
- Insufficient research, extension and capacity building
- Dearth of quality planting material
- Institutional finance and insurance coverage
- Weak market access for agroforestry produces
- Industry operations at a sub-optimal level

Thus, a national policy on agroforestry was launched in 2014 with the following major policy goals.

- Setting up a National Agroforestry Mission or an Agroforestry Board to implement the National Policy by bringing coordination, convergence and synergy among various elements of agroforestry scattered in various existing, missions, programmes, schemes and agencies pertaining to agriculture, environment, forestry, and rural development sectors of the Government
- Improving the productivity; employment, income and livelihood opportunities of rural households, especially of the smallholder farmers through agroforestry
- Meeting the ever-increasing demand of timber, food, fuel, fodder, fertilizer, fibre, and other agroforestry products; conserving the natural resources and forest; protecting the environment and providing environmental security; and increasing the forest / tree cover, there is a need to increase the availability of these from outside the natural forests

Major policy objectives

The National Agroforestry Policy was launched on February 10, 2014, on the occasion of World Congress on Agroforestry held in New Delhi and India became the first country in the world to have a National Agroforestry Policy. The basic objectives of this policy were

- Encourage and expand tree plantation in complementarity and integrated manner with crops and livestock to improve productivity, employment, income and livelihoods of rural households, especially the small holder farmers

- Protect and stabilize ecosystems, and promote resilient cropping and farming systems to minimize the risk during extreme climatic events
- Meet the raw material requirements of wood-based industries and reduce import of wood and wood products to save foreign exchange
- Supplement the availability of agroforestry products (AFPs), such as the fuel-wood, fodder, non-timber forest produces and small timber of the rural and tribal populations, thereby reducing the pressure on existing forests
- Complement achieving the target of increasing forest/tree cover to promote ecological stability, especially in the vulnerable regions
- Develop capacity and strengthen research in agroforestry and create a massive people's movement for achieving these objectives and to minimize pressure on existing forests

Strategies for implementation of the National Agroforestry Policy

1. **Establishment of institutional setup at national level to promote agroforestry:** An institutional mechanism, such as a Mission or Board is to be established for implementing the agroforestry policy. It will provide the platform for the multi-stakeholders to jointly plan and identify the priorities and strategies, for inter-ministerial coordination, programmatic convergence, financial resources mobilization and leveraging, capacity building facilitation, and technical and management support.
2. **Simple regulatory mechanism:** There is a need to create simple mechanisms / procedures to regulate the harvesting and transit of agroforestry produce within the State, as well as in various States forming an ecological region. There is also the need to simplify procedures, with permissions extended on automatic route as well as approval mode through a transparent system within a given time schedule.
3. **Development of a sound database and information system:** Security of land tenure is a critical issue. Due to the longer gestation period of trees stability and security of tenure rights is a necessary condition for farmers to take up agroforestry.
4. **Investing in research, extension and capacity building and related services:** The non-existent extension system for agroforestry may be the key reason for non-adoption of technologies. A common web-based platform may be setup to bring all research findings and available knowledge in the area of agroforestry for access of all stakeholders.

5. **Improving famers' access to quality planting material:** Certification of nurseries, seeds and other planting materials for agroforestry is required to make available good quality planting material at the required scale. Private sector can play an important role in augmenting supplies of improved planting stock.
6. **Providing institutional credit and insurance cover for agroforestry:** There is need for setting up of special purpose banking institutions to address specific needs of agroforestry sector. Agroforestry sector should also be benefitted with the provisions of interest subvention in the line of agricultural credit.
7. **Facilitating increased participation of industries dealing with agroforestry produce:** The role of agroforestry/biomass-based industries in the promotion of agroforestry is crucial. Therefore, such industry sector needs to be encouraged and facilitated.
8. **Strengthening farmer access to markets for tree products:** The marketing system for agroforestry produce is mostly unorganized. Marketing infrastructure similar to what is available for agricultural commodities, including market information be introduced with more private sector participation.
9. **Incentives to farmers for adopting agroforestry:** Regional and thematic differentiation in the agroforestry policy needs to be minimized. Incentive and support structure, such as the input subsidy, interest moratorium, etc. during the gestation period for promotion of agroforestry be provisioned to encourage farmers.
10. **Promoting sustainable agroforestry for renewable biomass-based energy:** Emphasis needs to be on raising fast growing trees/bushes/ grasses on marginal and degraded farmlands keeping in view their uses for meeting various energy requirements for making profitable agroforestry practices.

The policy is not only seen as crucial to India's ambitious goal of achieving 33% tree cover but also to mitigate greenhouse gas emissions from agriculture sector. After launching the National Agroforestry Policy in 2014, considerable progress has been made in terms of putting it into practice. Department of Agriculture Cooperation and Farmers Welfare under the Ministry of Agriculture and Farmers Welfare is also playing a significant role in the promotion of agroforestry. It has taken a policy decision to include trees in all its programmes, and this will significantly increase tree-planting on farms, especially under schemes funded by the National Mission on Sustainable Agriculture. Efforts are on to issue guidelines on the production and supply of high-quality planting material and accreditation of nurseries producing planting material (Dhyani, 2018).

28.427 million hectares, or 8.65% of the India's total land area, are thought to be covered by agroforestry (Arunachalam *et al.*, 2022). By 2050, there is still potential to add another 28.0 million hectares to the agroforestry sector. Since the beginning of time, the agroforestry system has been acknowledged as a crucial integrated agricultural technique for providing food, fodder, fuelwood, fibre, and lumber in addition to aesthetic and environmental benefits. Thus, the 2014 National Agroforestry Policy has the potential to revitalise agroforestry and significantly lower rural poverty in India.

LATEST DEVELOPMENTS IN AGROFORESTRY RESEARCH

Abhay Kumar *et al.* (2020) have discussed on several latest developments in agroforestry research.

Mitigation and Adaptation to climate change through agroforestry

Large volumes of carbon are known to be sequestered in the biomass of trees in agroforestry systems. Additionally, a number of studies have demonstrated the beneficial effects of agroforestry systems on soil organic carbon stocks, particularly following conversion from croplands. Additionally, these systems are encouraged as a way to diversify food production systems and adapt agriculture to climate change.

In the tropics, where agroforestry is widely seen as a means of adapting to climate change (CC), shade trees are frequently employed to shield crops from excessive heat and light. In temperate climates, growing crops under shade trees may be an interesting way to adapt agriculture to climate change. This can be achieved by buffering temperature at both the annual and daily scales, reducing yield loss from extreme temperatures (heat waves, frost), reducing soil evaporation through decreased radiation, wind speed, temperature, and air humidity, and lowering the risk of flooding. However, because of decreased rainfall and a higher danger of drought in the setting of CC, the rivalry between trees and crops for water may be intensified. Although the deeper rooting pattern in AF compared to forest settings can provide the plants access to greater water resources, tree production in AF circumstances could therefore be lower.

Agroforestry for combating land degradation

In the arid parts of the world, agroforestry contributes to improving soil restoration efforts to counteract desertification and land degradation. Arid and semi-arid regions are distinguished by intermittent droughts and poor, irregular rainfall that seldom surpasses 700 mm annually. Significant land degradation may be caused by human-induced variables, including overgrazing and other inappropriate land use practices including overexploitation of vegetation, excessive tillage, and crop residue clearance. A piece of land that has lost some of its natural productivity as a result of processes driven by humans is considered degraded. Negative alterations to the physical, chemical, and biological characteristics of the soil, together with plant deterioration, are commonly referred to as land degradation. Negative alterations in ecosystems' ability to provide products and services for society and the environment are also included in the concept of land degradation. Such deterioration might contribute to adverse microclimatic shifts at the landscape level, which would speed up the process of desertification. Degradation of land and desertification can have detrimental impacts on rural populations' lives by lowering food security and water availability and making them more susceptible to natural and biological risks as well as the effects of climate change. Land productivity may be restored with the use of several agroforestry methods. Trees, for instance, have the ability to stabilise the soil, particularly when utilised in contour farming and terracing techniques to prevent soil erosion. Additionally, they help shield soils from wind erosion. To improve the organic matter and nutritional condition of the soil, their branches can be clipped and used as mulch. presentations of scientific results on how well various agroforestry methods work to restore soil and prevent desertification and land degradation in arid places, taking into account socioeconomic factors and scalability.

Agroforestry for biodiversity conservation

It has been shown that agroforestry is a workable substitute for less biologically varied agriculture in striking a balance between productivity and biodiversity protection. Nevertheless, other research has also shown that the benefit of agroforestry systems for conserving biodiversity varies according to management choices. The availability of resources and the vegetative composition of animal species can be impacted by various management choices, which might have an overall impact on the conservation value of such species. The appropriate balance between production and conservation must be fostered by management decisions, and more data is required to precisely identify the trade-offs between the two. Since these are the best alternatives for protecting ecosystem functioning and the services required to provide sustainable production, the identification and quantification of these trade-offs is essential to the development of biodiversity-friendly management methods. An Agroforestry system for the preservation of biodiversity and the various ways that management decisions may impact an agroforestry system's worth.

Agroforestry for watershed restoration

Agroforestry techniques have the potential to help landowners worldwide in more social, economic, ecological, and environmental ways. Almost one billion hectares of agricultural land worldwide now have more trees than before because to these approaches. The incorporation of perennial trees into any given agricultural watershed may lead to the following benefits: improved water quality, reduced amounts of non-point source pollutants entering water sources, reduced amounts of carbon (C) sequestered at the watershed level, increased economic returns resulting from enhanced biodiversity, increased soil health and fertility, positive influence on stream geomorphology, and increased energy capture at different trophic levels at the landscape/watershed level. Riparian filter strips with woody plants are, however, scarce, particularly in places near streams and rivers where contemporary intensive agricultural operations are carried out. Therefore, the installation of riparian buffer strips or filter strips may be quite advantageous for these sites. Perennial vegetation (trees, shrubs, and perennial grasses) along the banks of streams found in both wooded and agricultural settings, or as continuous plant barriers incorporated in sensitive sections of the crop field itself, are known as riparian buffer strips or filter strips. These can arise spontaneously or be created purposefully. Despite making up a very small percentage of the landscape, they are extremely productive systems that regulate several significant biological processes disproportionately at various scales.

Agroforestry and the migration challenge

Migrations have always played a role in human history and can provide insight into the history of agricultural techniques and land usage today. Migrants carry with them customs, knowledge, and occasionally even seeds and animals when they journey. They alter land usage and have an impact on how farming has evolved where they dwell. In addition, they blend in with the community, taking on certain customs and changing others to fit their own experiences. With migratory fluxes, agroforestry techniques may be shared, learned, and transported from one location to another. Agroforestry is a type of forest conversion method that may be found on agricultural boundaries with forests. In isolated locations, their capacity to offer a wide range of goods for both rental and self-sufficiency is a significant benefit. They are also commonly used in savannah land occupation methods, either as a response to drought restrictions or because they are labour-intensive and can swiftly take a greater area of land. Migration can be caused by a number of factors, including harsh or recurring weather events that threaten crops, poverty, and insecurity. Agroforestry techniques contribute to the mitigation of climate change and can help make agricultural systems more resilient to climatic shocks. They might contribute to poverty reduction and food security, particularly in isolated locations. Degraded landscapes may also be restored via the application of agroforestry techniques.

Agroforestry encourages jobs and green growth

In order to ensure that natural resources continue to offer the resources and environmental services that are essential to our well-being, green growth entails promoting economic development and growth. Agroforestry is a globally applicable method and technique for sustainable land use that works with any kind of land cover. Because it promotes green growth and job creation, agroforestry may meet any of the three pillars of sustainability: economic, environmental, and social. Agroforestry is the best tool for eco-intensification because it can optimise the use of resources like light, moisture, and nutrients. Green growth is the result of diversifying the production of agricultural and forest systems, which should select the best agroforestry components (woody and agricultural production from the lower strata) not only to increase farm profitability but also. In order to support economic growth and development and, consequently, the creation of new employment globally, this seeks to produce the finest mixes of woody perennials and agricultural output from the lowest strata.

Scaling up of scientifically confirmed agroforestry innovations

Across a range of settings and scales, including technology, socioeconomics, service delivery, policies, and institutional frameworks, agroforestry research is integrated into development. This might hasten the effects of agroforestry technologies and systems to benefit the greatest number of people across the largest region in a better, equitable, and sustainable way. This not only optimises trade-offs between the current practices' agronomic, economic, and environmental performances, but it also helps bring about changes in institutions, policies, laws, rules, cultural values, and beliefs that may support the scaling up of agroforestry innovations with tangible and long-lasting effects.

Agroforestry and certification

The purpose of voluntary market instruments such as 'internal control systems' (ICS), 'participatory guarantee system' (PGS), and 'second and third-party certification systems' (STC) is to offer a reliable assurance to consumers looking for environmentally friendly or lawful items. This guarantee will be used as a tool to improve smallholder farmers' access to markets in agroforestry systems, particularly where smallholder farmers are engaged. It is, therefore, frequently viewed as a barrier and limited to industrial manufacturing on a big scale. To prove organic or excellent agricultural practices for food and sustainability for items derived from forests, there are several certification and verification procedures. It is recommended that professionals or researchers with certification in agroforestry transmit their knowledge and expertise to the forestry or food sectors. Agroforestry, or 'Trees outside Forests', is a successful and sustainable industry. They can encourage innovation and promote its development by citing instances of the benefits on livelihoods and current certification advances.

Agroforestry in practice

The interaction benefits of integrating trees and shrubs with crops and/or animals, both environmentally and productively, are leveraged by agroforestry approaches. It provides strategies for leveraging fresh, lucrative product markets while putting specialised knowledge and abilities to use in building dependable, resilient, and sustainable production systems. Long-term agroforestry land use methods and woodlot management that meet the three requirements-intensive, interactive, and integrated-are the focus of this workshop. In addition to achieving financial profitability, the goal of the agroforestry practice session is to implement a wide range of case studies of best management practices (BMPs) while producing noteworthy co-benefits, such as production diversification, landscape intensification, soil and water stewardship, and ecosystem restoration.

Agroforestry in public policies

All nations are urged to meet 17 social, environmental, and economic goals that advance prosperity while preserving the environment under the Sustainable Development Goals, which were agreed upon in 2015. Farmers continue to identify administrative and policy constraints as the main deterrents to agroforestry, despite the overwhelming evidence that agroforestry can help achieve these goals. At the regional, national, and worldwide levels, agroforestry may be effectively promoted by public policies, initiatives, strategies, and action plans, as shown by the Sustainable Development Goals. It is also encouraged for researchers, practitioners, and policy makers to contribute ideas and take part in an engaging round table discussion regarding the essential elements of policy that will draw farmers to agroforestry.

Economics of agroforestry: the link between nature and society

Both agroforestry methods and practices and environmental factors (biodiversity, soil and fauna effect) are comparatively well characterised. The reasons for some farmers' decision to switch to agroforestry are still largely unaccounted for in terms of methods, long-term economic rewards, and effects on farmer income. Social concerns are mostly seen through the lens of ethnobotany. The socio-economic component of agroforestry includes payments for ecosystem services, payments for income analysis (especially for vulnerable nations and people linked to food security), social components (farmers' perspectives, organisation, and policy), and other environmental challenges. Long-term practices are essential because of agroforestry techniques' favourable environmental effects, ability to secure revenue production, ability to improve resilience, and social vision of this paradigm. The goals are to

- close the knowledge gap between research and policy by contributing to the understanding of the social and economic aspects of agroforestry techniques
- to offer a more comprehensive perspective on the fundamentals of agroforestry development and practices within a multidisciplinary framework
- to showcase instances that connect research, regional communities, and policy

Agroforestry value chains

The whole set of steps necessary to take a product or service from its inception to its many stages of manufacture, distribution to end users, transformation, and disposal after use is known as a value chain. Value chain analysis looks to understand how value is produced and distributed among chain members as well as to describe how chain operations are carried out. It is well known that agroforestry systems enable farmers to have many revenue streams and to diversify their product offerings at the farm level. The goal of "Agroforestry and Value Chains" is to comprehend how implementing agroforestry techniques may raise the value of agricultural goods that are produced sustainably. Agroforestry should take into account scientific contributions that centre on

- Techniques and strategies utilised for value chain analysis, as well as instruments for involving value chain stakeholders
- The relationship between the various chain actors: the ways in which various stakeholders contribute to the formation and growth of the chain, as well as their respective relationships
- The value chain's potential and bottlenecks: identifying the innovations that need be implemented to raise the chain's value
- Assessing, comprehending, and enhancing value chain performance in terms of sustainability

Socio-economic and cultural controls on technology adoption in agroforestry systems

Adoption of technology has frequently been a major barrier to increasing farming productivity, income, and yields, especially in developing nations where family labour is the main source of production, market-based production systems are underdeveloped, the subsistence economy is still strong, and land is held under communal tenure. We define technology broadly to encompass both material and non-material technologies, such as new processing machines, inorganic fertiliser, labour arrangements, shifting gender relations, and land tenure regimes, as well as non-material technologies like these. Together, the talks are expected to give an overview of the main obstacles that smallholder production and the adoption of technology face in much of the developing world, as well as suggest possible approaches and solutions to overcome these obstacles.

Agroforestry landscape

Globally, there is a crisis affecting rural resource-based livelihoods, agricultural systems, and ecosystem health. Given the effects of climate change, population growth, and economic development, agriculture will need to undergo a remarkable transformation over the next fifty years in order to satisfy output demands in a sustainable manner. The conservation of biodiversity and ecosystems will need the development of new tactics to deal with the expansion of agriculture. In order to ensure that their landscapes provide the entire spectrum of products and services required, as well as to identify synergies and minimise trade-offs, populations and enterprises will need to devise new strategies for managing their natural resource base. A crucial component of that transformation will be integrated management of agricultural landscapes, or ILM. The Sustainable Development Goals and inclusive green growth will be implemented through partnerships and platforms at the landscape level. Agroforestry techniques and systems have the potential to significantly contribute to the transformation of agriculture in sustainable landscapes.

Urban agroforestry

Over the past several decades, the globe has seen unprecedented levels of urbanisation; this trend is anticipated to continue, making food and nutritional security one of the main issues in many cities. Regreening is another important issue since, far too frequently, the conversion of forest and agricultural land to urban area has been done with little to no regard for environmental quality, which results in the loss of ecosystem services and tree cover. As long as they are not affected by pollution, developing agroforestry in the remaining urban and peri-urban green spaces may provide an answer to these worries. By combining food production with the restoration of trees and plant cover and all of its ecosystem services, it would improve food and nutritional security for the urban poor and re-establish a connection between the urban population and the natural world. Studies on urban agroforestry can include a broad range of geographic settings and sizes, from local to worldwide, and can incorporate elements of agro-ecology, permaculture, agriculture, and forestry in an urban environment.

Agroforestry mapping

Considering the exponential rise in the need for current geospatial data to support informed decision making, mapping agroforestry systems at the landscape level and implementing associated interventions is a crucial strategy for developmental planning at the local, regional, and national levels. In order to address these issues, mapping organisations need to act more effectively and efficiently. Over the years, these issues have been successfully addressed with the use of remote sensing, GPS, and GIS approaches. However, there are innovative and ground-breaking developments in data collection and analysis. Unmanned aerial vehicles (UAVs) are unquestionably among the newest and most inventive instruments available, providing the benefits of using greater reach and agility in data collecting. Applications of remote sensing in agroforestry; mapping land suitability for agroforestry; modelling and mapping agroforestry parameters; development of geospatial information systems to support efficient management of agroforestry systems, etc. Agroforestry mapping and monitoring applications using Lidar and Radar remote sensing, as well as newly developing and revolutionary developments in UAV applications, will receive particular attention and focus.

Agroforestry and business

Beyond the realm of great science, feeding the world's growing population at the necessary scale requires the participation of all societal sectors, but especially the corporate sector. This is made up of hundreds of millions of businesses, from small-scale farmers in developing nations to the massive conglomerates that manage a large portion of the world's commodities flows. One method is to use agroforestry to restore damaged landscapes and increase the productivity of the forestry and agricultural landscapes.

Opportunities and trade-offs of agroforestry with perennial crops

Agroforestry techniques have been extensively marketed as having advantages for perennial crops, especially tropical ones, in terms of sustainability, revenue diversification, and climate change mitigation. They are not, however, universal and are reliant on the institutional and socioeconomic background, as well as the local ecological parameters (such as soil, temperature, and species of companion trees). The difficulties, possibilities, and trade-offs associated with temperature in tropical agroforestry systems using perennial crops, with a focus on the following problems

- Perennial agroforestry systems encourages ecological intensification
- The growth and intensity of tree crops are associated with both deforestation and reforestation.
- Trade-offs between ecosystem service delivery and productivity (soil fertility, biodiversity preservation, carbon sequestration, pest and disease management, and a buffer against extreme weather).
- Climate-smart techniques to help perennial agroforestry systems adjust to changing climate conditions value chains, eco-certification, and enduring agroforestry systems
- Analysing costs and benefits and diversifying income streams for enduring agroforestry systems
- Knowledge of local agroforestry and collaborative innovation
- In agroforestry systems based on tree crops, timber production is facilitated by productivity, risk and vulnerability management, and supportive frameworks (legal, political, financial, and cultural).

Rural development through tree-animal-pasture relationships

Worldwide, silvipastoral systems have been and continue to be prevalent. Large agricultural landscapes with less productive soils and/or climates are dominated by traditional systems; however, in recent decades, a variety of new silvipastoral schemes have emerged to address various environmental issues related to livestock farming, enhance the forage autonomy of extensive pastures, and show the value of these systems for both livestock farmers and the provision of public ecosystem services. Improved pastures for shade conditions, woody forage banks, animal carrying capacity, assessment of the value of tree shelter for animal production and welfare, tree regeneration, monitoring and promotion, control of nutrient leaching and greenhouse gas emission, and carbon sequestration are all results of tree-pasture management and interaction in the silvopastoral system. Additionally, silvipastoralism supports climate change adaptation and mitigation.

Dynamics of farmer managed natural regeneration

In response to the problems posed by population increase and climate change, farmers in regions such as sub-Saharan Africa, South Africa, and Southeast

Asia have made investments in the preservation and care of naturally regenerating trees and shrubs within their fields. This method differs from the majority of other agroforestry techniques in that it is only focused on managing and protecting woody plants that naturally regenerate on agricultural land. Nurseries and subsequent planting are not necessary. Based on available data, farmer-managed natural regeneration (FMNR) has mostly developed in high population density environments where farmers were compelled to increase agricultural productivity.

Renewal of tree pollarding

Farmers in various climes adopted the global practice of pollarding trees, which involves harvesting all of the branches at regular intervals. In addition to providing wood, pollards may be used for energy, mulching, bedding, animal food during dry or cold seasons, and much more. Because of their lower canopy size, which reduces rivalry with crops, their complementarity with pastures and crops for light and subsurface resources, and their mature trees' well-established root systems, pollards are particularly adapted to agroforestry systems. Numerous wildlife species depend on pollarded trunks for vital room for survival, feeding, and hiding. Cave-dwelling species, insectivorous passerines, and nocturnal predatory birds like owls find them appealing. They also serve as homes for martens, bats, squirrels, and other pollinating insects that are important to agricultural output, such as beetles and bees. Because pollards resemble ancient trees that are extinct in managed forests, they are frequently valuable historical or cultural trees and can serve as the only home for endangered species. Recently, exciting advancements in pollarding management have surfaced, such as the automation of tree pollarding to save manpower costs. Examining the customary applications of pollards and showcasing inventive methods to management will provide advantages of incorporating pollarding into contemporary agricultural techniques, such as its contribution to improving biodiversity, sequestering carbon dioxide, generating biomass, and conserving water.

Ever green agriculture

In order to develop more sustainable and fruitful agricultural systems, Ever Green Agriculture (EVA) is a type of more intensive farming that incorporates certain tree species into crop and animal production systems at the field, farm and landscape sizes. Opportunities for long-term food crop production that are bolstered by low to medium concentrations of trees with particularly advantageous characteristics have been brought to light by EVA. Scientists are now debating whether or not current 'successes' can be generalised and repeated elsewhere.

Agroforestry and agroecology

An agroforestry approach integrates the long-term protection of natural resources as a component of food, fuel, and fibre production, and it fits in well with agroecological food production systems (organic, biodynamic, permaculture, etc.). Agroforestry involves applying ecological principles and practices to the design and management of agroecosystems. Agroecological farmers are accustomed to handling more complicated systems; they rely on system variety to provide ecosystem services like fertility maintenance and pest control. The management of agroforestry systems using an agroecological approach is not without its difficulties, though, especially when it comes to competing for scarce resources like fertilisers and controlling weeds in the tree rows.

Tree genetics resources for agroforestry systems

Trees are the "first resource" of agroforestry systems because they comprise varied species, diverse populations, diverse DNA, and diverse land races. Numerous trees play a vital role in enhancing the standard of living for millions of smallholder farmers by providing essential goods including fuel, food, medicine, wood, and fodder, as well as environmental benefits like carbon sequestration and soil fertility and health. In order to create agroforestry landscapes that are both productive and sustainable, as well as to promote their restoration, proper management of them is essential.

- Preserving tree genetic resources in human-interfered farming and forest systems; investigating the best ways to maintain and enhance variety; and guaranteeing its continuous accessibility for utilisation.
- Tree domestication, which includes bringing potential wild tree species into cultivation and enhancing the growth of existing domesticated trees.
- Delivery of tree planting materials: looking into ways to guarantee that farmers receive the appropriate material

Pests, diseases and weeds in agroforestry systems

Using a variety of interrelated mechanisms, such as improved soil quality, altered host plant physiology, permanent shelter for invertebrates (beneficial pests), and alternative sources of disease inoculum, agroforestry systems can influence the growth of pests, diseases, and weeds at various scales. Depending on their particular needs, these processes may result in the control of weeds, diseases, and pests, or on the other hand, in a rise in their numbers. As a result, agroforestry systems structure and management are crucial to maximising system performance by strengthening positive processes and reducing negative ones. Quantifying the yield losses and quality losses of the products in the system is one way to evaluate how well agroforestry systemsare doing in terms of controlling weeds, diseases, and pests.

Roots matters in agroforestry

The success of co-existing crops in agroforestry systems depends on the interactions between agricultural root systems and trees, which are not well understood. While a larger extension of fine roots in space and time might improve access to water and nutrients, there can be significant competition between crop and tree roots for the same resources. In addition to fostering complimentary and facilitating connections, appropriate management can assist to lessen this competition. Appropriate species selection, planting spacing, scheduling, mulching, shoot trimming, and root or ploughing pruning are examples of management methods. Different species have different patterns of root activity surrounding individual plants; by using appropriate spacing and fertiliser distribution, it may be possible to prevent excessive competition and wasteful nutrient losses in agroforestry systems, the relationships between various agricultural and tree species' root systems.

Biophysical modelling of interactions in agroforestry

In agroforestry, models are being used more often for policy guidance, farmer decision assistance, and scientific research. Such models need to be able to project field, farm, and food system performance across a broad variety of climatic and socioeconomic variables in light of the changing climate. The complexity of agroforestry systems, including the interactions between different parts, the vast spatial domains that trees and crops cover in both horizontal and vertical directions, the long life spans of trees, and the variations in their management, remain among the many challenges that modelling these systems still faces. To further complicate matters, growth resources are shared by system components simultaneously above (light and water) and below (nutrients and water) the ground for a variety of outputs (such as providing, sustaining, and regulating services). All of these interactions should be simulated in order to gain a thorough understanding of the overall impacts on the system.

INTERNATIONAL INSTITUTES WORKING IN FOREST CONSERVATION AND RESEARCH

Food and Agriculture Organization (FAO)

The Food and Agriculture Organization (FAO) is a specialized agency of the United Nations that leads international efforts to defeat hunger. Its goal is to achieve food security for all and make sure that people have regular access to enough high-quality food to lead active and healthy lives. FAO has

194 member nations, one member organisation (European Union) and two associate members.

FAO plays an important role in coordinating and implementing forestry genetic resources policy within its overall aim of providing technical assistance. Emphasis is paid, in technical terms, on the use of multipurpose species. The species in the arid and semi-arid regions of developing countries are given priority as they are subjected to high human and biotic stress. Efforts are done to explore, use and conserve gene resources of forest trees. It disseminates the information on forest tree seed supplies, seed collection, handling, storage, testing and certification.

International Board for Plant Genetic Resources (IBPGR)

The International Board for Plant Genetic Resources (IBPGR) is an international scientific-organization under the aegis of the Consultative Group on International Agricultural Research (CGIAR) and established by them in 1974. The basic function of IBPGR is to advance the conservation and use of plant genetic resources for the benefit of present and future generations. It is working on agricultural crops, forest species, particularly fuelwood species, for their conservation and improvement. It has raised general awareness of the issues related to genetic resources, which in turn has stimulated the demand for training and research, for technical publications, and for scientific assistance to national plant genetic resources programmes. The institute has been involved in the collecting of 200,000 samples of crops in 120 countries.

International Tropical Timber Organization (ITTO)

The International Tropical Timber Organization (ITTO) is an intergovernmental organization promoting the sustainable management and conservation of tropical forests and the expansion and diversification of international trade in tropical timber from sustainably managed and legally harvested forests. The major activities of ITTO are

- Develops internationally agreed policy guidelines and norms to encourage sustainable forest management (SFM) and sustainable tropical timber industries and trade.
- Assists tropical member countries to adapt such guidelines and norms to local circumstances and to implement them in the field through projects and other activities.
- Collects, analyses and disseminates data on the production and trade of tropical timber.

- Promotes sustainable tropical timber supply chains.
- Helps develop capacity in tropical forestry.

ITTO is an action and field-oriented organization with more than 30 years of experience. It has funded and assisted in the implementation of more than 1000 projects and other activities addressing the many aspects of SFM, such as forest restoration, wood-use efficiency, the competitiveness of wood products, market intelligence and transparency in the tropical timber trade and tropical timber supply chains, forest law enforcement and governance, illegal logging, biodiversity conservation, climate-change mitigation and adaptation, the contributions of non-timber forest products and environmental services, and the livelihoods of forest-dependent communities. Its headquarters is in Yokohama, Japan.

International Union for the Conservation of Nature (IUCN)

The International Union for Conservation of Nature (IUCN) is a unique international agency composed of both government and nongovernmental civil society organisations with a mission to influence, encourage and assist societies throughout the world to conserve the integrity and diversity of nature and to ensure that any use of natural resources is equitable and ecologically sustainable. It provides public, private and non-governmental organisations with the knowledge and tools that enable human progress, economic development and nature conservation to take place together. By facilitating these solutions, IUCN provides governments and institutions at all levels with the impetus to achieve universal goals, including on biodiversity, climate change and sustainable development, which IUCN was instrumental in defining.

IUCN is concerned with species level conservation and has paid increasing attention to plant genetic resources. It produces the IUCN plant Red data book, a series that gives detailed case histories on rare and threatened plants in all parts of the world; for each species, data are given on conservation status, threats to survival, distribution and habit, together with a short description and an evaluation of its interest or potential value to humankind.

IUCN's headquarters is in Gland, Switzerland. It has a World Conservation Monitoring Centre (WCMC) located at Cambridge, UK for data storage and processing and also provides information on international trade in endangered plants and animals.

International Union of Forestry Research Organization (IUFRO)

The International Union of Forest Research Organizations (IUFRO) is a non-profit, non-governmental international network of forest scientists. It unites more than 15,000 scientists in almost 700 Member Organizations in over 110 countries with its headquarters in Vienna, Austria. Scientists cooperate in IUFRO on a voluntary basis. IUFRO's vision is of science-based sustainable management of the world's forest resources for economic, environmental and social benefits. It is the global network for forest science cooperation.

IUFRO promotes global cooperation in forest-related research and enhances the understanding of the ecological, economic and social aspects of forests and trees. It disseminates scientific knowledge to stakeholders and decision-makers and contributes to forest policy and on-the-ground forest management. IUFRO has placed strong emphasis on industrial species and works on provenance testing, progeny testing and breeding of specific species. It is also working on conservation and on population genetics for the species of temperate zone and Mediterranean conifers, as well as *Quercus* spp., *Eucalyptus* spp., *Populus* spp.

With the 'Strategy 2015-2019', IUFRO addresses five research themes and associated emphasis areas, and three institutional goals. The following five themes aim to guide the science collaboration within IUFRO's global network in the current Strategy period.

- Forests for people
- Forests and climate change
- Forest and forest-based products for a greener future
- Biodivesity, ecosystem services and biological invasions
- Forest, soil and water interactions

The three institutional goals are

- Research excellence: Strive for quality, relevance and synergies
- Network cooperation: Increase communication, visibility and outreach
- Policy impact: Provide analysis, insights and options

United Nations Environment Programme (UNEP)

The United Nations Environment Programme (UNEP), an agency of the United Nations, coordinates the organization's environmental activities and assists developing countries in implementing environmentally sound policies and practices. UNEP's activities cover a wide range of issues regarding the atmosphere, marine and terrestrial ecosystems, environmental governance and green economy. UNEP's mission is to provide leadership and encourage partnership in caring for the environment by inspiring, informing, and enabling nations and peoples to improve their quality of life without compromising that of future generations. It has played a significant role in developing international environmental conventions, promoting environmental science and information and illustrating the way those can be implemented in conjunction with policy, working on the development and implementation of policy with national governments, regional institutions in conjunction with environmental non-governmental organizations. Its headquarters is in Nairobi, Kenya.

Central America and Mexico Coniferous Resources Cooperative (CAMCORE)

An organization called Central America and Mexico Coniferous Resources Cooperative (CAMCORE) was formed in 1980 and made part of the Department of Forestry & Environmental Resources at North Carolina State University, USA. The driving force behind its formation was the private forest sector. CAMCORE demonstrates that the private sector, in collaboration with government agencies and universities, can make substantial contributions to tree breeding, gene conservation and environmental stewardship, while simultaneously providing economically sound forest management. This private organization has been able to directly affect the development of genetic resources by the cooperative efforts.

CAMCORE makes seed collections in threatened forest stands and then plants these on members' land in more protected areas in genetic field trials (called progeny tests) and conservation areas (called *ex situ* conservation banks or parks) with similar climates in countries around the world. CAMCORE professionals at North Carolina State University analyse the data from the progeny trials and produce annual summaries to help members decide on what to grow. Early results from CAMCORE genetic tests indicate that height gains of approximately 20% are possible in the tropics by simply using alternative species and more-productive seed sources. CAMCORE is now working with 50 different forest species and has sampled 11,000 trees in 500 locations. It has more than 2,500 hectares of genetic trials and conservation areas. It has the largest database on tropical and subtropical pines and non-Australian eucalypts in the world.

Tropical Agricultural Research and Higher Education Centre (CATIE)

The Centre Agronomics Tropical de Investigation Ensenanza (Tropical Agricultural Research and Higher Education Centre, CATIE) is an international institute for agricultural development and biological conservation and sustainable use of natural resources in Central America and the Caribbean, combining science, education and innovation. CATIE is the first graduate school in Agricultural Sciences in Latin America. It works for the renewal of natural resources by the renewable natural resources development which has four programmes such as agroforestry, silviculture, wild lands and watershed management. It also works on *ex-situ* conservation.

International Development Research Centre (IDRC)

International Development Research Centre was established by an act of Canada's parliament in 1970 with a mandate to initiate, encourage, support, and conduct research into the problems of the developing regions of the world and into the means for applying and adapting scientific, technical, and other knowledge to the economic and social advancement of those regions. The IDRC funds research in developing countries through various projects which work for forestry research in developing countries to promote growth, reduce poverty, and drive large-scale positive change. The IDRC headquarters is located in Ottawa, Canada. It has four regional offices in Montevideo, Uruguay; Nairobi, Kenya; Amman, Jordan; and New Delhi, India. IDRC is governed by a board of up to 14 governors, whose chairperson reports to Parliament through the Minister of International Development.

Commonwealth Scientific and Industrial Research Organization (CSIRO)

The Commonwealth Scientific and Industrial Research Organisation (CSIRO) is an independent Australian federal government agency with its role to improve the economic and social performance of industry for the benefit of the community. The centre works on Australian forest tree resources. The Tree Seed Centre collects and distributes high quality and source identified seeds of commercially promising Australian woody plants for research purposes. It provides professional advice on the choice of species and seed supply, and also provides technical information on species of value. The CSIRO headquarters is located in Canberra

U.S. Department of Agriculture's Forest Service and State Programmes

The activities of the U.S. Department of Agriculture's Forest Service include research on the biological, physical, and social sciences related to the diverse forests and rangelands. The genetic improvement programmes are also conducted in national forests. Research is primarily aimed at species of high commercial value, but there is growing emphasis on maintaining the diversity of forests. Individual states have independent conservation programmes. Conifer germplasm conservation project is designated to provide information and resources needed for long term protection of the diversity of forests. It is working on *Pinus taeda*, *Pinus ponderosa*, *Pseudotsuga* spp., etc.

14

Agroforestry for Climate Change Adaptation and Mitigation

Human societies over the ages have depleted natural resources in different ways and degraded their local environments. Indiscriminate cutting down of trees and accelerating the construction works lead to global warming and climate change. Climate scientists believe that human-induced deforestation is responsible for 18-25% of climate change. Accumulation of greenhouse gases in the lower atmosphere is the main cause of global climate change. The concentration of these gases in the earth's atmosphere is increasing, mainly due to deforestation and the combustion of fossil fuels which release carbon dioxide.

The earth receives heat energy constantly from the sun through radiation. Some of the heat is absorbed and some is reflected. Carbon dioxide works like an invisible blanket that wraps around the earth and traps the heat inside, similar to the function of a greenhouse. A greenhouse is a structure that is used in colder climates to grow plants. Even when outside temperatures drop below freezing point, greenhouses are still warm enough inside to grow plants. Greenhouses are made of glass, which allows solar radiation to enter. The heat is trapped inside the greenhouse, allowing plants to grow all the year round. Increasing concentrations of carbon dioxide in the earth's atmosphere insulate it like a greenhouse, leading to a gradual warming of the earth's atmosphere. Carbon dioxide is the major greenhouse gas, but there are others as well, including methane and nitrous oxide.

Mitigation means addressing the causes of climate change, while adaptation attempts to tackle its effects. The Intergovernmental Panel on Climate Change defines mitigation as an anthropogenic intervention to reduce the sources or enhance the sinks of greenhouse gases and adaptation as the adjustment in natural or human systems to a new or changing environment (IPCC, 2007).

EFFECTS OF CLIMATE CHANGE

The term 'climate change' refers to significant changes in the average weather parameters, such as temperature, wind and rainfall, experienced in a region over a long period of time. The earth is facing one of the greatest environmental, economic and social threats through climate change. Today, climate change is a global concern especially in the agriculture sector. Altering cropping seasons, increasing incidence of pests and diseases, rationing water supply and limiting food supply are all evidences of climate change. These factors force farmers to adapt using meagre resources. A significant effect of global climate change is the altering of rainfall patterns, which certainly affects agriculture. Extended drought can cause the failure of small and marginal farms with resultant economic, political and social disruption. Stronger and more frequent typhoons, warmer nights, and longer and hotter days are some other most common evidences of climate change.

Ecological imbalance

Climate change can make summer months warmer or cold months cooler than usual. These changes in temperature can cause animals to migrate to more suitable places or force them to adapt with adverse effects on their physical conditions. Some may even die and become extinct because of their inability to adapt to extreme changes in environmental conditions. Biodiversity is thus at more risk now than ever because of climate change.

Extreme climate variability

Climate variability refers to short term changes in climate such as longer dry or rainy season, intense heat during summer, more or scanty rains during rainy months.

Increasing temperature

Global warming is the biggest threat being faced by mankind today. It refers to the observed and projected increase in the earth's average temperature due to natural or anthropogenic climate change. The earth's average temperature rose about 0.7° Celsius in the 20th century. Gases created through industrial and agricultural practices increase the heat-reflecting potential of the atmosphere, thereby raising the planet's average temperature.

Limited food supply

Climate change can alter the schedule and duration of cropping seasons. Farmers can become confused as to when they should plant crops, thereby affecting length of cropping season, time of harvest and food supply. Water shortage during dry months can also affect crop growth and overall food production.

Increased occurrence of pests and diseases

Climate change can vary the life cycle of pests by increasing their population at a different time. Farmers may be unaware of or caught unprepared for these changes. Diseases can also become prevalent based on the environmental conditions resulting from climate change.

More powerful typhoons

Global warming increases ocean temperature and rate of evaporation. As water vapour in the atmosphere increases, the typhoons carry more rains. Climate change, through global warming has increased the frequency and intensity of typhoons over the years.

Rise in sea levels

Increasing global temperature causes glaciers and polar icecaps to melt, thereby making sea levels rise. If sea levels continue to increase, more land could be submerged under water permanently.

AGROFORESTRY FOR ADAPTATION TO CLIMATE CHANGE

Climate change threatens the suitability and productivity of crops and livestock and jeopardizes wood and non-wood production. Agroforestry may be promoted to combat the threats of soil erosion and desertification due to climate change. Adaptation activities can also aim to improve management practices and establish sustainable agroforestry systems. A wide range of agroforestry systems now exist with a potential to improve productivity, favourably influence microclimate, prevent soil degradation and restore soil fertility and diversify income generating opportunities. Agroforestry can both sequester carbon and produce a range of economic, environmental and socioeconomic benefits.

As adaptation emerges as a science, the role of agroforestry in reducing the vulnerability of agricultural systems to climate change or climate variability needs to be assessed more effectively. Rainfall variability is a major constraint in the semiarid regions and also to the upland farms that do not have access to irrigation. Climate change may translate into reduced total rainfall or increased occurrence of dry spells during rainy seasons in many semiarid regions. Successful and well managed integration of trees on farms and in agricultural landscapes often results in diversified and sustainable crop production, in addition to providing a wide range of environmental benefits such as erosion control and watershed services. Therefore, optimising the use of increasingly scarce rainwater through agroforestry practices could be one way of effectively improving the capacity of farmers to adapt to drier and more variable conditions. In low rainfall years, water availability to crops is the most important factor that decides the extent of crop failure and food production. In this case also agroforestry can play an important function by buffering agricultural crops against water deficiencies. There are other mechanisms such as improved microclimate and reduced evapotranspiration through which agroforestry practices may improve the adaptive capacity of farmers. Insect pests, diseases and weeds are considered as major obstacles to crop production in many tropical agroecosystems. Integrating tree and herbaceous species in agricultural landscapes can produce positive interactions that could contribute towards controlling pest and disease outbreaks. The agroforestry practices can also control weeds through litter mulching and shading effects.

Agroforestry options may provide a means for diversifying production systems and increasing the sustainability of smallholder farming systems. Tree based cropping systems have some obvious advantages for maintaining production during wetter and drier years.

1. The deep root systems of trees are able to explore a larger soil volume for water and nutrients, which will help during droughts.
2. Increased soil porosity, reduced runoff and increased soil cover lead to increased water infiltration and retention in the soil profile which can reduce moisture stress during low rainfall years.
3. Tree based systems have higher evapotranspiration rates than agricultural crops and can thus maintain aerated soil conditions by pumping excess water out of the soil profile more rapidly than other production systems during excess rainfall years.

Thus, diversifying the production system to include a significant tree component may buffer against income risks associated with climatic variability.

AGROFORESTRY IN MITIGATING CLIMATE CHANGE

Sustainable diversification of agricultural systems

Diversification of agricultural enterprises is one of the best practices to reduce the risks associated with variable climate. It is an adjustment of the farm enterprise pattern to increase farm income or reduce income variability by reducing risk and exploiting new market opportunities. At the farm level it is the adoption of multiple production activities that are complementary in economic and/or ecological dimensions involving crops, trees and livestock. Agroforestry systems are a suitable pathway for sustainable diversification of agricultural systems. Agroforestry ensures food security by generating direct benefits to farmers such as food, fodder, feed, fuelwood, live fences and other products. The diversity of crops provides multiple harvests at different times of the year, thereby reducing the risk of crop loss and food shortage. Quality of life of farmers is improved by increasing income due to multiple harvests and sale of products from different components, thereby providing regular income throughout the year.

Improvement in rain water use efficiency

Climate change has both direct and indirect impacts on water availability. The direct impacts include changes in precipitation patterns while the indirect ones are increases in losses through runoff and evapotranspiration. The presence of trees has positive effect on water budget of the soil and crops growing between or beneath them. There are several mechanisms whereby agroforestry may use available water more effectively than the annual crops. Agroforestry systems with a perennial tree component can make use of the water remaining in the soil after harvest of annual crops and the rainfall received outside the crop season. A mulch or litter layer increases the infiltration of rain water while simultaneously reducing evaporation from the soil. The changes in microclimate also reduce the evaporative demand and make more water available for transpiration.

Soil and water conservation

There are several ways by which climate change manifests soil degradation. Higher temperatures and drier conditions lead to lower organic matter accumulation in the soil resulting in poor soil structure, reduction in infiltration of rain water and increase in runoff and erosion. The occurrence of extreme rainfall events has adverse impacts on the severity, frequency and extent of erosion. Trees improve soil fertility by processes of increasing additions to the soil, reduction of losses from soil, and improving soil physical, chemical and biological conditions. Trees have a different impact on soil properties than annual crops in terms of their longer residence time, larger biomass accumulation and more extensive root systems. Trees increase or at least maintain the organic matter levels of the soil. This is mainly through litter fall and continuous degeneration or sloughing-off of roots of standing trees. Pruning materials from the woody perennials used in agroforestry systems add a huge quantity of organic matter to the soil and this is generally high in nitrogen.

Potential to limit carbon emissions and sequester carbon

The tree component of the agroforestry systems can be a significant sink for atmospheric carbon. An efficient agroforestry system not only maximizes the benefit it provides but also ensures the link to climate change mitigation. Trees are important carbon warehouses that filter massive quantities of carbon from the atmosphere, trapping it in their biomass. Through the natural process of photosynthesis, plants absorb carbon dioxide from the atmosphere and release oxygen. They store this carbon in their leaves, branches, trunks and roots. An average tree can remove about 23 kg of carbon dioxide from the atmosphere annually. In agroforestry systems, the amount of carbon sequestration is further increased. The interactions of the different components of agroforestry systems can help absorb and sequester carbon dioxide and other greenhouse gasses from the atmosphere. Thus, trees in an agroforestry system make it a potential strategy in mitigating climate change.

Microclimate amelioration

The full genetic potential of many crops can only be exploited under optimum environmental conditions. Any change in these conditions, especially during the reproductive stage, will have a direct impact on the production and economic viability of certain crops. In an agroforestry system, the presence of trees reduces heat and light. Trees bring about favourable changes in the microclimatic conditions by influencing radiation flux, air temperature, wind speed all of which will have a significant impact on modifying the rate and duration of photosynthesis and subsequent plant growth, transpiration and soil water use.

Interventions for drought mitigation

Tree canopies lower the temperature and increase relative humidity in an agroforestry system which leads to a lower transpiration rate. Tree litter and pruning materials act as mulch on ground surface that reduces the evaporative loss of soil moisture. The water storage capacity of soil is also increased due to improved soil organic matter. Higher survival and biomass generation under agroforestry is possible due to greater efficiency for photosynthesis, tapping of nutrients and water from deeper layers, efficient utilization of off-season rainfall by the perennial tree component and better environmental conditions.

Protection and stabilization of ecosystem

Trees give protection from adverse effects of different kinds of pollutants such as dust, dirt and other physical air pollutants. The dust level in the air can be as much as 75 percent lower on the sheltered side of the tree compared to the windward side. It is estimated that one hectare of a close forest filters about 50 t of dust and dirt every year. Trees also absorb some of the chemical air pollutants.

CARBON SEQUESTRATION POTENTIAL OF AGROFORESTRY SYSTEMS

Greenhouse gas emission is likely to rise by 25-90% by 2030 relative to 2000 and the earth could warm by 3° C at the end of this century. Even with a temperature rise of 1-2.5° C, the Intergovernmental Panel on Climate Change predict serious disastrous effects, including reduction in crop yields in tropical and subtropical areas leading to increased risk of food shortage, spread of climate responsive diseases such as malaria, and an increased risk of extinction of 20-30% of all biodiversity present on earth (IPCC, 2007). Carbon dioxide is the most important among all greenhouse gases. Carbon sequestration may be defined as the process of removing C from the atmosphere and depositing it in a reservoir. It entails the transfer of atmospheric carbon dioxide, and its secure storage in long-lived pools (UNFCCC, 2007).

From the agroforestry point of view, carbon sequestration involves primarily the uptake of atmospheric CO_2 during photosynthesis and transfer of fixed carbon into vegetation and soil pools for 'secure' storage. It occurs in two major segments of agroforestry systems; aboveground and belowground. The total amount sequestered in each part differs greatly depending on a number of factors including the region, the type of system, site quality, and previous land-use (Nair, 2012). On average, the soil and aboveground parts are estimated to hold major portions, roughly 60 and 30%, respectively, of the total carbon stored in tree-based land-use systems (Lal, 2010). Sequestering CO_2 from atmosphere through natural techniques such as afforestation, reforestation, natural regeneration of forests and the adaptive agriculture is more economically and ecologically sound in increasing the carbon storage capacity of the terrestrial ecosystems (Dhyani, 2014). Although carbon sequestration through afforestation and reforestation of degraded natural forests has long been considered useful in climate change mitigation, agroforestry offers some distinct advantages (Murthy *et al.*, 2013). The carbon in the aboveground and belowground biomass in an agroforestry system is generally much higher than the equivalent land use without trees. An agroforestry system has higher carbon sequestration potential than pastures or field crops (Kirby and Potvin, 2007).

CARBON SEQUESTRATION POTENTIAL OF AGROFORESTRY SYSTEMS IN INDIA

Majority of the agroforestry systems have the potential to sequester carbon, which may vary according to tree species (Prasad *et al.*, 2012) and management practices (Newaj *et al.*, 2001). Carbon sequestration in terrestrial pools include the above ground plant biomass, such as timber, fuelwood and belowground biomass such as roots, soil microorganisms, and all the forms of organic and inorganic carbon in soils including deep root zone. In agroforestry systems, two major components (trees and crops) are mainly responsible for CO_2 sequestration. The total amount sequestered in each component differs greatly and is dependent largely on a number of factors that includes the type of system, nature of components, age of plant, site quality, and previous land-use (Newaj and Dhyani, 2008).

The carbon sequestration potential of various trees in AFS in different parts of the country is presented in table 14.1. The CSP of trees varies with species, structure, age and spatial distribution. In agroforestry systems, although trees sequester more carbon, but crops also fix and store carbon in considerable amounts. Crop improves the organic matter in soil, which is a significant component of the terrestrial carbon pool (Ciampitti *et al.*, 2011). An increment in carbon pool in soil can be achieved through adoption of appropriate crop rotations (Wright and Hons, 2005), integrated soil fertility management (Srinivasarao *et al.*, 2012), precise use of fertilizers and organic amendments (Majumder *et al.*, 2008), and adoption of conservation agriculture (Lal, 2009). However, the choice of crop species, cropping system, timing of fallowing and quality and quantity of residue returned to the soil have considerable impact on soil organic carbon stocks (Mandal *et al.*, 2007). Maize, sorghum and pearl millet have higher potential to carbon sequestration as compared to rice, finger millet and soybean (Kushwah *et al.*, 2014).

Table 14.1 Carbon sequestration potential of trees in India

Region	Agroforestry system	Tree species	Trees/ha	Age (year)	Carbon sequestration potential (Mg C/ha/ year)	References
Himalaya	Block plantation	*Eucalyptus tereticornis*	2500	3.5	4.40	Dhyani *et al.*, 1996
		Tectona grandis	570	10	3.74	Negi *et al.*, 1995
		Cedrus deodara	100	19	2.47	Wani *et al.*, 2014
	Agrisilviculture	*Dendrocalamus hamiltonii*	1000	7	15.91	Kaushal *et al.*, 2014
		Populus deltoides	500	8	12.02	Singh and Lodhiyal, 2009
	Farm forestry	*Acacia catechu*		30	1.5	Hooda *et al.*, 2007
		Pinus spp.		30	7.1	
		Mixed plantation		30	5.9	
		Mangifera indica		30	1.7	
Indo-Gangetic	Agrisilviculture	*Populus deltoides*	400	7	1.98	Rizvi *et al.*, 2011
	Block plantation	*Acacia nilotica*	1250	7	2.81	Kaur *et al.*, 2002
		Dalbergia sissoo	1250	7	5.37	
		Prosopis juliflora	1250	7	6.50	
Humid & subhumid	Agrisilviculture	*Gmelina arborea*	592	5	3.23	Swamy and Puri, 2005
	Silviculture	*Tectona grandis*	444	20	3.32	Negi *et al.*, 1990
		Gmelina arborea	452	20	3.95	

Region	Agroforestry system	Tree species	Trees/ha	Age (year)	Carbon sequestration potential (Mg C/ha/ year)	References
Arid & semi-arid	Block plantation	*Albizia procera*	312	10	1.79	Rai *et al.*, 2000
		Albizia amara	312	10	1.00	
		Acacia pendula	312	10	0.95	
		Dalbergia sissoo	312	10	2.55	
		Dichrostachys cinerea	312	10	1.05	
		Emblica officinalis	312	10	1.55	
		Hardwickia binata	312	10	0.58	
		Leucaena leucocephala	2500	9	10.32	Rao *et al.*, 2000
		Eucalyptus camaldulensis	2500	9	8.01	
		Dalbergia sissoo	2500	9	11.47	
		Albizia lebbeck	625	9	0.62	
		Acacia albida	1111	9	0.82	
		Acacia tortilis	1111	9	0.39	
		Acacia auriculiformis	2500	9	8.64	
		Eucalyptus tereticornis	320	2	13.86	Pragason and Karthik, 2013
		Acacia farnesiana	170	2	2.42	
		Cassia montana	154	2	1.84	
		Prosopis juliflora	138	2	1.16	
		Acacia nilotica	106	2	5.70	

Region	Agroforestry system	Tree species	Trees/ha	Age (year)	Carbon sequestration potential (Mg C/ha/ year)	References
Arid & semi-arid	Agrisilviculture	*Albizia procera*	312	7	3.70	Newaj and Dhyani, 2008
		Leucaena leucocephala	4444	4	14.42	Prasad *et al.*, 2012
			10000	4	15.51	
		Casuarina equisetifolia	833	4	1.57	Viswanath *et al.*, 2004
Tropical	Homegarden	Mixed tree species	667	6.5	1.60	Saha *et al.*, 2009
		Acacia mangium	5000	6.5	12.59	Kunhamu *et al.*, 2011
			2500	6.5	9.94	
			1250	6.5	9.51	
			625	6.5	6.37	

Source: Newaj *et al.*, 2014; Dhyani, 2016

Soils are the largest terrestrial carbon pool having approximately two-thirds of the carbon in ecosystems. Soil organic carbon is the prime component of soil organic matter and it has been recognized as a key component of soils both in natural and in managed cultivated lands. Carbon sequestration in soil is the process of transferring CO_2 from the atmosphere into the soil through crop residues, tree roots and other organic solids, and in a form that is not immediately reemitted. The sequestered carbon in soil helps in reducing CO_2 present in the atmosphere, while enhances soil quality and a sustained productivity. Soil carbon sequestration can be accomplished by management practices that add high amounts of biomass to the soil and cause minimal soil disturbance. Agroforestry systems with better management have greater amounts of carbon sequestration potential in and out of the soil. Below ground biomass of trees in the form of roots comprises about 1/5th to 1/4th of the total living biomass and there is constant addition of organic matter to the soil through decaying dead roots, which leads to improvements in the carbon status of the soil (Dhyani and Tripathi, 2000). Besides tree, there are several other factors like precipitation, temperature, evapotranspiration, decomposition rates, litter fall, decay rates of tertiary tree roots, crop residue, species of tree/crop, chemical properties of the litter, etc. also influence soil carbon sequestration under agroforestry systems (Dhyani, 2016).

FACTORS AFFECTING CARBON SEQUESTRATION IN AGROFORESTRY SYSTEMS

Silvicultural practices

Silvicultural practices ranging from site preparation to stand management operations can alter tree growth and productivity and, thus, improve above- and belowground carbon sequestration, but sometimes can also contribute to carbon emissions (Nair *et al.*, 2010). The silvicultural practices such as tree and stand management, choice of species, site quality and rotation length have favourable effects on carbon sequestration. Similarly, harvesting and burning have unfavourable effects on carbon sequestration. Practices such as weed control and fertilization may increase stand growth and promote carbon sequestration (Johnsen *et al.*, 2001), but may also emit CO_2. Thus, agroforests should be managed with appropriate silvicultural and agronomic practices to maximize carbon gain.

The tree density has a great influence on the stand structure and the way trees grow is dependent on stand density. Therefore, designing appropriate regimes of density to meet specific objectives is an important aspect of sound stand-density management (Kumar *et al.*, 1995). High density stands enhance the vegetation carbon pools. But this may adversely affect the productivity of understorey arable crops in an agroforestry system. On the other hand, excessively high stand densities may also adversely affect tree growth and productivity through severe competition, resulting in lower vegetation carbon pools. Thinning improves the growth of stands and thus leads to more carbon assimilation. If the thinned plants are burnt or plant parts are decomposed there will be a net release of carbon to the atmosphere (Lavigne, 1991). Thinning/pruning of trees also brings about changes in understorey light, air/soil temperature, and soil moisture regimes and accelerate detritus turnover rates, further reducing soil carbon sequestration (Nair *et al.*, 2010). Wider tree-spacing favours understorey crops, but spreading crowns may require adoption of tree management practices such as tree-canopy pruning in agroforestry systems. Crown pruning generally favours stem development (Kunhamu *et al.*, 2010) and thus promotes carbon sequestration in the stem wood, besides promoting understorey production.

Choice of species and site-specific factors

The tree growth and carbon sequestration potential are dependent on species, age and rotation length, site quality, and tree management (Nair *et al.*, 2009). The choice of species for agroforestry may be based on a rigorous species-site matching. Trees which are leafless during the active growth phase of understorey crops are particularly preferred (Nair *et al.*, 2010). Fast-growing species may accumulate more carbon before they are 10 years old than slower-growing species; however, the slower-growing species accumulate more carbon in the long-term (Redondo-Brenes, 2007). The wood of slower-growing species also possesses higher specific gravity, which further increases the carbon sequestration potential in the long-term (Balvanera *et al.*, 2005). The more valuable, high-specific gravity wood used for construction, furniture, woodcrafts constitute a longer-term sink for fixed carbon than low specific-gravity wood used for short-lived purposes such as packaging cases and poles.

Agroforestry practices differ considerably in their carbon sequestration potential depending on a number of site-specific factors such as ecology, species composition, and management. Organic agriculture may lead to higher soil organic matter storage than conventional agriculture, due to the recirculation of animal manures and crop residues (Foereid and Hogh-Jensen, 2004). Longer tree rotations increase the stored carbon stock and would reduce carbon emissions from harvesting and processing operations per unit of wood harvested. The largest amount and most permanent form of carbon may be sequestered by increasing the rotation age of trees and by manufacturing durable products from them upon harvesting (Markewitz, 2006).

Agronomic practices

Fossil fuels utilized for silvicultural activities such as fertilization and site preparation, intended to increase carbon sequestration, may emit CO_2 and can play a significant negative role in the carbon balance of forestry and agroforestry systems. Carbon emissions occur due to fossil fuel combustion in operating heavy equipment in forestry activities (Lorenz and Lal, 2010). Although addition of inorganic fertilizers may accelerate both aboveground and soil carbon sequestration potential, fertilizer manufacture is an energy intensive process (Nair *et al.*, 2010). Carbon emissions due to increased activities such as use of herbicides or water pumping for irrigation may counter-balance any potential gains in carbon sequestration (Schlesinger, 2000). Although herbicides and other agrochemicals may promote tree growth and aboveground carbon sequestration especially by suppressing competitors, parasites and predators, the effects are seemingly more complex. The associated gains in aboveground carbon sequestration may be offset by carbon emissions on account of the fossil energy consumption in the manufacture and application of such chemicals (Nair *et al.*, 2010).

Crop rotation has significant effect on soil organic matter and their relationship with soil physical attributes. Total carbon level in soil increases due to rotation of crops. Soil carbon storage can be increased within an ecosystem. Cropping systems with higher soil nitrogen contents influence greater carbon processing and storage capacity than systems with lower soil nitrogen levels. Recalcitrant carbon derived from root biomass are stabilised in presence of sufficient available nitrogen.

Soil management

The three main anthropogenic factors (Jarecki and Lal, 2003) that influence soil carbon sequestration are

- Disturbance of the soil, such as tilling
- Amendments, such as fertilization or irrigation
- Incorporation of organic matter

Soil organic C dynamics is controlled in large part by climate, soil physical properties and the amount of plant litter input. Soil management also influences the change in carbon stock efflux in soils. The maintenance of soil carbon in soil is governed by annual temperature, precipitation and many interacting factors such as soil types, tillage, application of fertilizers, quality and quantity of organics returned to soil and the method of residue management. Due to ploughing of soil, poor return of crop residues to the soil results soil degradation and retarding soil health. Change in land use results alteration of soil carbon at various depth. In the agricultural practices for higher yields, return of crop residues and conservation tillage system increased soil carbon. This organic carbon management is considered to be the key environmental management system for sustainable soil health as well as agricultural productivity. Soil carbon pools change rapidly in response to land-use pattern. Globally, soil contains three times as much as carbon as plant biomass. Land-use change has large impacts on soil carbon stock which decreases when forests are converted to agricultural land (Chapman *et al.*, 2012).

Conventional tillage practices decrease carbon sequestration in agricultural soils (Alvaro-Fuentes *et al.*, 2009). No-till agriculture is considered as the management practices that preserves soil and increases soil organic carbon pools compared with conventional tillage practices. Changes in agricultural management from conventional tillage to no-till and increased crop-rotation diversity can increase accumulation of soil carbon (West and Post, 2002).

The effect of fertilization practices on soil organic carbon depends on the nature of the fertilizer, climate and other site-specific factors. Inorganic fertilization on tree plantations may be positive, negative, or neutral with respect to carbon sequestration depending on a number of local factors such as site fertility, species, fertilizer dose, stage of stand development, etc. Incorporation of crop residues and pruned materials from the trees and use of organic manures in agroforestry for nutrient management, may promote soil organic carbon sequestration (Kundu *et al.*, 2007).

The herbicides and pesticides affect soil ecology, disrupting the processes that lead to the formation of soil macroaggregates. The roots of understorey plants play an important role in aggregate formation, especially in sandy soils where physical protection is required for aggregate formation. Use of herbicides in sandy soils restricts plant residue inputs from the surface and root inputs in soil (Sarkhot *et al.*, 2007). As a result, soil organic matter is not physically protected and it breaks down more quickly which adversely affect soil carbon sequestration.

AGROFORESTRY AS A TOOL TO MITIGATE AND ADAPT TO CLIMATE UNDER LULUCF ACCOUNTING

Utilising the fact that atmospheric CO_2 may store as carbon in plants and soils in terrestrial ecosystems can slow the pace at which CO_2 builds up in the atmosphere. According to the United Nations Framework Convention on Climate Change, a ‘sink’ is any procedure, action, or device that absorbs greenhouse gases (GHGs) from the atmosphere. Through land use, land-use change, and forestry (LULUCF), human activities affect terrestrial sinks. As a result, the exchange of CO_2 (the carbon cycle) between the terrestrial biosphere and the atmosphere is changed.

According to the sixth assessment report of the Intergovernmental Panel on Climate Change (IPCC), the "Agriculture, Forestry and Other Land Use (AFOLU)" sector contributed, on average, 13-21% of all human greenhouse gas emissions worldwide between 2010 and 2019. Based on accounting models, the estimated anthropogenic net CO_2 emissions from AFOLU between 2010 and 2019 result in a net source of +5.9±4.1 Gt CO_2eq/yr with an uncertain trend. Net AFOLU CO_2 emission fluxes are driven by changes in land use, with deforestation accounting for 45% of total AFOLU emissions. Land affects climate through evapotranspiration, albedo effects, and aerosol loading from emissions of volatile organic compounds, in addition to being a net carbon sink and source of GHG emissions.

According to the IPCC report, the LULUCF sector provides food, timber, and other renewable resources, as well as biodiversity protection, and offers substantial near-term mitigation potential. For the most part, mitigation strategies in forests and other natural ecosystems offer the greatest potential for LULUCF mitigation between 2020 and 2050. Reducing deforestation has the most potential to lower anthropogenic greenhouse gas emissions among LULUCF operations. This is followed by carbon sequestration in agriculture and ecosystem restoration, which includes afforestation and replanting.

A significant influence is also played by other terrestrial systems. Soil and subsurface plant organic matter contain the majority of croplands' and grasslands' carbon stores. According to the IPCC, this means that 1.8 - 4.1 Gt CO_2eq/yr may be reduced with the help of better and sustainable crop and animal management as well as soil carbon sequestration in agriculture, which includes soil carbon management in croplands and grasslands, agroforestry, and biochar.

The Sustainable Development Goals (SDGs) and mitigation efforts in the LULUCF sector may complement one other in a number of ways, including improving jobs, local livelihood, biodiversity, and ecological functioning. In addition, the IPCC report acknowledges that certain actions, like planting trees on naturally barren land, may exacerbate the risks that climate change poses to livelihoods, water and food security, and biodiversity. This is particularly the case when such actions are taken without careful consideration for the local environment, which includes the land's current use. The best way to handle such risks is to pursue LULUCF mitigation in response to the requirements and viewpoints of many stakeholders in order to produce results that optimise synergies while minimising trade-offs.

But the primary obstacles to LULUCF operations are the possibility of reversibility and the transience of carbon reserves. The sequestration of CO_2 via vegetation and soil management is susceptible to the effects of climate change and can be undone by human activity, natural disturbances, or a combination of the two. Severe effects like as wildfires, widespread tree death, and a reduction in the capacity of natural terrestrial carbon sinks are already evident and are expected to worsen with each degree of global warming. The IPCC estimates that losses and costs associated with climate change in ecosystems would be significantly decreased by coordinated, swift, and persistent efforts to keep global warming to almost 1.5°C.

Kyoto Protocol

Adopted in Japan at the close of 1997, the Kyoto Protocol (KP) obliges industrialised nations to stabilise greenhouse gas emissions in accordance with the Conventions' tenets. Adopted in 2005, the overarching goal of KP was to lower emissions by an average of 5% between 2008 and 2012 when compared to 1990 levels. Subsequently, the Kyoto Protocol's Doha Amendment was approved, and it would go into effect on January 1, 2013, and last until 2020. Parties agreed to cut GHG emissions in the eight-year period from 2013 to 2020 by at least 18% below 1990 levels; however, the participating nations differed in the first and second periods.Since soils store 81% of the carbon in terrestrial ecosystems, maintaining soil carbon in these ecosystems and encouraging soil carbon growth are crucial components of LULUCF operations (IPCC 2000). Because woody plants have deeper roots, their presence on terrestrial biomasses is directly correlated with the quantity of C in the soil.

Role of Agroforestry within the LULUCF Accounting

The most recent IPCC (Intergovernmental Panel on Climate Change) assessment made multiple references to agroforestry as an agricultural activity. It defines agroforestry (AF) as an integrated system that includes mixed systems and explains how "AFOLU (Agriculture, Forestry and Other Land Use) mitigation measures linked to increases in food production (e.g., agroforestry, intensification of agricultural production, or integrated systems) can increase food availability and access especially at the local level." A variety of AF activities are acknowledged as being carried out by the LULUCF and can help to reduce climate change.The first one relates to paragraph 3 of Article 3 and is about enhancing protection against natural disturbances like fire, pests, and storms. AF practices can lessen the likelihood that areas designated as afforestation and reforestation (AR) will be reclassified as deforestation (D), which is a pool that remains permanent for the duration of the commitment period. Using silvopasture on forest areas and silvoarable farming on arable land is the best way to prevent this change. Because the biomass is transformed into animal products with excellent safety standards and lowers the danger of fire, forest grazing is the most affordable and environmentally friendly method of clearing understory from forests.Simultaneously, regeneration (McEvoy *et al.* 2012) and biodiversity (Rigueiro *et al.* 2006) are encouraged. Furthermore, silvoarable practices, such as planting a combination of annual arable crops in the early years of the plantation when the canopy cover of the trees does not decrease crop productivity, encourage the development of deeper root systems in trees, which improves their anchoring and increases their resistance to natural disturbances like storms, strong winds, flooding, or significant snow events.

The second one relates to paragraph 4 of Article 3 and addresses the administration of various land management practices, such as

1. Not considering the areas incorporated in AR or D, forest management (FM) is a set of methods for stewardship and use of forest land, including plantations and natural forests.
2. Grazing land management (GM), which is a system of practices on land used for livestock production with the goal of controlling the quantity and kind of vegetation and livestock produced,
3. Crop management (CM), which is a system of practices on land on which agricultural crops are grown, as well as on land that is set aside or temporarily not being used for crop production, and
4. Revegetation (RV) is a direct, human-induced activity to raise carbon stocks on sites by the creation of vegetation that covers a minimum area of 0.05 ha. RV is not to be confused with afforestation or reforestation.

Identifying agroforestry within agricultural land and agricultural activity, agroforestry has been recognised as an indicative measure that may be included in the information on LULUCF initiatives presented in accordance with article 10(2)(d) as part of the CM. The EU has also included some preliminary measures related to pasture enhancement and management of grazing land. The EU's proclaimed GM and grassland development policies support agroforestry-related activities such as nitrogen management, productivity enhancement, and the introduction of deeply rooted species.Woody vegetation increases productivity and improves fertiliser management (Rigueiro *et al.*, 2009). Woody plants are often deeply rooted. As was previously indicated in paragraph 3 of Article 3, agroforestry practices, primarily those pertaining to the protection of soil C, as well as increased agricultural use of forest land, increasing harvest wood forests and production in existing forests—all contribute to the promotion of forest activities. Agroforestry also helps to reduce deforestation and increase defences against natural disturbances like fire, pests, and storms. If trees are planted in arable land, for example, and cut branches are connected to composting, this will also support the idea of using wood products and contribute to the bioeconomy.

Table 14.2: EU indicative measures that may be included in the information on LULUCF actions submitted pursuant to article 10(2)(d) (Decision 529/2013/EU) and may relate to agroforestry

Measures related to	**Examples**
Cropland management	Agroforestry
Grazing management and pasture improvement	• Preventing grassland to cropland • conversion to native vegetation increasing productivity • Improving nutrient management • Introducing more appropriate • Species, in particular deep rooted species
Forest activities	• Afforestation and reforestation • Conservation of C in existing forest • Enhancing production in existing • Forests Increasing harvested wood products • Enhancing forest management • Optimize species composition, tending, thinning and soil conservation
Preventing deforestation	
Strengthening protection against natural disturbances such as fire, pest and storms	
Substitution GHG intensive energy feedstock and materials with harvested wood products	

Source: Mosquera *et al.* 2016

15

Management Practices of Multipurpose Trees

Multipurpose trees and shrubs are defined as all woody perennials that are purposefully grown to provide more than one significant contribution to the production and/or service functions of a land-use system that they occupy. In agroforestry systems, different species of trees and shrubs can be planted with many types of crops in a variety of patterns. Thus, it is required to know about the methods of propagation and nursery raising, planting and other silvicultural management practices along with the diverse uses of multipurpose trees and shrubs of tropical and subtropical areas before their selection for forestry and agroforestry activities. It is important to select the most suitable tree species since it is not easy to replace them once they have been planted.

MANGIUM (*Acacia mangium*)

Acacia mangium is a single-stemmed evergreen tree that grows to 25-35 m in height and up to 60 cm in diameter. The bole is usually straight, often fluted near the base, free of branches for up to half its height. Mangium is native to Australia, Indonesia and Papua New Guinea. However, it tolerates varied site conditions and has adaptability to different planting objectives. Mangium shows most vigorous growth on well-drained, fertile soils in high rainfall areas in the humid tropics. It is valued for its rapid growth and has been planted throughout the humid tropics and is a major plantation species in the Asia-Pacific. Provenances from Papua New Guinea consistently show better growth in height and diameter.

Propagation and nursery

Mangium starts to flower after 2 years of planting. Pods are broad, linear and irregularly coiled when ripe. Pods can be collected from the trees when the pods turn very dark-green to light-brown in colour. Seeds are extracted manually after sun-drying. Pods and seeds should not be left to dry in the sun for long. The average number of seeds in one kg of pure seed comes about 125,000. The seed can retain its viability for several years when stored at room temperature in a dark place in a sealed container.

Propagation by seedling is best although direct sowing is also possible. To break dormancy of mangium seeds, hot water treatment is recommended. Seeds are treated before sowing by immersing them in boiling water for 30 seconds then soaking them in cold water for 24 hours. Seed inoculation with appropriate rhizobial strain is recommended before sowing. Mangium seedlings are ready for pricking out in 6-10 days after sowing. Polythene bags are the most common containers used in the tropics for pricking out. Mangium seedlings attain a target size of 25-40 cm height in about 12 weeks. Seedlings are hardened by progressively reducing watering and removing shade in the nursery. If the seedlings have grown larger than the target size in the nursery, they may be lopped.

Planting and tree management

Planting is usually done in pits of 20 cm depth and 10-12 cm diameter. In pure stands, spacing of 2 m × 2 m or 2.5 m × 2.5 m is common. In agroforestry situations, spacing within rows and between rows must consider the effect of shade and root competition on the yield of associated crops. *Acacia mangium* demands full light for good development and its growth in shade is stunted. In the 1st year, the plantation should be protected from livestock, as they browse the trees.

First weeding must be carried out two months after planting and thereafter at regular intervals depending on weed growth. Fertilizers may be applied @ 30-40 g N, 15-20 g P_2O_5 and K_2O per seedling per year from the second year to the fifth year. Mangium needs regular pruning and thinning if the plantation objective is to produce quality sawlog on a 15 to 20 year rotation. These operations in general are not required for pulp wood production on a 6 to 8 years rotation. In agroforestry systems, frequent pruning is required to allow light for the associated crops. In pruning, branches are carefully removed in one or more steps along the bottom trunk up to about 6-7 m height. For sawlog production, the following silvicultural schedule is recommended (Table 15.1).

Table 15.1 Silvicultural management schedule for mangium sawlog regime

Age	Activity	Remarks
Once at 4 months after planting and again 2 months thereafter	General slashing	All climbers within 45 cm radius of each plant are uprooted. All branches at height less than 30 cm from the ground are removed.
12 months after planting	General slashing and first pruning	All branches up to 1.5-2.0 m height are removed.
2 years after planting	First thinning and high pruning	600 trees/ha are maintained and excess plants are removed which may be used as poles. Branches up to 6 m height of the 200 selected trees (to be retained till end) are pruned.
4-5 years after planting	Second thinning	Another 200 trees/ha are harvested retaining 400 trees/ha.
8-9 years after planting	Final thinning	Another 200 trees/ha are harvested and only 200 trees/ha are allowed for final harvest at 15-20 years after planting.

Protection

Although root rots disease caused by *Ganoderma* sp., *Phellinus* sp. and *Rigidoporus lignosus* are major problems in mangium stands, there are no specific control recommendations against these fungi. Signs of the disease are evident on the roots after the tree has fallen or upon excavation. The usual method of controlling root rot caused by fungi that spread by root contact is to remove and destroy all diseased roots and woody debris. Chemical protection against pink disease (*Corticium salmonicolor*), especially in endemic areas, can be achieved by using copper fungicides. The best way to prevent pink disease, however, is to plant tolerant varieties.

Progressive decay of the heartwood (heart rot) is another malady afflicting mangium trees. Normally, fungi that decay heartwood do not attack sapwood. Thus, such trees continue to grow to maturity and may outwardly appear healthy and vigorous. However, since heart rot is progressive, there will be considerable decay to the heartwood at the end of the rotation. A variety of basidiomycete fungi have been associated with this malady. At present there are no control measures against mangium heart rot. The best way is to avoid injury to trees and wound dressing.

Although about 30 insect species are reported to be pests of mangium, only a few such as root feeders, branch and stem borers and the red coffee borer are considered economically important. Root feeders (*Sternocera aequisignata*) can be controlled by carbofuran and chlorpyrifos application to the soil or seedbeds. For controlling termites, water solution of chlordane 1% or dieldrin 0.5% is applied around the affected area. To prevent branch and twig borer (*Sinoxy lonanale*) occurrence, all broken branches in which breeding takes place are removed and burnt. The only effective method to control red coffee borer (*Zeuzera coffeae*) damages is to inject insecticide into the holes where larvae push out their frass.

Uses

Timber is used for a variety of purposes like wood-based panels, pulp and paper industry, etc. Mangium wood gives attractive furniture, cabinets, moulds and door/window components. However, the presence of flutes and incidence of rots and termite attack will detract both the quality and quantity of sawn timber from mangium logs. Therefore, mangium has greater potential as a component of composite wood products such as veneer and plywood, laminated veneer lumber, fibreboards, etc. and for chemical uses such as pulp, paper and tannin production, besides fuel-wood. The pulp is readily bleached to high brightness levels and is excellent for papermaking. With a calorific value of 4800 kcal/kg, *Acacia mangium* provides good quality charcoal and is suitable for the manufacture of charcoal briquettes and artificial carbon.

Mangium bark has high tannin content (18-39%), which can be commercially exploited. Mangium trees form a symbiosis with soil bacteria of the genus *Rhizobium*, leading to root nodules, in which the bacteria transform free nitrogen into organic and inorganic compounds containing nitrogen.

The species is suitable for many agroforestry systems like alley cropping, agrisilvicultural, silvipastoral, hortisilvicultural, etc. due to its quick growth and low branching habit. It has potential in some intercropping combinations, such as with maize, pulses, groundnut, turmeric, pineapple, arrowroot, etc.

SISSOO (*Dalbergia sissoo*)

Dalbergia sissoo is a medium to large-sized deciduous tree, growing up to 30 m in height and 80 cm dbh (diameter at breast height, i.e., 1.37 m) under favourable conditions. The bole is often crooked and branchless for up to 8 metres, occasionally for as much as 20 m. It develops a long taproot with numerous lateral ramifying roots from an early age. It is native to Afghanistan, Bangladesh, Bhutan, India, Malaysia and Pakistan, and cultivated as a forest tree in south Asia and tropical Africa. Abundant moisture and adequate sunlight are required for desired growth of this plant.

Propagation and nursery

Propagation of *Dalbergia sissoo* takes place most commonly through seed and vegetatively through suckers arising from the root system. There are approximately 45,000-55,000 seeds/kg. Sissoo rarely regenerates under the parent canopy. Ripe pods may be collected manually by climbing trees and picking the fruits or by shaking the branches and picking the pods from the ground. Usually, the seeds are not extracted from the pods, but the pods are broken into 1-seeded pieces. Seeds have no dormancy and remain viable for only a few months. Pre-treatment of seeds is not necessary, but soaking in water for 24-48 hours accelerates germination up to 80%. Germination of fresh seed takes 1-3 weeks. Shading is recommended during the hottest hours of the day during the germination period. Propagation by root suckers is done best by cutting stems just below the soil surface. While it is difficult to propagate using stem and branch cuttings without hormone treatments, exogenous application of auxins has been found to improve survival and growth rates.

Planting and tree management

Dalbergia sissoo plantations are established in block or strip plantations at 1.8 m × 1.8 m to 4 m × 4 m. Closer spacing is used for straight timber of good quality. When the canopy closes, at about 6 years, 30-40% of the stems are thinned to selectively remove suppressed, diseased and badly formed trees. Thinning is recommended every 10 years where the rotation is 30-60 years. There is evidence that the stumps begin to lose vigour after 2 or 3 rotations when managed as a coppice crop. It coppices vigorously up to about 20 years of age.

Uses

After teak, *Dalbergia sissoo* is the most important cultivated timber in India. The heartwood is very hard and close grained with a specific gravity of 0.62-0.82. It seasons well and does not warp or split. It is extremely durable and is one of the timbers least susceptible to dry-wood termites in India. Wood offers resistance to sawing and cutting but is excellent for turnery, takes a good polish and finishes to a smooth surface. It is used for high-quality furniture, cabinets, decorative veneer, marine and aircraft grade plywood, ornamental turnery, carving, engraving, tool handles and sporting goods.

Dalbergia sissoo is reported to have pesticidal properties. Aqueous extracts from the leaves, stems and roots inhibit the reproduction, growth and development of the insect pest *Utethesia pulchella.* Methanol extract from the roots has insecticidal properties, especially against *Diacrisia obliqua, Spodoptera litura* and *Argina cubrania.* Oil obtained from the sissoo seeds is used to cure skin diseases. The powdered wood, applied externally as a paste, is reportedly used to treat leprosy and skin diseases.

Young branches and foliage form an excellent fodder with a dry-matter content of 32.5%. The foliage has normally been used as emergency feed when other fodder sources fail. A useful source of honey but the flowers are only lightly attached to the flower branch and fall easily. The bees are therefore not able to take full advantage of the large number of flowers. The species is fast growing, hence suitable for firewood. Sapwood and heartwood have calorific values of 4900 and 5200 kcal/kg, respectively.

The species is suitable for many agroforestry systems like agrisilvicultural, silvipastoral, hortisilvicultural systems. It may be planted as one component of a multitiered homegarden system, where it contributes several products. It is also used as a windbreak in mango, coffee and tea plantations. These shade-tolerant crops also benefit from improved soil fertility under *Dalbergia sissoo*. The tree nodulates and thus improves soil fertility.

KHAIRA (*Acacia catechu*)

Acacia catechu is a small or medium-sized, deciduous tree up to 15 m tall with a light feathery crown and dark brown, glabrous, slender, thorny, shining branchlets, usually crooked. It is especially common in the drier regions on sandy soils of riverbanks and watersheds. It is native to India, Myanmar, Nepal, Pakistan and Thailand. In India, three varieties, namely var. catechu, var. catechuoides and var. sundra are recognized.

Propagation and nursery

Acacia catechu can be raised from direct sowing, coppice, planting out nursery-raised seedlings or by stump planting. The seeds adhere to the light pod valves after the pods dehisce and are often blown to a considerable distance from the trees. Seed fall takes place in the month of January and February. The seeds can be collected by lopping small pod bearing branches in January and spreading them in the sun for a few days. The seeds are separated from pods by thrashing. Seed viability is lost within 1 year at room temperature at 11-15% moisture content. There are 15,000-40,000 seeds/kg. Seeds are sown in the nursery in the month of April or May. It is recommended, but not necessary, to put the seeds in boiling water and then leave for 24 hours them to cool. Germination commences from about the 4th day after sowing and its completion may linger on up to 4-5 weeks.

Irrigation is essential in the nursery till the outbreak of monsoon. The seedlings require daily irrigation with a precaution that the water does not accumulate at the roots of the plants. The nursery should be kept free of weeds as these are liable to kill seedlings by suppression. Khair seedlings are comparatively resistant to damping off disease in the nurseries, however, water logging may sometimes predispose the seedlings to damping off in the early stage of development.

Direct sowing gives good results and it is very easy. Under optimum conditions, khair can also be propagated by stumps. The stumps should be made from seedlings about 15 months old raised in nurseries from seed sown in April of the previous year. Cuttings should be made from well-developed seedlings. The root and shoot should be 23 to 31 cm and 2.5 to 5.0 cm, respectively. The best size of stumps at the root collar is 10 to 15 mm in diameter. Planting of stumps should be done soon after the break of rains.

Planting and tree management

Acacia catechu plantations are established in block or strip plantations at 1.5 m × 1.5 m to 3 m × 3 m. Since early growth is slow, weeding is essential, especially when the plants are still young. Protection from grazing by animals is required. Early thinning is very important for the proper development of the crop. The first thinning is done at the age of 3 years. Subsequently thinnings are done at the ages of 5, 10, 15, 20 and 25 years. In coppice crops, it becomes necessary to reduce the number of the several coppice shoots sprouting from a single stump to one or two within 3-5 years.

Rotation regimes depend upon the intended use, for fuelwood production, felling is usually at 10-15 years of age. Trunks with a diameter of 30-35 cm are considered the most economic. For extracting the tanning agent cutch, this size is achieved only after 30 years.

Protection

The fungus *Ganoderma lucidum* causes serious mortality due to root rot in reforested stands. Khair is susceptible to the attack of pathogen at all ages. Young plants are killed soon after infection while the mature trees die when most of the roots become affected. The disease can effectively be checked by extraction of old stumps and cleaning of debris from the site, digging of isolation trenches in young plantations, planting of resistant species like *Bombax ceiba* and *Ailanthus excels* and mixed cropping (50:50) with resistant species.

Fomes badius causes heart rot in khair and is common in all khair forests, both natural and planted. The fungus causes decay in the heart wood only. Though sapwood remains healthy and free from infection the heart rot increased with age of tree and mature trees become unfit for extraction of cutch and katha due to its complete disintegration. The disease can be managed to some extent by avoiding injuries to the trees.

Uses

A substance called catechu (chemically, catechutannic acid), which is marketed as a solid extract, can be isolated from the heartwood khair. Depending on the way of processing, several products can be obtained from crude catechu. Dark catechu or cutch, which is mainly obtained as a by-product of the catechu industry is used for dyeing cotton and silk and preserving of fishing nets, sailing ropes and mail bags. About 63,000 tonnes of khair wood in India is annually consumed for manufacture of cutch and catechu.

Khersal, a crystalline form of cutch sometimes found deposited in cavities of the wood is used for the treatment of coughs and sore throat. The bark is said to be effective against dysentery, diarrhoea and in healing of wounds. The seeds have been reported to have an antibacterial action.

Though khair is chiefly used as a source of cutch and catechu, it is also a useful timber. The timber is comparatively heavy with a density of 880-1000 kg/cubic m at 15% moisture content. The wood is also very strong, durable and resistant to white ants. Timber is used for house posts, agricultural implements and wheels. The wood is excellent firewood. The calorific value of sapwood and heartwood is estimated to be 5142 and 5244 kcal/kg, respectively. Dry wood on destruction gives 38.1% charcoal of very good quality.

It is considered to be a good fodder tree and is extensively lopped to feed goats and at times cattle also. For leaf fodder, finger-thick branches are lopped usually before main leaf fall occurs. The leaves contain 13.0-18.7% crude protein.

EUCALYPTUS (*Eucalyptus* spp.)

Eucalyptus is a genus of over seven hundred species of flowering trees. Most species of Eucalyptus are native to Australia. A few species are native to other countries. Eucalyptus have been grown in plantations in many countries because they are very fast growing and have valuable timber and can be used for pulpwood, for honey production or essential oils. In India Eucalyptus is second most widely planted species after teak. Mainly five species of eucalypts are grown in different agroclimatic zones of the country. While *Eucalyptus camaldulensis*, *E. globulus* and *E. grandis* have restricted adaptability, *E. tereticornis*has been planted in the country both in the forests and outside in the agricultural fields on a large scale, except in the north-eastern states. Performance of species varies considerably with the site and climatic conditions. The suitability of the species in different agroforestry zones has been given in the following table. Trees are tall up to 50 m high. Bole is straight and clean with whitish mottled bark.

Table 15.2 Suitability of eucalypts species in different agroclimatic zones of India

Species	Climatic zone	State
Eucalyptus camaldulensis	Tropical and sub-tropical	Himachal Pradesh, Uttrakhand
Eucalyptus globulus	Mild temperate to cool tropical climate	Kerala, Karnataka
Eucalyptus grandis	Lower sub-tropical areas	Kerala
Eucalyptus teriticornis	Dry tropical to moist tropical areas	Punjab, Haryana, Uttar Pradesh, Rajasthan, Gujarat, Madhya Pradesh, Maharashtra, Bihar, Odisha, West Bengal, Andhra Pradesh, Telengana, Karnataka and Tamil Nadu
Eucalyptus citriodora	Mild temperate areas	Hilly areas

Propagation and nursery

Natural regeneration is very poor due to destruction of seeds by ants. *Eucalyptus* leaves also contain germination inhibitor. Seeds should be collected from mature trees of 10 years or more old before the capsules dehisce. In a year seeds may be collected twice, February-March and October-November. Capsules should be dried in shade for one day and then gently shaken to clean the shells. Seeds are stored in tins at cool dry places. Seeds are viable up to three months. About 3,67,400 seeds weigh one kg.

Pre-treatment of seeds is not required. The optimum temperature for germination is 32° C, but a wide range is tolerated. Seed should be sown in raised nursery bed in the month of October-November or February-March. Nursery soil should be sterilized with Aldrex or BHC against termites. Seed should be sown 20 g/m^2 of bed in lines 10 cm apart and 2.5 mm deep. The germination rate is generally high and can reach almost 100%. The seeds germinate in 4-15 days. 60-75 seedlings are obtained from one gram of seeds. Seedling can also be raised in seed trays. Trays should be kept moist with a fine spray of water until germination begins. The seedlings should be pricked out when they have two pairs of leaves into poly-bags of size 22 cm × 10 cm or root trainers. Polythene bags initially should be placed in shade for few days and then shifted in the open in sunken beds. The seedlings are hand watered for some time and later the beds may be irrigated.Three-month-old containerized stock is recommended for planting. *E. camaldulensis* is also suited to mass vegetative propagation. Cuttings from juvenile shoots root readily. Eucalyptus is a very good coppicer.

Planting and tree management

Though planting is done at the break of monsoon, eucalyptus can be planted throughout the year, provided irrigation is available. Planting is usually done in 20 cm × 20 cm × 20 cm pits at 3 m × 3 m spacing. Poor competition ability with weeds and the development of an open crown necessitates frequent weeding, until the canopy closes 3-5 years after planting. Depending on site conditions, *E. grandis* and *E. tereticornis* may respond to mineral fertilization with accelerated growth. Fertilizers may be applied at the rate of 30 g N, 30 g P_2O_5 and 15 K_2O per sapling per year during the second, third and fourth years. For production of pulpwood and fuelwood, 6-10 year rotations are used without thinning. A thinning of 50% plants at 5 years provides posts, poles, fuelwood and pulpwood, leaving the better trees for the production of other products, such as sawn timber after 15 years.

Protection

Polyphagous insects seem to attack the nursery stock. Quinalphos or malathion 0.05% is recommended against these insects. Drenching the containers with chlorpyrifos is a preventive measure against termite attack in plantations. Quinalphos 0.2% solution is recommended to control stem borer attack. *Cylindrocladium* leaf blight and pink diseases are common in eucalyptus trees. To control *Cylindrocladium* leaf blight, the nursery is drenched with carbendazim 0.05%. Bordeaux paste is recommended against pink disease. Using disease tolerant clones is a sure means of preventing the incidence of both diseases.

Uses

Eucalyptus grandis wood is pink to pale reddish brown in colour. It has good bending properties. It is used for house construction, floors, furniture, crates and veneers. *E. tereticornis* produces dark red wood. It is hard, strong, tough, heavy, very durable and resistant to termite attack. It is used for a wide range of construction applications. The timber requires special care in sawing and drying because of high incidence of spiral grain and has a tendency to split. *Eucalyptus globules* is good for papermaking and is widely used for pulp. The density of the wood is 900-980 kg/m^3 at 12% moisture content. The firewood is suitable for industrial use in brick kilns but is not preferred for domestic use because it is too smoky and burns too fast. However, it makes good-quality charcoal.

E. camaldulensis is a major source of honey, producing heavy yields of nectar in good seasons. The honey is light gold and of reasonable density with a distinctive flavour. The oils extracted from eucalyptus leaves are used as an inhalant with steam and other preparations for relief of colds and influenza symptoms. Because of its refreshing odour and its efficiency in killing bacteria, the oil is also used as an antiseptic. It helps to treat lung infections, gastrointestinal ulcers and angina. The leaves are also used for the extraction of commercially important eucalyptus oil known as eucalyptol.

With its light crown, *Eucalyptus* is well suited for agroforestry systems. It is a valuable tree for windbreaks and shelterbelts.

CASUARINA (*Casuarina equisetifolia*)

Casuarina equisetifolia is a large evergreen tree with a straight bole and numerous, long, slender, drooping, jointed, leafless branchlets arising from rough woody branches. The jointed branchlets, which are partly deciduous, are green and perform the functions of leaves. Leaves are minute scale like and arranged in the form of a cup at the joints of the branchlets. Bark is brown, rough, fibrous and exfoliating in longitudinal strips. Although the tree is evergreen, it usually sheds a large number of twigs throughout the year. Casuarina is native to Australia, the Indian subcontinent, south east Asia, and islands of the western Pacific Ocean. The tree is widely planted throughout the tropics, especially for being able to provide timber whilst growing on poor sandy soils near the coast and thus also providing shelter and protection for the soil. The tree is commonly grown as an ornamental, especially near the sea where it can also offer shelter from the wind. In cultivation, *Casuarina equisetifolia* hybridizes with *Casuarina glauca* and *Casuarina junghuhniana.*

Propagation and nursery

Propagation is by seeds or through vegetative means. The fruit is a woody, oval structure superficially resembling a conifer cone, made up of numerous carpels, each containing a single seed. Seed viability is often low. For seedling production, about half kg seeds are sown on raised nursery beds of 10 m × 1 m. This will produce about 10,000 good quality seedlings. In sandy soils, farmyard manure is mixed with the topsoil. After sowing the seeds, a thin layer of sand is sprinkled to cover the seeds. Usually sowing is done in November-December. The nursery beds are kept moistened and proper shading is provided to facilitate rapid seed germination. Germination takes about 10 days and seedlings attain a height of 10-15 cm in 6 weeks. They are then pricked out into polythene bags or transplanted into beds of size 1 m × 10 m in January-February.

Vegetative propagation is done by branch cuttings, stump cuttings and layering. For vegetative propagation by rooting of branch cuttings, 5-7 cm long cladode cuttings are treated with rooting hormones. The hormone-treated cladodes are transferred to pre-soaked vermiculite and kept in a mist chamber. About 100% rooting is obtained within 15 days. The rooted cuttings are then transferred to a mixture of sand, soil and farm yard manure (2:1:1) for hardening. After 15 days, the hardened propagules can be transferred to the field. Air-layering is sometimes practised but is too costly for large-scale operations.

Planting and tree management

Casuarina has a wide environmental adaptability and hence occupies sites ranging from arid regions to coastal zones. Root nodules containing the actinorhizal symbiont *Frankia* enable *Casuarina equisetifolia* for biological nitrogen fixation. Therefore, it thrives best on sandy soils low in nitrogen and has the potential to improve the nitrogen capital of impoverished sites. It requires huge sunlight and a well-drained alkaline to neutral soil for proper growth. Plants can tolerate inundation by sea water at extremely high tides. Young plants are susceptible to drought until their roots reach the groundwater table, which may take up to 2-3 years after planting, but then they become very drought resistant.

Site preparation includes ploughing the land 2-3 times and making 30 cm × 30 cm × 30 cm pits before the onset of monsoon. The pits are filled with farm yard manure and topsoil. Planting is done immediately after the first rains. Casuarina can be planted in block, row and line or strips. Spacing varies depending on the objective and the end product. Usually, a spacing of 75 cm × 75 cm is adopted. Young trees are susceptible to competition from weeds, especially grasses. Thus, one or two weedings may be required immediately after the rains. Timely thinning is essential as *Casuarina* species trees demand light. When the trees are about 3 m in height, the lateral branches are pruned to a height of about 2 m. Pruning is usually done at the end of the second year or after the beginning of the third year. In plantations established at close spacing, one thinning in the second year or third year depending on tree growth is desirable, where 25-50% of the trees are felled. In mixed species systems such as agroforestry, spacing and thinning practices are mainly dependent on the cropping systems and the nature of the associated species. If the associated crops are shade intolerant generally wider spacing and/or intensive thinning are recommended. Fertilizers may be applied at the rate of 20-25 g N, 15-20 g P_2O_5 and 15-20 g K_2O per seedling per year from the second year to the fifth year. The rotation period ranges from 4 - 5 years for fuelwood and 10 - 15 years for poles.

Protection

The common diseases in *Casuarina* nurseries are damping off, seedling blight, stem canker and seedling rot nurseries. Emisan 0.01% is effective against these diseases. Stem-wilt or bark blister disease caused by *Trichosporiumvesiculosum*is a serious disease in the plantations. The disease affects trees of 3-4 years and causes mortality up to 80%. Maintaining a soil *p*H of 6.5 to 6.8 and treating the plantation with fungicidal sprays can control this disease. Other diseases include stem canker and dieback caused by *Phomopsis casuarinae*, pink disease caused by *Corticium salmonicolor*, root rot disease caused by *Ganoderma lucidum* and heart rot caused by *Polyporus glomeratus*, *Fomes fastuosus* and *Fomes senex*. Stem canker and dieback can be controlled by carbendazim @ 0.01%. There are no serious insect pest infestations inflicting extensive economic losses to casuarina trees.

Uses

Yield of high-density fuelwood plantations varies from 10-20 tonnes per ha per year on 7-10 years rotations. Higher yields are reported from irrigated and fertilized sites. The highly regarded wood ignites readily even when green, and ashes retain heat for long periods. It has been called 'the best firewood in the world' and also produces high-quality charcoal. Calorific value of the wood is 5,000 kcal/kg and that of the charcoal exceeds 7,000 kcal/kg.

Wood is hard to very hard and strong. Heartwood is also resistant to termites. The wood is used for house posts, rafters, electric poles, tool handles, oars and wagon. The wood is used to produce paper pulp, and as a raw material for rayon fibres.

Casuarina plants are reported to have some medicinal values. Root extracts and barks are used for the treatment of dysentery, diarrhoea and stomach-ache. The powdered bark is used for treating pimples on the face. The cambium layer beneath the bark is squeezed and used to sedate a mentally ill or aggressive patient.

Since it is salt tolerant and grows in sand, the plant is used to control erosion along coastlines, estuaries, riverbanks and waterways. It is a nitrogen fixing species and therefore, often planted for reclaiming and improving the land. In the tropical cyclone-prone areas, casuarina is used as a protective planting due to its tolerance to strong winds. The abundance of highly branched twigs absorbs wind energy well. In areas with hot, dry winds the tree protects crops and animal herds. It is remarkably suited for boundary planting as it does not intercept much of the incoming solar radiation and yields substantial quantities of green leaf manure on lopping. With high productivity and properties that enhance soil fertility, it shows promise as an agroforestry species for arid and semi-arid areas. The tree is also grown as an ornamental, especially near the sea where it can also offer shelter from the wind.

TEAK (*Tectona grandis*)

Teak (*Tectona grandis*) is a large tropical tree reaching over 30 metres in height with an open crown that has many small branches. The bole, which can be unbranched for up to 15 m, is up to 1 metre in diameter. The species is indigenous to India and the Southeast Asian region. But it is naturalised and cultivated in many countries in Africa and the Caribbean. In India teak is distributed naturally in the peninsular region. It prefers a warm moist tropical climate with mean annual precipitation of 1100 -2000 mm and a well-drained fertile soil. Being a strong light demander, it does not tolerate overcrowding and does not withstand waterlogging.

Propagation and nursery

Generally, stumps or seedlings are used as planting material. For stump preparation, the seedlings have to be maintained in the nursery for about one year.

Seeds should be collected from vigorously growing middle-aged trees characterized by straight boles, desirable branching habit, good form and less fluting. Freshly fallen intact fruits with inflated calyx from such trees are collected during December-February. Fruit is fleshy and bears 1-4 seeds which are enclosed in a stony covering. After cleaning and drying the seeds should be safely stored in gunny bags or sealed containers. Seeds of diameter greater than 9 mm are usually collected. For convenience in storage and transport, the bladder like calyx of the fruit is removed. This is done by half-filling a bag with the fruits and vigorously rubbing and shaking it or by beating with sticks, after which the remains of the calyces are separated from the nuts by winnowing.

Due to hard seed coat, germination of one-year old seeds is better than that of fresh seeds. Thus, it is necessary to break dormancy of the seed. Seed germination in the untreated seeds is totally absent or very in-significant. Several methods are in practice to break the dormancy of teak seeds.

- Seeds are spread over raised platform in August. The seeds get soaked in rains and then with sun they get dried thus getting natural weathering.
- Freshly collected seeds are put in gunny bags which are then submerged under flowing water for four days. The gunny bags are taken out and spread over in strong sun for four days. It is again submerged in water for 3-4 days following drying. It is repeated for 3-4 times until endocarp and mesocarp gets easily cracked.
- Seeds are put in alternate layers of seeds and straw and daily watered for seven days then dried and stored till the time of sowing.

- Soaking the seed in concentrated H_2SO_4 for 20 minutes and thorough washing in running water hasten germination.
- Soaking the seeds in the mixture of cowdung and water enhance germination.
- Stored seed germinates best if soaked for 24 - 48 hours in warm water, prior to sowing, changing the water frequently.
- Another method of encouraging germination is to char (or half burn) the fruits by covering them with a thin layer of grass and lighting it.
- The freshly collected teak fruits are spread on the ground in a 5 cm layer immediately after collection. After about five weeks the termites remove the exocarp and subsequent germination after alternate wetting and drying is found to be better.

Raised seedbeds of about 30 cm high, 10 m length and 1 m width are prepared. A mixture of sand, soil and farmyard manure forms the top layer of the seedbed. Sowing is done after the bed is watered. Usually the sowing is done by broadcast method or dibbling in April-May. After sowing, the seeds are pressed into the beds and a thin layer of soil is sprinkled to cover the seeds. The beds are also mulched with green leaves to reduce evaporation losses. The bed is then dusted with carbaryl 10% to prevent insect attack. The germination rate is low, usually 60-70%. Germination starts after 10 days but may take 2-3 months. The seedlings can be transplanted to polythene bags after 3-4 months or it can be maintained in the nursery beds for 10 to 12 months for preparation of stumps. Seedlings should be given a little shade. One year old seedlings of 1-2 cm at the thickest portion below the collar are uprooted and all the leaves and secondary roots are removed. Stumps with 15-20 cm of tap root and 2-3 cm of stem are prepared with sharp knives. Stump planting is generally preferred and it is easy for transport. For seedling plantation, young seedlings are shifted to polythene bags containing soil mixture and maintained in the nursery for 3 to 6 months.

Planting and tree management

With the pre-monsoon showers, stump planting is done in crowbar holes. The site must be cleared of stubble or other competing vegetation. If seedlings are used, then optimal time of planting is after the onset of southwest monsoon. They are usually planted in pits of size 30 cm × 30 cm × 30 cm. Spacing recommended for monospecific woodlot is 2 m × 2 m. However, if intercrops are to be raised, then row-to-row distance can be altered. For one or two row strip plantings at farm boundaries, a closer plant-to-plant spacing of 1 m is recommended. Teak is very susceptible to weed competition and frequent weedings may be required during the first two years. Fertilizers may be applied @ 30-40 g N, 15-20 g P_2O_5 and 15-20 g K_2O per plant per year from the second year to the fifth year and thereafter once in three to four years for 10-12 years. In agroforestry situations, if the intercrops are fertilized, the quantities of chemical fertilizers applied to teak can be proportionately reduced or even skipped. Providing life-saving irrigation during the summer season favours teak growth.

For a fifty-year rotation, monospecific teak plantation with initial spacing of 2 m × 2 m, thinning may be carried out at 4, 8, 12, 18, 26 and 36 years after planting. Thinning in short rotation (25-30 years) high input plantations can be at 4, 8, 12 and 16 years. The thumb rule governing thinning is that trees should not be allowed to compete with each other for site resources, as intense competition may depress teak growth. Therefore, considering the site characteristics, tree growth rate and merchantability of the thinned-out materials, a flexible thinning schedule can be adopted.

Teak can be integrated to various agroforestry systems like, agrisilvicultural (teak + casuarina with agricultural crops maize, cotton, turmeric, tomato and chilli), agrisilvihorticultural (teak + guava/mango with agricultural crops pineapple, turmeric, vegetables, maize and cotton), silvihorticultural (teak + guava/annona) and silvipastoral (teak as tree component and napier and guinea as pasture components). Fruit/spice/medicinal trees also can be successfully intercropped with teak throughout its growth. Inclusion of nitrogen fixing trees such as *Gliricidia* or *Leucaena* either in alternate rows or every third row improves teak growth. However, the nitrogen fixing trees should be pruned regularly so that these do not compete with teak for light.

Protection

Hyblaea puera and *Eutectona machaeralis* are considered to be the major pests in teak. These insects are known to occur on seedlings in nurseries and also in grown up trees in plantations. *H. puera* feeds on tender foliage during the early part of the growth season and *E. machaeralis* feeds on older foliage towards the end of the season. These pests can be controlled by spraying of the foliage with endosulfan 0.05-0.075% or quinalphos 25 EC 0.05% spray.

Vascular wilt disease (*Burkholderia solanacearum*) is noticed in nursery and young plantations. As preventive measures against this disease, proper drainage should be ensured and root injury is to be avoided. Leaf spot disease (*Phomopsis* sp. and *Colletotrichum gloeosporioides*) in nursery and young plantations can be controlled by mancozeb 0.05% or carbendazim 0.05% application. Against pink disease (*Corticium salmonicolor*) in young plants, Bordeaux paste is applied.

Uses

Teak is one of the most important timbers in the world. A rare combination of superior physical and mechanical properties makes it a paragon of timber. Teak is recognized as the best timber for the manufacture of door, window frames and shutters, wagon and carriage, furniture, cabinets, ships, agricultural implements, decorative flooring and wall panelling because of its moderate weight, appropriate strength, dimensional stability and durability, easy workability and finishing qualities and most appealing grain, texture, colour and figure.

Various parts of the tree, including the wood are credited with medicinal properties. A wood tar paste is made from the powdered wood by putting it into hot water. It promotes digestion, relieves headaches and toothaches and reduces inflammations of the skin. Oil extracted from the seeds promotes hair growth and the oil extracted from the roots is used to treat eczema, ringworms and inflammation. Flowers are considered useful against a number of diseases such as biliousness, bronchitis and urinary discharges. Leaves are used in indigenous medicine and their extract indicates complete inhibition of pathogenic bacteria *Mycobacterium tuberculosis*, the causative agent of tuberculosis. The leaves also contain yellow and red dyes, which have been recommended for dyeing of silk, wool and cotton. The leaves are occasionally used as plates for dining purposes, for making cheap umbrellas and for thatching temporary huts in some places. The bark is regarded as an astringent and considered useful in bronchitis. Activated charcoal can be prepared from its saw dust.

It is reported that the fine dust produced in machining operations may cause irritation of the skin or bronchial asthma and rhinitis after inhalation. Thus, a dust extractor fan is recommended. The substance responsible for the allergic reaction is probably the naphthoquinone desoxylapachol.

NEEM (*Azadirachta indica*)

Azadirachta indica, commonly known as neem, is a fast-growing medium to large sized tree that can reach a height of 15-20 metres. The branches are wide and spreading. The fairly dense crown is roundish and may reach a diameter of 20-25 metres. The trunk is relatively short, straight and may reach a girth of 1.5-3.5 m. It is native to the Indian subcontinent. It is typically grown in tropical and semi-tropical regions. The neem tree is noted for its drought resistance. Neem is one of a very few shade-giving trees that thrive in drought-prone areas. The bark is hard fissured or scaly and whitish-gray to reddish-brown. The sapwood is greyish-white and the heartwood reddish.

Propagation and nursery

Neem regenerates naturally by seeds, coppice and root suckers. Fruit ripening coincides with rainy season. The fallen fruits germinate within a fortnight giving a thicket of seedlings under mature tree. It is easily raised in the nursery and planted out as potted plants or seedlings. Seeds are collected in June-July. Seed do not require any scarification or pre-sowing treatment, however, de-pulping and cleaning the seed considerably improve germination percentage. Germination is commonly 75-85% for cleaned seeds sown within a week of collection. Seeds do not retain their viability for more than 2 months and should be sown within 2 weeks of collection. Germination begins from 4 to 10 days after sowing and is usually complete within 3 to 5 weeks.

Seeds may be sown either in June-July in seedling containers or in open beds at a spacing of 2.5 cm in lines 15 cm apart. Depth of sowing is to 2.5 cm. The seeds should be lightly covered with soil and watered. Nursery soils should be loose and freely drained. Shade is not necessary but may be useful under extremely hot conditions. Seedlings grown in open beds should be thinned to a 15 cm × 15 cm spacing when they are approximately 2 months old. Pricking out of seedlings or wilding is done at three leaves stage into polythene bags. Seed can also be sown directly in poly bags.

Planting and tree management

Seedling growth in the nursery is moderate and is greatly hindered by competition. Plantations can be established at the beginning of the rainy season by direct sowing, by planting container-grown seedlings, or by using stump plants. In India, direct sowing has been found to be the most economical and successful method provided that seeds are sown as soon as possible after collection. Seedlings raised in the nursery can be planted when they reach 7 to 10 cm in height and have a taproot that is usually about 15 cm long. On very dry sites, larger seedlings of at least 45 cm in height are often preferred. Stumps from 1 to 2-year-old seedlings prepared by cutting back the stem to 2.5 cm and the taproot to 25 cm are commonly planted in India. Planting should be done in July-August in the pits of 1 m × 1 m × 1 m dimension. Cleaning around the pits should be done at the time of planting. Neem also responds well to chemical and organic fertilizers.

Neem can also be propagated vegetatively by air-layering, root and shoot cuttings, grafting and tissue culture. Successful air-layering can be achieved by treatment with 0.1% indole-butyric acid or naphthalene acetic acid.

Neem is a light-demanding species, but it tolerates moderate shade during the early stages of growth. It is intolerant of grass competition during the seedling and sapling stages, requiring weeding to ensure its survival, particularly in dry areas. It can withstand drought better than excessive rainfall. It is intercropped with a variety of field crops, usually cotton, sesame, and pigeon pea, greengram for up to 4 or 5 years.

Protection

Neem has few serious pests, but several scale insects have been reported to infest it. *Aonidiella orientalis* and *Pulvinaria maxima* feed on sap of young branches and young stems. In India, *Psuedocercos porasubsesessilis* is the most common fungus attacking the leaves. The bacterium *Pseudomonas azadirachtae* may damage leaves.

Uses

Almost every part of neem tree is useful from its roots, trunk, bark, leaves, flowers, fruits and seeds in day-to-day life. Neem timber grain is rough and does not polish well. The wood is used to make wardrobes, bookcases and closets, as well as packing cases because its insect repellent quality helps to protect the contents from insect damage. The main stem of the tree is also widely used to make posts for construction or fencing because the wood is termite resistant. Charcoal made from neem wood is of excellent quality and the wood has long been used as firewood. Its oil is burned in lamps throughout India.

Azadirachtin is the principal active compound of neem plant. Extracts can be made from leaves and other tissues, but the seeds contain the highest concentrations of the compound. In India, many neem-based pesticides are available. These act as insect repellents, inhibiting feeding, and disrupting insect growth, metamorphosis and reproduction. Azadirachtin affects egg production and hatching rates. It can inhibit moulting, preventing larvae from developing into pupae. Many foliage-feeding species avoid plants treated with neem compounds or cease eating after ingesting the neem. Azadirachtin as an antifeedant is effective on about 100 insect species. Thus, the extracts work especially well to protect plants from defoliation without affecting beneficial pollinating insects like honeybees. A traditional agricultural practice involves the production of 'neem tea'. The seeds are dried, crushed and soaked in water overnight to produce a liquid pesticide that can be applied directly to crops. Crushed seed kernels are also used as a dry pesticide application, especially to control stem borers on young plants. These homemade remedies are often very effective in repelling pests or acting on insects as a feeding deterrent. The active compounds break down quickly, so an application of 'neem tea' generally provides protection for only about one week. Neem extracts may have toxic effects on fish and other aquatic wildlife and on some beneficial insects. Therefore, care should be taken that any unused extracts are disposed of by exposing them to heat or sunlight to break down the active compounds.

Neem has proved effective against certain fungi that infect humans. Various parts of neem tree have antiseptic, diuretic and purgative actions, and are also used to treat boils, pimples, eye diseases, hepatitis, leprosy, rheumatism, scrofula, ringworm and ulcers. Leaf teas are used to treat malaria. People use the twigs as toothbrushes, and dentists find twigs effective in preventing periodontal disease. Neem oil has been used traditionally as a typical treatment for skin symptoms in both humans and livestock. The young twigs and flowers are occasionally consumed as vegetables.

The large crown of neem tree makes it an effective shade tree, planted widely as an avenue tree in towns and villages and along roads in many tropical countries. Tree bark contains 12-14% tannins. Neem oil is used for preparation of soaps, cosmetics, pharmaceuticals and other non-edible products. An exudate can be tapped from the trunk by wounding the bark. This high-protein material has potential as a food additive and is widely used in Southeast Asia as 'neem glue'.

Farmers in India use neem oilcake (the residue left after extracting oil from the seeds) as an organic manure and soil amendment. It is believed to enhance the efficiency of nitrogen fertilizers by reducing the rate of nitrification and inhibiting soil pests including nematodes, fungi, and insects. Leaves and small twigs are used as mulch and green manure.

TAMARIND (*Tamarindus indica*)

Tamarindus indica is a large beautiful fruiting tree, growing up to 15 m tall with a dense, spreading crown. The trunk is short and the lower branches are borne almost horizontal. Fruit, a pod is indehiscent, sub-cylindrical, 10-18 cm × 4 cm, straight or curved and rusty-brown. The shell of the pod is brittle and the seeds are embedded in a sticky edible pulp. Tamarind is believed to be indigenous to tropical Africa, probably introduced long back to India by the Arabs.

Propagation and nursery

Tamarind can be raised easily through direct sowing, planting out of entire plants or stump planting or poly bag container plants. Plant begins to yield seed when about 8-10 years of age. Ripe pods are collected in the month of March-April. Seed is separated from edible pulp by washing with water then dried and stored. Dry seeds retain viability for several years at ambient temperatures.

Seeds are soaked for 24 hours in warm water and sown in an irrigated nursery seedbed or containers in April-May. Germination starts in about a week and about 80% germination is achieved in about a month. Germination is best when seeds are covered by 1.5 cm sandy loam soil. The seedlings quickly develop a taproot and so should not be allowed to grow in a nursery seedbed for more than 4 months before being transplanted into containers. Seedlings attain plantable size by July-August when they are about 3-4 months old and attain a height of 30-40 cm. The pit size should be 30 cm × 30 cm × 30 cm for final planting. Tamarind can be propagated through sowing in lines or patches directly in the main field. Depth of sowing should be about 1.5 cm. The lines are spaced at 4-5 m apart and the seeds are sown 10 cm apart. About 20 kg of seed is required to sow one hectare area. It gives 80% survival against 60% in transplanting.

Planting and tree management

Outstanding mother trees are propagated asexually through soft terminal cuttings, grafting, budding, or air-layering. The cuttings are prepared with current year's shoots and it is collected with new leaves flush early in the morning in turgid condition. Cuttings are treated with 1000 ppm indole-butyric acid.

Growth is generally slow. Seedling height increases by about 60 cm annually. The juvenile phase lasts up to 4-5 years. Young trees are pruned to allow 3-5 well-spaced branches to develop into the main scaffold structure of the tree. After this, only maintenance pruning is required to remove dead or damaged wood. Trees generally require minimal care. The species is strong light demander but resistant to drought. A long, well-marked dry season is necessary for fruiting. Size-control measures include close spacing of about 500 trees/ha and pruning to rejuvenate the fruiting wood. Spacing should be at least 13 m × 13 m in a pure plantation. Spacing may be reduced with vegetatively propagated plants, as they usually do not attain the same size as seeded trees. Smaller trees are easier to harvest. The trees also respond to coppicing and pollarding. Trees commence bearing fruit at 7-10 years of age, with maximum yields being obtained from about 15 years onwards. Trees can continue yielding for 200 years.

Uses

Sapwood is light yellow, heartwood is dark purplish brown; very hard, durable and strong, and takes a fine polishes. The wood is used for carving. It is used for general carpentry, sugar mills, wheels, agricultural tools, boat planks, toys, panels and furniture. Tamarind tree provides good firewood with calorific value of 4850 kcal/kg. It also produces an excellent charcoal.

The chief value of this tree lies in its fruits which are used for various types of food preparations and sherbets. The fruit pulp, mixed with a little salt, is a favourite ingredient of the curries and chutneys popular throughout India. The pulp consists of 8-14% tartaric acid. Acidity is caused by the tartaric acid. The ripe fruit of the sweet type is usually eaten fresh, whereas the fruits of sour types are made into juice, jam, syrup and candy. The juice has a high content of vitamin B (thiamine and niacin) as well as a small amount of carotene and vitamin C. The flowers, leaves and seeds can be eaten and are used in preparation of a variety of dishes. Tamarind seeds are also edible after soaking in water and boiling to remove the seed coat. Flour from the seed may be made into cake and bread.

The bark may be used to relives sores, ulcers, boils and rashes. It may also be administered as a decoction against asthma and amenorrhea. Leaf extracts exhibit anti-oxidant activity in the liver, and is a common ingredient in cardiac and blood sugar reducing medicines. A sweetened decoction of the leaves is good against throat infection, cough, fever, and even intestinal worms. Filtered hot juice of young leaves is used for conjunctivitis. The pulp may be used as a massage to treat rheumatism. Powdered seeds are used to cure dysentery and diarrhoea.

Flowers are a good source for honey production. Both leaves and bark are rich in tannin. The bark tannins can be used in ink or for fixing dyes. Leaves yield a red dye, which is used to give a yellow tint to cloth.

Tamarind is not very compatible with other plants because of its dense shade, broad spreading crown and allelopathic effects. Thus, it is not suitable for agroforestry systems. It makes an ideal avenue tree by virtue of its shade, ornamental flowers and longevity. The deep roots make it very resistant to storms and suitable for windbreaks.

PHALSA (*Grewia* spp.)

Genus *Grewia*. has several species out of which *Grewia asiatica* (known as *phalsa* in India and *falsa* in Pakistan) and *Grewia optiva*(known as beul or dhaman) are the most important and cultivated species in India. *Grewia asiatica* is extensively cultivated in India and in some other areas of the subtropics and tropics for its edible fruit, which fetches good prices in local markets. The fruit is an edible drupe, 5-12 mm diameter, purple to black when ripe. Itis a deciduous small tree growing up to 8 m tall. *Grewia optiva* is a very popular tree of the farmers of the sub-Himalayan tract for its fodder and fibres. A full-grown tree is moderate sized with spreading crown, reaching a height up to 12 m with a clear bole of 3-4 m and a girth of about 80 cm. The fruit is a fleshy drupe, 2-4 lobed, olive green when immature and black when ripe, and edible.The phalsa is indigenous to India and Southeast Asia.

Propagation and nursery

Seeds are the usual means of propagation of *Grewia asiatica* and *Grewia optiva*and they germinate in 15-20 days. Seeds do not require any special pre-sowing treatment but soaking in water for 12 hours may be required as seed coat is hard. Ground-layers, treated with hormones, have been 50% successful and air-layers up to 85%. Cuttings are difficult to root. Only 20% of semi-hardwood cuttings from spring flush, treated with 1,000 ppm naphthaleneacetic acid (NAA), and planted in July may root and grow normally. Long taproot is formed during the first year which is longer than the shoot length

In *Grewia optiva* fruiting starts in June-July and fruits ripen in October-November. The fruits are rubbed and washed in water to separate out the seeds. Each fruit contains 2-4 seeds. The seeds have a hard testa and can be stored well for at least a year without any appreciable drop in vitality. The dibbling method of sowing with twice a day irrigation proved to be the best for germination. The seeds are sown in March-April, about 2 cm deep in lines 15 cm apart. The seedlings are spaced about 10 cm apart in lines. Watering is done regularly till germination is over. The seedlings attain a plantable height of 30 cm by July.

Planting and tree management

Seedlings uprooted from the nursery with balls of earth are wrapped in moist gunny bags and transported to the planting site safely. Planting is done in 30 cm × 30 cm × 30 cm pits at a spacing of 4 m × 4 m for block planting and 4-5 m for single row planting along the field bunds. For stump planting 15-month aged seedlings are used.

Phalsa grows in both tropical and subtropical climates but will tolerate other climates, except at high altitude; however, it does best in regions having distinct summer and winter seasons. In India, it grows well up to an elevation of 1,500 m. The temperature range for growth is reported to be 10-44° C with the optimum between 23-34° C. Phalsa is a strong light demander and requires complete overhead light. The plant tolerates short periods of light frosts. Established plants are drought resistant. As seedlings are suppressed by weeds, weeding may be required at regular intervals at early growth period.

When grown commercially, *Grewia asiatica* plants are cut back hard annually to encourage the growth of new shoots. Flowers develop only on the new shoots of the current growing season. The period between flowering and fruit maturity is 45-55 days. Fruit yields are improved if the plant is cut back to the ground each year, thus encouraging plenty of new growth upon which the fruit is produced. Phalsa begins to bear fruit after only 2 years. The annual yield is

3-5 kg/plant or 4.5-6 tonnes per hectare. Summer is the fruiting season. Only a few fruits in a cluster ripen at any one time, so continuous harvesting is necessary.

Protection

Leaf-cutting caterpillars attack the foliage at night. A blackish caterpillar causes galls on the growing shoots. Termites often damage the roots. In some areas, leaf spot is caused by *Cercospora grewiae.*

Uses

The fruits of *Grewia asiatica* are eaten fresh as dessert or are made into syrup, and extensively employed in the manufacture of soft drinks. The squash is prepared from the fruit pulp by mixing it with sugar and used as an astringent, stomachic and cooling agent. The juice ferments so readily that sodium benzoate must be added as a preservative. The fruits are a cooling tonic and cure inflammation, heart and blood disorders and fevers. The fruit is also good against throat trouble. The bark cures urinary troubles. An infusion of the bark is used in treatment for diarrhoea. The root bark is employed in treating rheumatism. The leaves are applied on skin eruptions and they are known to have antibiotic action.

The wood of *Grewia asiatica* is medium-weight to heavy hardwood with a density of 730-900 kg/cubic m at 15% moisture content. Heartwood is pale grey to pale brown, not sharply differentiated from the sapwood; grain interlocked; texture fine; wood with some silver grain. The wood seasons well, is moderately soft to moderately hard, tough and moderately strong; it works satisfactorily with hand and machine tools. Non-durable when exposed to the weather or in contact with the ground, but durable for interior use. Under cover, the heartwood is moderately resistant to dry-wood termites. Wood is generally used for small articles where toughness is required, such as tool handles, spades, shafts of golf sticks, shoulder poles for carrying small loads, pestles, bows, billiards cues and shingles.

The wood of *Grewia optiva* is white, heavy, hard, elastic, strong and fine-textured. It is used for oar-shafts, axe-handles, shoulder poles, cat frames, bows and several other purposes, where strength and elasticity is required. The wood is difficult to saw when green but also difficult to work by hand after seasoning. It is reported to be suitable for paper-making.The shoots obtained after annual prunings are used for making quite strong baskets that are used to transport fruits and vegetables. Though used as a fuelwood, not liked very much because of the foul foetid smell it emits on burning. The bark yields a fibre of inferior quality used for ropes. The mucilaginous extract of the bark obtained after pounding in water is used to clarify sugarcane juice during the preparation of jaggery.

MULBERRY (*Morus alba*)

There are three species of mulberry of economic importance, namely white mulberry (*Morus alba*), black mulberry (*Morus nigra*) and red mulberry (*Morus rubra*). *Morus alba*, is a fast-growing small to medium-sized mulberry tree which grows to 10-20 m tall. It is generally a short-lived tree. It is native to northern China and is widely cultivated and naturalized in the United States, Mexico, Australia, Kyrgyzstan, Argentina, Turkey, Iran, India, Cambodia, Indonesia, Japan, Laos, Myanmar, Pakistan, Thailand, Vietnam, Zanzibar, etc. The white mulberry is widely cultivated to feed the silkworms employed in the commercial production of silk. Its berries are poisonous when unripe, but are otherwise edible.

Morus nigra is a deciduous tree, slender but with numerous branches, grows to 6-9 m in height. Black mulberry is thought to have originated in the mountainous areas of Mesopotamia and Persia and are now widespread throughout Afghanistan, Iraq, Iran, India, Pakistan, Syria, Lebanon, Jordan, Palestine and Turkey. It has long been cultivated for its edible fruit in Europe and eastern China.The black mulberry was imported into Britain in the 17th century to be used in the cultivation of silkworms. It was unsuccessful because silkworms' preference to the white mulberry.

Morus rubra is a small to medium-sized deciduous tree, growing to 10-15 m in height and up to 50 cm in diameter. It lives up to 125 years.Red mulberry is native to eastern and central North America. Though it is common in the United States, it is listed as an endangered species in Canada. It is mainly cultivated for its large, soft, juicy and sweet with a good flavour fruits used to make delicious confection, pastries and wine.

Propagation and nursery

Morus alba is propagated either by planting out nursery raised seedlings or through rooted branch cuttings. Nursery raised seedlings may be planted out either as entire plants or as stumps. The stumps give better results than the seedlings. Direct sowing does not produce good results. Mulberry plants flower in March-April and fruits ripen in May-June. Trees of about five years age start producing viable seed. Ripe fruits are collected from the trees and are heaped in the shade, rubbed and washed in water to separate out the seed which is dried in sun for a few days before storage. Seed stored in gunny bags lose vitality after one year's storage. Thus, it should be stored in sealed tins. The seed stratified in layers of fine dry sand or ash keeps well for over two years.

For germination, the seed requires moist and well-drained soil. Light shade is favourable for germination and seedling establishment. The seedlings can establish under canopy of trees having light crowns. Sowing in the nursery is done in May-June, soon after seed collection. Sowing is done in lines about 20 cm apart. Soaking of the seed in cold water for about a week hastens and ensures uniform germination. Pre-sowing treatment of the seed with kerosene oil or camphor water is also recommended to protect it from being carried away by ants. Since seeds are very small in size usually these are sown by mixing with sand or ash to ensure uniform sowing. Germination commences in about a week and is completed in another 10 days.

Planting and tree management

White mulberry or *Morus alba* grows in areas with a subtropical or mild temperate climate. The plant's water requirement is high. The shade-tolerant tree is highly susceptible to drought. It can withstand light frost. *Morus alba* grows on a variety of soils ranging from sandy loam to clayey loam. The tree cannot tolerate alkalinity and grows best on soils with *p*H ranging between 6.0 and 7.5.

The seedlings are transplanted in main site when about 10 cm tall at a spacing of about 60 cm × 60 cm. Transplanting is done in winter. All but a few terminal leaves are stripped before seedlings are planted in the main field. For the production of stumps, the seedlings may be retained in the nursery for one or two years depending on their growth rate. About 2 cm collar diameter is considered to be the most suitable size for stumps. Stumps with about 22 cm root and 8 cm shoot are prepared with a sharp tool so that these do not spilt. These are wrapped in moist gunny bags during transport. Planting is done either in crow bar holes or in 30 cm × 30 cm × 30 cm pits. The spacing depends upon the objectives of raising the plantations. Close spacing may suffice if the trees are to be pollarded for leaf production. Wider spacing of 4 m × 4 m or 5 m × 5 m may be necessary if timber and leaf production are to be combined. The field is to be weeded when plants are young. Prunning may be required to regularize the shape of the plants and allow the growth of new shoots. The plants respond very well to coppicing and pollarding.

Uses

Morus alba yields a medium-weight hardwood. Heartwood texture is moderately straight with attractive silver grain. It is easy to turn, bend and finish, and it also seasons well. Wood is in chief demand for sports industry especially used for hockey sticks, tennis and badminton rackets, cricket bats, etc. It is also suitable for house building, boats, beams, posts, flooring, bridge building, agricultural implements, cabinet work, furniture, poles, shafts and bent parts of carriages and carts. It makes medium quality fuelwood with a calorific value of 4370-4770 kcal/kg. Wood yields sulphate pulp with satisfactory strength for white writing and printing paper.

The leaves are used for silk worm rearing. Leaves are also used as very good quality fodder for livestock. Dairy cows can be fed up to 6 kg of leaves a day to improve milk yield. Shade-dried leaves incorporated into feed enhance health and egg production in poultry.

Bark is said to be good in the treatment of stomach-ache, neuralgic pains and dropsy. The leaves and young branchlets are used for treating heavy colds, cough, red eye, insect bites and wounds, and the fruits are used in the treatment of sore throat, dyspepsia and melancholia. Fruit juice may be fermented and used to make liquor.

POPLAR (*Populus deltoides*)

Poplars are amongst the fastest growing tree species under appropriate agroclimatic conditions. Genus *Populus* has 25-35 species, mostly native to North America. Few species like *Populus ciliata*, *P. laurifolia*, *P. gamblei*, *P. alba* and *P. glauca* are indigenous to the Himalayan region of India. However, an exotic species *Populus deltoides*, indigenous to United States of America became very popular in the north-western plains of Uttar Pradesh, Haryana and Punjab in the last few decades. Certain clones of *Populus deltoids* have been found to be eminently suitable for afforestation and agroforestry plantation. *Populus deltoids* is a fast-growing tall tree with a fairly straight and slim trunk, rather open crown composed of a few large branches and attaining a height of 30 m and girth of 2 m. The branches are more or less angled or almost winged. Outer bark forms early, furrowed by cork-like ridges and deep fissures.

Populus deltoids can survive on soils varying from sandy loam to fairly stiff clay, but it makes its best growth on moist, well-drained, deep, medium-textured, alluvial soils that are fertile and well-aerated. Coarse sands and heavy clayey soils deficient in organic matter are unsuitable. It tolerates frost and waterlogging. *Populus deltoids* clones have been cultivated in a number of countries. In fact, *Populus deltoids* and its hybrids together make about 90% of the total cultivated poplars of the world. In India, it has been successfully cultivated as a forest crop or agroforestry crop in the Punjab plains and in the Terai region of Uttar Pradesh.

Propagation and nursery

The tree produces seed at intervals. Poplar seeds are very minute and 14,000 seeds weigh one kg. The seeds remain viable for about three weeks and should be used for sowing soon after collection. But it can be stored for a long period after properly drying. The viability of seeds can be maintained for several years by storing in cold storage at seed moisture content of 4-5% in sealed containers. Germination of *Populus deltoides* seeds is less than 20% under laboratory conditions. Thus, the plant is rarely raised by seeds due to its low germination rate. The seedbed is prepared with two parts soil and one part river sand duly sterilized to give higher survival of seedlings. Nursery beds are often fumigated with methyl bromide to control damping off. Seeds should not be covered or pressed into the soil of the seedbed. Young seedlings are very susceptible to drying out, and the seedbed must be kept water saturated for germination and at least one month thereafter.

Poplar is commercially propagated vegetatively through rooted cuttings and stumps. Cuttings should be collected from one year old shoots from lower 2/3rd portion of the shoots or one year old plants from existing nurseries for better results. Cuttings from the main stem of the *Populus deltoides*give better results than those obtained from branches. Diseased, dying off-type and suppressed plants must be culled from the beds where stock is reserved for preparation of cuttings for the new nurseries.The plants can be cut at 2 cm from ground level. The length of the cuttings should be around 20 to 22 cm with thickness of about 1 to 2 cm having at least four buds. The upper cut of the cutting should be slightly above from the active bud. Cuttings should be prepared with a very sharp and fairly heavy tool to obtain a very clean and smooth cut and care should be taken to prevent splitting of cuttings. All cuttings must be submerged under fresh water in drums for 24 hours immediately after preparation of cuttings. Cuttings obtained from young, healthy and vigorous plants perform better. When cuttings are taken from the trees, the age of trees and of the parts of trees from which the cuttings are taken markedly affect rooting.

Cuttings treated with fungicidal solution for 10 minutes should be planted in nursery beds at a spacing of 80 cm × 60 cm by inserting them into holes made in beds. The best time of planting cuttings in the nursery is January to February. Cuttings made earlier than January and later than March do not perform well in open conditions. Each cutting with its thinner end up should be planted in the planting hole in such a way that the upper portion is just 2 cm above the ground level. After planting, the soil around each cutting should be compacted. The soil all around the cuttings should be drenched with Aldrex emulsion 0.2% immediately after planting the cuttings. Irrigation should be provided as soon as the planting of cutting in the bed is completed. The first irrigation should have about 5-7 cm water uniformly above ground level. Soil moisture in the nursery bed should be kept high during rooting of cuttings. Subsequent irrigation should be light and the interval may vary between 8 to 12 days depending upon the type of soil. The nurseries should be irrigated till the onset of monsoons and the topsoil should not be allowed to develop cracks and become absolutely dry. Proper and effective drainage of excess water during rainy season is essential to prevent water logging and collar rot. After the rainy season one to two irrigations per month is adequate.

As the poplar plants grow very fast, the nursery soil has to be enriched frequently. Application of farm yard manure @ 20-25 t/ha nursery area should be applied during bed preparation. Nursery beds are depleted of fertility after producing plants for one year, if no fertilizer is applied. After onset of the monsoon, urea @ 5 g/plant should be applied. Phosphorous application to nursery has also a good effect on root development. Weeding and hoeing may be required depending upon the incidence of weeds.Regular debudding is done by gently rubbing the newly formed buds with gunny bags up to 2/3rd height of plant from the base from June to October. But care must be taken not to damage the young leaves.

The nursery raised plants called 'entire transplants' (ETPs) attain a height of about 3 to 4 m in one growing season. These are utilized for planting in the main field. While removing these plants, the root is cut at 25 cm depth and all side roots more than 10 cm long are also trimmed. The plants are taken out with naked root and without foliage. ETPs can be prepared from one- or two-year-old plants raised from cuttings in the nursery. This is the best method for transplanting poplars. Poplar propagation can also be made by stumps prepared from one year old nursery plants keeping 25 cm long root and 3 cm of shoot.

A lot of good quality clones of *Populus deltoides* have been developed and launched commercially. Some of them are G-3, G-48, D-61, D-67, D-121, L-34, L-49, L-52, WS-39, WSL-22, WSL-27, WSL-32, PL-1, PL-2, PL-3, PL-4, PL-5, PL-6, PL-7, etc.

Planting and tree management

Selection of proper sites for poplar plantations is of utmost importance because survival and growth rates of plants depend largely on the site quality. The land should be well drained as poplars do not withstand water logging. Assured irrigation facilities should also be ensured. Fertile loam or silt loam soils rich in organic matter should be preferred for poplar plantation. Areas affected with salinity or alkalinity should be avoided.

One deep ploughing up to 25-30 cm depth is desirable before final planting of poplars. A light preparatory irrigation may be required to bring the soil to proper moisture to facilitate ploughing. The field should be levelled after ploughing. Phosphatic and potassium fertilizers should be added during land preparation. As poplars are very sensitive to zinc deficiency, zinc sulphate @ 25 kg/ha should be applied in the area at the time of site preparation.

The best time for planting poplar is during the months of January-February before the opening of new buds. One year old ETPs of 3 m length containing 25 cm root portion give the best results. Nursery beds should be given a light irrigation 7-10 days before uprooting so that there will be no damage to the roots or any injury to the stem while uprooting. Bundles of ETPs should be placed horizontally in the pits filled with fresh water to be replenished from time to time. Storage under fresh water for minimum 48 hours is desirable. If the ETPs suffer dehydration and damage to phloem and cambium layers the establishment and survival of the plants will be difficult.

ETP should be held vertically in the planting pit which should then be filled completely with soil. In block plantation a spacing of 4-5 m × 2 m is maintained. The best spacing under agroforestry systems is 5 m × 4 m. The depth of planting will depend upon the soil type, depth of water table, size of planting stock, etc. As poplar is a fast-growing species, the crop should be adequately fertilized for proper growth. A basal dose of 2 kg farm yard manure, 50 g single super phosphate and 25 g muriate of potash per plant are applied. Nitrogenous fertilizer should be applied in split doses, first dose of nitrogen as 75 g urea should be applied during the first week of June, second dose of 150 g urea during first week of July and third dose of 250 g urea during second and third week of August. Application of fertilizer must be followed by light irrigation.

Sprouting starts by the end of March and is completed by end of April. Certain clones sprout fairly late up to the middle of May. Some of the plants may not sprout at all and this can be ascertained by inspecting the buds of the plants. If the buds do not show signs of swelling, the plants should be cut back at the ground level in early May. Special care by way of irrigation and soil working should be taken for the cut back plants which will sprout if the same were green and alive at collar level at the time of cutting back. Precautions should be taken to ensure that wet or very moist soil is not heaped around the stem of the plants as that may encourage collar rot.

Timely irrigation holds the key to the successful establishment and growth of poplar plants. As the poplars grow and root system develops and spreads over large parts of the plantation area, flood irrigation during summer months in beds to cover the entire area will be very helpful. However, flooding may result in lodging if there is a strong wind storm.

Pruning is essential for ensuring a large clear bole with concentration of maximum volume in the main stem for improving yield and quality of commercial timber. During the first year of growth debudding operations should be carried out in the lowest one third part of the stem during June-July. Pruning of branches at the lower portion of the stem should be carried out as per requirements. However, between second and third years of age pruning should be restricted to a maximum of lower one-third part of the total height of the tree. During subsequent years pruning could be carried out up to a maximum of half of the total height of the tree. Main stem should not be damaged during pruning. Bordeaux paste should be applied immediately after carrying out pruning. After one year pruning should be carried out only during winter and not in summer or rainy season.

Poplars can be harvested at short rotations of 8 to 10 years. Poplars can attain 90 cm girth at breast height and mean annual increment of 20 m^3/ha at 8 years rotation under good care.

Protection

Major insect pests causing damage to the poplars are leaf defoliator, stem and shoot borer, leaf webber, bark eating caterpillar, case worm and leaf hopper. Leaf defoliators are active during the rainy season while leaf webbers are active from April to November with peak period from July to October. Chlorpyriphos 500 ml + cypermethrin 200 ml/ha should be sprayed to control these pests. Stem and shoot borer can be controlled by applying phorate 10 G @ 12 kg/ha up to second year of plantation. After 2 years, these can be controlled by pushing a small wick of cotton dipped in any liquid fumigant in the holes through which frass is being pushed out by the borer. All holes must be closed with mud paste after such treatment. Bark eating caterpillar and case worm can be controlled by spay of 2.5 kg carbaryl 50 WP/ha at feeding sites and the infested branches should be heavily pruned. The leaf hoppers are active from April to November with peak period of their activity from July to October. A spray of oxydemeton-methyl 25 EC @ 750 ml or dimethoate 30 EC @ 625 ml/ha is recommended to control this pest. If mite infestation is observed metasystox @ 2 ml/l of water should be sprayed 2-3 times at an interval of 15 days. Termites can be controlled by applying chlorpyriphos @ 3.75 l/ha.

The cuttings are dipped in 0.5% solution of emisan-6 for 15 minutes before planting as preventive measure against collar rot. If blight infestation is observed the crop should be sprayed with carbendazim @ 2 g/l of water. Wilt can be controlled by spraying of wettable sulphur powder @ 1.5 kg/ha.

Uses

Populus deltoids hasa medium hardwood, light, free from knots, easy to saw and work. The wood has good nail holding power and can be used for making cases for fruit and food stuffs. Wood obtained from poplars is suitable for manufacture of match splints, veneering products, artificial limbs, interior panelling, sports goods, and cheap furniture, etc. It is also suitable for structural uses such as false ceilings, partition and almirah shelves, etc. However, most of the poplar wood is used in the match and plywood industry. The species is suitable for making general purpose plywood, marine plywood, concrete shuttering plywood, etc. Young poplars of 2-3 years age group are an excellent source of cellulose fibre for making various grades of fine paper, packing paper, newsprint, etc. Wood is used for fuel.

Any traditional crops except paddy can be grown reasonably well in between the lines of poplars during the first 3 years. Sugarcane is preferred as an intercrop for the first two years as it is more profitable. Subsequently shade tolerant crops like ginger, turmeric, colocasia, pineapple, arrowroot, etc. can be raised successfully. Short duration winter vegetables or *rabi* crops like wheat, oat, potato, coriander, soybean, lentil, chickpea, toria, cabbage, tomato, etc. can also be raised successfully as most poplar clones are leafless during autumn. Third year onwards cultivation of these crops can be raised throughout the rotation. However, intercrop yield decreases with the increase in age of poplars.

OAK (*Quercus* spp.)

The oak tree is one of the most loved trees in the world. It is a symbol of strength, morale, resistance and knowledge. The oak tree belongs to the genus *Quercus* which has about 800 species all over the world, especially in the northern hemisphere where they are native. It is a long-lived tree which can live more than 1000 years.

Oak is moderate to large sized tree with almost rounded crown. It attains a height of 20 m and diameter of 60 cm. It is an evergreen tree, old leaves fall after the appearance of new leaves and thus trees are never leafless. New leaves appear in the month of March-April.

Propagation and nursery

Oak can be propagated easily through direct sowing or planting out of nursery raised seedlings. Air layering is also possible and application of auxins encourage rooting. In oak, male catkins and female spikes appear on new shoots in April-May. The fruit takes 6-18 months to mature, depending on their species. The oak fruit is a nut called an acorn or oaknutborne in a cup-like structure known as a cupule. Each acorn usually contains one seed and rarely contains two or three. Acorns are collected in December-January preferably from tress and are dried under shade only. The seeds can be stored for one year at low temperature and high humidity.

Oak can be successfully raised by direct sowing either dibbling the seed in cultivated line or patches. The seeds are sown about 2 cm deep before the onset of the rains. About 5 kg of acorns are sufficient for one hectare area plantation. Oak seeds germinate in heavy or moderate shade but fail to germinate in places exposed to direct sun. Drought is the most adverse factor causing seedling mortality. Fresh seed gives 60-70% germination. Germination starts in about 10-12 days and takes to 4-5 days to complete. Seedling develops a long taproot and the growth of root is generally more than that of shoot during the first few years.

In case of nursery, beds should be prepared under light shade, and not under direct sunlight. Seeds are sown in lines about 20 cm apart with 5 cm spacing within the lines at a depth of 2 cm in raised beds. Deep sowing delays the germination and also reduces germination percentage.

Planting and tree management

Regular weeding is required in nursery beds. Seedlings soon develop a taproot and become intolerant of root disturbance. So they should be planted into their permanent positions while young. Seedlings of about 15-20 cm height are suitable for planting out as taller seedlings are difficult to handle due to their massive roots. Planting should preferably be completed by July. Seedlings should be planted at a spacing of 3-4 m ×2-3 m. Oak species are slow growing and thus these need weeding and cleaning for several years.

Oak is moderate light demander and can tolerate to shade in early age. But older trees benefit from full light. Seedlings are sensitive to frost and must be protected during the winter, but older trees are more frost-resistant.

Uses

The major use of oak is for fuelwood and charcoal making, for which it is in great demand. Calorific value of fuelwood is 4600 k cal/kg. Tree is lopped for fodder and a mature tree can yield 20-25 kg leaves annually. The wood is very hard, strong and durable. It has a good grain if properly cut, but does not season well, being rather apt to warp and shrink. The wood is used for light construction purposes, agricultural implements, axe-handles, poles, toys, etc.

Oak seeds, raw or cooked have a sweet taste. It can be eaten whole, though it is more commonly dried, then ground into a powder and used as a thickening in stews or mixed with cereals for making bread. The roasted seed of many

Quercus species has been used as a coffee substitute. The oak seeds are also used in the treatment of diarrhoea, indigestion and asthma. The leaves of most species in this genus are more or less rich in tannins. The bark of oak trees is also usually rich in tannins.

BAMBOO (*Dendrocalamus strictus*)

Bamboos are tall grasses consisting of large canopies. Different species of bamboo are found both naturally and cultivated in farms. *Dendrocalamus strictus* is the most common bamboo species which is widely distributed throughout India, Myanmar and Thailand. This is the hardiest of all Indian bamboos. *Bambusa bambos* is another species occurring throughout India, except in the Himalayas and sub-Himalayan region and the valleys of Ganges and Indus. It is very common in Odisha, the Western Ghats, throughout south India. Besides these two dominant species, several other bamboo species and sub-species are widely found in India and other Southeast Asian countries. These include *Bambusa balcooa, B. glaucescens, B. longispiculata, B. nutans, B. polymorpha, B. tulda, B. vulgaris, Dendrocalamus giganteus, D. hamiltonii,* etc. Different species have different growth rates, development patterns, and different requirement of environmental conditions.

Dendrocalamus strictus also known as male bamboo, solid bamboo or Calcutta bamboo is a tropical and subtropical clumping species. It is native to India where it occupies 53% of the total bamboo area. *Dendrocalamus strictus*is a medium-sized bamboo with culms of about 8-20 m tall and 2.5-8 cm in diameter. The internodes are 30-45 cm long and thick-walled. Culms are hollow under humid conditions, but nearly solid under dry conditions. This species has pale blue-green culms when young, and dull green or yellow culms on maturity. Its nodes are somewhat swollen and basal nodes are often rooting.

Propagation

Dendrocalamus strictus prefers a low relative humidity and mean annual temperatures between 20° - 30° C, but can withstand extreme temperatures, as low as -5° C and as high as +45° C. The optimum annual rainfall is 1,000-3,000 mm, with 300 mm per month during the growing season, but is very drought resistant. The species does not grow well on water-logged or heavy soils. It prefers sandy loam soils with good drainage and a *p*H 5.5-7.5.

Bamboos can be propagated through seed or by vegetative means. Bamboo seeds should be sown within 2-3 months of harvest. Seeds are sown in containers in a lightly shaded position. Germination usually takes place readily. Seedlings are pricked out into individual pots as soon as these are large enough to handle. Seedlings are then planted out into permanent positions when 20 cm tall.

However, some species such as *Bambusa balcooa* and *Bambusa vulgaris* do not produce seed and some other species such as *Bambusa bambos* and *Dendrocalamus strictus* do flower at long intervals varying from 30 to 70 years. Some species flower gregariously and then the parent plants all die, some flower sporadically with or without parent plant deaths, and some species combine both patterns. Only a few species flower more frequently and set seeds. Moreover, bamboo seeds are short-lived and difficult to store outside research institutions. Thus, bamboos are mostly propagated by vegetative means like rhizomes, culm cuttings and branches. However, for this high air humidity, appropriate rooting medium, moderate to full light intensity, protection from pests and diseases, waterlogging and strong winds are required. Of these, maintenance of high air humidity and lack of water logging are the most critical.

Rhizomes

Natural regeneration of bamboo occurs through rhizomes. The rhizome is an underground stem, closely similar in structure to the above ground portion of the stem. The buds on rhizomes, which usually develop are generally one year old. Separated out rhizomes can be planted in the rainy season. Basically, there are three ways; use of rhizome with roots, rhizome with culm and roots or rhizome with culm-stock and roots.

1. *Rhizome with roots*: Rhizomes should be healthy not more than 2-3 years old; not damaged; and should have roots. Rhizomes without culms are cut 50-60 cm long with roots.
2. *Rhizome with culm and roots*: Generally, young culms along with associated rhizomes and roots are used. If the culm is young, it is maintained intact, if it is large the upper part can be removed, or branches trimmed. It is essential that the rhizome attached bears buds; other rhizomes can be cut away.
3. *Rhizome with culm-stock and roots*: This is essentially a rhizome with roots but only 30-60 cm of the basal portion of the connected culm.

February to March appears to be the best time for both collecting and transplanting. If the region has a cold climate the best planting time is April, if it is warm then it should be the cooler month. If rhizomes alone are used then they need a period in the nursery during which they are laid horizontally to root. The two other types are transplanted directly into the field. In a rooting bed the medium should be 3: 1 soil: sand and rhizome planted 10-15 cm deep, and 25 cm apart. Watering is necessary and straw mulch is helpful. Shade is needed when young shoots emerge. If many plants are to be produced from one piece of rhizome, they are to be separated and usually l-2 sprouts may be kept per piece. The piece should be at least 15-20 cm long, with three or four nodes, each with an intact bud. The length of the rhizome affects the survival of the shoot. The longer the rhizome, the more nutrient it contains to support the new shoot. Rhizomes are planted in pit of 60 cm^3 and at spacing of 6-8 m × 6-8 m.

Culm cuttings

The culm is the above ground stem which grows from the underground rhizomes. It is segmented with nodes and internodes. The length of internodes and their texture vary according to species. The nodes bear one or more buds, their shape and size also depend on species. The buds occur on alternate sides of the culm.

Culm cuttings are segments of the culm usually 1 or up to 2-3 nodes with buds or branches. They are suitable for clump-forming species but usually not for non-clump-forming species. Generally,the culm selected for the cuttings should be not more than 2 years old and buds should be healthy. Segments are selected from the lower to mid zone of the culm. The upper part and the lateral branches of the upper culm are discarded. Mid-March to May is the best time for cuttings. The branches on the selected part of the culm are pruned to a length of 10-30 cm, care being taken not to injure existing buds.

A segment is cut with a sharp knife or saw. There must not be any splitting at the cut. A length of 5-10 cm stem should be kept on either side of the node. Immediately after cutting, the cut ends are waxed or segments are wrapped with moist gunny bags to minimise water loss from cut ends. If cuttings are 2 or more noded, an opening of 2 cm long and 1 cm wide is made, or 2 holes each 7 mm diameter are drilled in the centre of the internode. A rooting hormone solution is poured into the segment cavity and the holes are closed by wrapping and tying a polythene strip.

Branch cuttings

Propagation through branch cuttings is one of the most practical methods due to ease of handling. Like culms, branches are stem material. Thick-walled species with stout branches are ideal, e.g., *Bambusa* spp. or *Dendrocalumus* spp., but survival sometimes is only about 50%. However, branches require 6-12 months for rooting, and 12-30 months for rhizome development which is essential for production of new culms.

Branching patterns vary among bamboo species. Each bud or culm node can potentially produce a branch; however, in many species the primary branch remains dominant and stout. An offset is the lower part of a single culm (usually with 3-5 nodes, i.e., about l-2.5 m) with the rhizome axis basal to it and its roots. Planting of these is the most conventional way of propagating bamboo. The culm is between 1 and 2 years old. The culm is cut with a slanting cut and the rhizome to which it is attached is dug up and cut off to a suitable length to include well developed buds. Offsets are normally obtained just before the rainy season or after a pre-monsoon shower. Collecting of offsets is done 2-3 months before planting; in this case they should be kept in a temporary nursery near the site.

Propagation bed

Nursery beds are 1 m wide and range from 5-15 m depending on space available. They should be built on level ground and should be 20-30 cm deep. At the base, medium to coarse sand should be laid for 10-15 cm. Over that a top layer of fine sand should be laid for 10-15 cm. Before planting the cuttings, their branches are trimmed. Culm cuttings are placed horizontally in the bed with the bud upward, spaced 15-30 cm apart and so that branches if present are arranged away from those of another cutting. The cuttings are placed in the top fine sand layer about 3-5 cm above the coarse sand and covered by 3-6 cm of the fine sand. The bed is kept moist. Partial shading is required.

A new technology for mass production of bamboo seedling has been developed by the Forest Research Institute, Dehradun. Seeds are sown in July in the germination trays and when the seedlings reach 3-4 leave stage, they are planted in polybags containing equal proportion of soil, sand and FYM. At the age of eight months, seedlings are removed from polybags. Proliferated tillers of these seedlings are separated by cutting rhizome to act as propagules. Each propagule consists of a tiller along with rhizome and roots.

Management of established clumps

Bamboos have an interesting method of growth. Each plant produces a number of new stems annually. These stems grow to their maximum height in their first year of growth, subsequent growth in the stem being limited to the production of new side branches and leaves. Bamboo responds well to fertilizer application. Organic fertilizers are recommended if grown for edible shoots. Irrigation may be essential during the first two years to ensure better establishment and quicker culm production. If grown for edible shoots watering ensures enhanced sprout production. Regular thinning and cleaning should be carried out from the 4th year of clump establishment. All dry, dead and drying culms are to be removed from the clump so as to create sufficient space in the clump for new sprouts to grow up straight. As a regular practice these operations are to be carried out every year in the months of November-February. If managed properly with routine pruning, thinning and cleaning, bamboo usually escapes pest infestations. Proper sanitation measures should also be adopted for the control of fungal infections.

No current year culms are to be harvested. About 60% of the 3 year old culms and almost all of the 4th year culms can be cut and removed. However, it is always better to retain a few older culms in the clump to serve as support for the younger newly emerging culms. Harvesting should never be done during the growing season. It is recommended to cut the culms lower than 30 cm above the ground level, but not below the 2nd node.

If the stand is managed for edible shoots, they are to be cut either in the early morning or late evening when the sprouts attain 35-45 cm in height. Only 60% of the sprouts from all portions of the clump should be harvested while 40% are retained in the clump for culms production.

Uses

Dendrocalamus strictus is extensively used as raw material in paper mills and also for a variety of purposes such as light construction, furniture, musical instruments, bamboo board, mats, walking sticks, agricultural implements, rafts, baskets, woven wares and household utensils. Young shoots are edible and used as food. Leaves are used as forage. Many seasonal field crops like toria, sesame, pulses, vegetables, etc. can suitably be intercropped in agroforestry systems with bamboo.

Selected References

Abhay Kumar, Yadava, M. S., Shabuam, S., Oraon, P. R., Singh, B. K and Kumar, S. 2020. Recent Trends in Agroforestry. In: Advances in Agricultural Sciences. AkiNik Publications, New Delhi. pp. 35-52.

Abrol, I. P. and Dhruva Narayana, V. V. (Eds.). 1990. Technologies for wasteland development. ICAR, New Delhi.

Ahmed, P. 1991. Agroforestry: a viable land use of alkali soils. Agroforestry Systems 14: 23-37.

Akachuku, A. E. 1985. Cost-benefit analysis of wood and food components of agrisilviculture in Nigerian forest zone. Agroforestry Systems 3: 307-316.

Alavalapati, J. R. R., Luckert, M. K. and Gill, D. S. 1995. Adoption of agroforestry practices: a case study from Andhra Pradesh, India. Agroforestry Systems 32: 1-14.

Anderson, D. 1987. The Economics of Afforestation.The John Hopkins University Press, Baltimore, USA.

Andow, D. A. 1991. Vegetational diversity and arthropod population response.Annual Review of Entomology 36: 561-586.

Armson, K. A. 1977. Forest Soils: Properties and Processes. University of Toronto Press, Toronto, Canada.

Arnold, J. E. M. 1983. Economic considerations in agroforestry projects. Agroforestry Systems 1: 299-311.

Arnold, J. E. M. 1987. Economic considerations in agroforestry. In: Steppler, H. A. and Nair, P. K. R. (Eds.) Agroforestry: A decade of development. ICRAF, Nairobi, Kenya

Arunachalam, A., Rizvi, R. H., Handa, A. K. and Ramanan, S. S. 2022. Agroforestry in India: area estimates and methods. Current Science 123(6): 743-744.

Baker, F. W. G. 1992. Rapid propagation of fast-growing woody species. CAB International, Wallingford, U.K.

Balasubramaniyan, P. and Palaniappan, S. P. 2001. Principles and Practices of Agronomy. Agrobios India, Jodhpur.

Balvanera, P., Kremen, C. and Martinez-Ramos, M. 2005. Applying community structure analysis to ecosystem function: Examples from pollination and carbon storage. Ecological Applications 15: 360-375.

Bandopadhyay, A. K. 1997. A Text Book of Agroforestry with Application. Vikash Publishing House Pvt. Ltd., New Delhi.

Bastin, J. F. 2019.The global tree restoration potential. Science 365: 76-79.

Beer, J., Muschler, R. G., Somarriba, E. and Kass, D. 1998. Shade management in coffee and cacao plantations- a review. Agroforestry Systems 38: 139-164.

Beer, J. 1987. Advantages, disadvantages and desirable characteristics of shade trees for coffee, cacao and tea. Agroforestry Systems 5: 3-13.

Bellefontaine, R., Petit, S., Pain-Orcet, M., Deleporte, P. and Bertault, J. G. 2002. Trees outside forests: towards better awareness. FAO Conservation Guide 35. FAO, Rome.

Bellow, J. G., Hudson, R. F. and Nair, P. K. R. 2008. Adoption potential of fruit-tree-based agroforestry on small farms in the subtropical highlands. Agroforestry Systems 73: 23-36.

Belsky, A. J., Amundson, R. G., Duxbury, J. M., Riha, S. J., & Ali, A. R. 1993. The effects of trees on their physical, chemical, and biological environments in a semi-arid savanna in Kenya. Journal of Applied Ecology 30(2): 451-474.

Bene, J. G., Beall, H. W. and Cote, A. 1977. Trees, food and people: Land Management in the Tropics. IDRC, Ottawa, Canada.

Betters, D. R. 1988.Planning optimal economic strategies for agroforestry systems. Agroforestry Systems 7: 17-31.

Bhatt, B. P. and Todaria, N. P. 1990. Studies on the allelopathic effects of some agroforestry tree crops of Garhwal Himalaya. Agroforestry Systems 12: 251-255.

Blanford, H. R. 1958. Highlights of one hundred years of forestry in Burma. Empire ForestryReview 37(1): 33-42.

Bowen, G. D. and Nambiar, E. K. S. (Eds.). 1984. Nutrition of plantation forests. Academic Press, London.

Brady, N.C., 1984. The Nature and Properties of Soils. Macmillan Publishing, New York, pp. 350-354.

Budelman, A. 1989.Nutrient composition of the leaf biomass of three selected woody leguminous species. Agroforestry Systems 8: 39-51.

Buresh, R. J. and Cooper, P. J. M. 1999.The science and practice of short-term fallows. Agroforestry Systems 47: 1-358.

Buresh, R. J. and Tian, G.1997. Soil improvement by treesin sub-Saharan Africa. Agroforestry Systems 38: 51-76.

Burley, J. and Von-Carlowitz, P. 1984. Trees and Their Economic Importance in AgroforestrySystems. In: J. A. Franz and P. L. S. Rao (Eds.) Agroforestry: A Decade of Development (pp. 19-52). International Council for Research in Agroforestry, Nairobi, Kenya.

Burley, J. and Wood, P. J. 1991. Agroforestry in the UK. Forestry Commission Bulletin 95.

CAFRI. 2015. Vision 2050. ICAR-Central Agroforestry Research Institute, Jhansi, India.

Carter, E. J. 1994. The Potential of Urban Forestry in Developing Countries: A Concept Paper. Food and Agriculture Organization, Rome.90 pp.

Chauhan, S. K., Singh, R. V. and Singh, K. P. 2010. Potential of agroforestry systems in carbon sequestration: An Indian perspective. Journal of Agricultural Physics 10(1): 48-54.

Chavan, S. 2014. Agroforestry for adaptation and mitigation of climate change. Popular Kheti 2(3): 214-220.

Chavangi, A. H. and Zimmermann, R. 1987. A Guide to Farm Forestry in Kenya, Ministry of Environment and Natural Resources, Government of Kenya, Nairobi.

Chundawat, B. S. and Gautam, S. K. 1993. Textbook of Agroforestry. Oxford and IBH Publishing Company Pvt. Ltd., New Delhi.

Ciampitti, I. A., García, F. O., Picone, L. I. and Rubio, G. 2011. Soil carbon and phosphorus pools in field crop rotations in Pampean soils of Argentina. Soil Science Society of America Journal 75(2): 1-10.

Clark, K. H. and Nicholas, K. A. 2013. Introducing urban food forestry: A multifunctional approach to increase food security and provide ecosystem services. Landscape Ecology 28(9): 1649-1669.

Cnswell, E. T., SaJapongse, A., Howlett, D. J. B. and Dowling, A. J. 1997. Agroforestry in the management of sloping lands in Asia and the Pacific. Agroforestry Systems 38: 121-137.

Cohen, M. and Garrett, J. 2009. The food price crisis and urban food insecurity.International Institute for Environment and Development, London. 39 pp.

Cooper, P. J. M. and Denning, G. L. 2000. Scaling up the Impact of Agroforestry Research. ICRAF, Nairobi, Kenya.

Crawford, M. 2010. Creating a Forest Garden: Working with nature to grow edible crops. Green Books, Devon, U. K.

Dadhwal, K. S., Narain, P. and Dhyani, S. K. 1989. Agroforestry systems in the Garhwal Himalayas of India. Agroforestry Systems 7: 213-225.

Dagar, J. C. 2009. Opportunities for alternate land uses in salty and water scarcity area. International Journal of Ecology and Environmental Science 35: 53-66.

Danks, C. and Fortmann, L. 2004. Forest and Tree Tenure and Ownership. In: Burley, J., Evans, J. and Youngquist, J. A. (Eds.) Encyclopaedia of Forest Sciences. Elsevier Academic Press, Oxford, U.K. pp. 1157-1162.

Das, T. and Das, A. K. 2005. Inventorying plant biodiversity in homegardens: A case study in Barak Valley, Assam, North East India. Current science, 89(1): 155-163.

Davis, L. 1989. Analysis of Agroforestry Systems.Winrock International, Morrilton, Arkansas, USA.

Denning, G. L. 2001. Realising the potential of agroforestry: integrating research and development to achieve greater impact. Development in Practice 11(4): 407-416.

Dhyani, S. K., Puri, D. N. and Narain, P. 1996. Biomass production and rooting behavior of Eucalyptus tereticornis on deep soils and riverbed bouldery lands of Doon Valley, India. Indian Forester 122(2): 128-136.

Dhyani, S. K., Ram, Asha and Dev, Inder. 2016. Potential of agroforestry systems in carbon sequestration in India. Indian Journal of Agricultural Sciences 86(9): 1103-1112.

Dhyani, S. K., Sharda, V. N. and Sharma, A. R. 2003. Agroforestry for water resources conservation: issues, challenges and strategies. In: Pathak, P. S. and Ram Newaj, (Eds.) Agroforestry: Potentials and Opportunities. Agribios (India), Jodhpur.

Dhyani, S. K. 2014. National Agroforestry Policy 2014 and the need for area estimation under agroforestry. Current Science 107(1): 9-10.

Dhyani, S. K. and Tripathi, R. S. 2000. Biomass and production of fine and coarse roots of trees under agrisilvicultural practices in north-east India. Agroforestry Systems 50: 107-121.

Divya, M., Parthiban, K. T., Vanangamudi, K., Srinivasan and Govindarao, M. 2008. Social forestry and agroforestry. Satish Serial Publishing House, Delhi.

Dixon, J. A. Gibbon, D. P. and Gulliver, A. 2001. Farming systems and poverty: improving farmers' livelihoods in a changing world. FAO, Rome.

Dunn, W. W., Lynch, A. M. and Morgan, P. 1990. Benefit-cost analysis of fuelwood management using native alder in Ecuador. Agroforestry Systems 11: 125-139.

Dwivedi, A. P. 1992. Agroforestry Principles and Practices. Oxford and IBH Publishing Company Pvt. Ltd., New Delhi.

Etherington, D. and Mathews, P. 1983. Approaches to the economic valuation of agroforestry farming systems. Agroforestry Systems 1: 347-360.

Evans, J. 1992. Plantation Forestry in the Tropics, 2nd edition. Clarendon Press, Oxford, U. K.

FAO. 1981. India and Sri Lanka: Agroforestry. FAO, Rome.

FAO. 1985. Tree growing by local people. FAO Forestry Paper 64, Rome.

FAO. 1989. Arid zone forestry, a guide for field technicians. FAO Conservation Guide No. 20, Food and Agriculture Organization, Rome.

FAO. 2009. Livestock's long shadow: Environmental issues and options. Food and Agriculture Organization, Rome.

FAO. 2010. Global Forest Resources Assessment 2010: Main Report. Food and Agriculture Organization, Rome.

FAO. 2017. The future of food and agriculture-Trends and challenges. Food and Agriculture Organization, Rome.163 pp.

Fernandes, E. C. M. and Nair, P. K. R. 1986. An evaluation of the structure and function of tropical homegardens. Agricultural Systems 21: 279-310.

Filius, A. M. 1982. Economics aspects of agroforestry. Agroforestry Systems 1: 29-39.

Foereid, B. and Hogh-Jensen, H. 2004. Carbon sequestration potential of organic agriculture in northern Europe - A modelling approach. Nutrient Cycling in Agroecosystems 68: 13-24.

Fonzen, P. F. and Oberholzer, O. 1984. Use of multipurpose trees in hill farming systems in Western Nepal. Agroforestry Systems 2: 187-197.

Forster, P., Ramaswamy, V., Artaxo, P., Berntsen, T., Betts, R., Fahey, D. W. and Plattner, G. K. 2007. Changes in Atmospheric Constituents and in Radiative Forcing. In Climate Change 2007: The Physical Science Basis. Contribution of Working Group I to the Fourth Assessment Report of the Intergovernmental Panel on Climate Change (pp. 129-234). Cambridge University Press.

Garg, V. K. and Jain, R. K. 1992. Influence of fuelwood trees on sodic soils. Canadian Journal of Forest Research 22: 729-735.

Gholz, H. L. (Ed.). 1987. Agroforestry: Realities, Possibilities and Potentials.MartinusNijhoff, Dordrecht, the Netherlands.

Ghosh, S. P., Kumar, B. M., Kabeerathumma, S. and Nair, G. M. 1989. Productivity, soil fertility and soil erosion under cassava-based agroforestry systems. Agroforestry Systems 8: 67-82.

Gill, M. S., Singh, J. P. and Gangwar, K. S. 2009.Integrated farming system and agriculture sustainability. Indian Journal of Agronomy 54: 128-139.

Godsey, L. D. 2008. Economic budgeting for agroforestry practices. Agroforestry in Action. Centre for Agroforestry, University of Missouri, Columbia, USA.

Golley, F. B. 1986. Chemical plant-soil relationships in tropical forests. Journal of Tropical Ecology 2: 219-229.

Grainger, A. 1980.The development of tree crops and agroforestry systems. International Tree Crops Journal 1: 3-14.

Grewal, S.S. and Grewal, P. S. 2012. Can cities become self-reliant in food? Cities 29(1): 1-11.

Grey, G.W. and F.J. Deneke. 1986. Urban Forestry. Wiley, New York.299 pp.

Gupta, J. P. and Sharma, B. M. (Eds.) 1997. Agroforestry for Sustained Productivity in Arid Region.Scientific Publishers, Jodhpur, India.

Gupta, R. K. 1993. Multipurpose trees for agroforestry and wasteland utilization. Oxford and IBH Publishing Company Pvt. Ltd., New Delhi.

Gupta, T. 1982. The economics of tree crops on marginal lands with special reference to the hot and arid region in Rajasthan, India. International Tree Crops Journal 2: 155-192.

Hasan, M. K. and Ashraful Alam, A. K. M. 2006. Land degradation situation in Bangladesh and role of agroforestry. Journal of Agriculture and Rural Development 4:19-25.

Hegde, N. G. and Kelkar, V. D. 1989. Agroforestry- selected readings. BAIF Publications, Pune.

Hoekstra, D. A. 1987. Gathering socio and economic information for agroforestry projects. ICRAF Working Paper No. 50. ICRAF, Nairobi, Kenya.

Hoekstra, D. A. 1990. Economics of agroforestry.In: Mac-Dicken, K. G. and Vergara, N. T. (Eds.) Agroforestry Classification and Management. John Wiley & Sons, New York, USA.

Hoekstra, D. A. and Kuguru, F. M. (Eds.). 1982. Agroforestry Systems for Small-Scale Farmers. ICRAF, Nairobi, Kenya.

Huxley, P. A. (Ed.). 1983. Plant Research and Agroforestry. ICRAF, Nairobi, Kenya.

Huxley, P. A. 1983. The Tree/crop Interface. ICRAF Working Paper No. 13. ICRAF, Nairobi, Kenya.

Huxley, P. A. 1984. The basis of selection, management and evaluation of multipurpose trees. ICRAF Working Paper No. 25. ICRAF, Nairobi, Kenya.

Huxley, P. A. 1985. Systematic designs for field experimentation with multipurpose trees. Agroforestry Systems 3: 251-266.

Huxley, P. A. 1986. Rationalizing research on hedgerow intercropping: An overview.ICRAF Working Paper No. 40. ICRAF, Nairobi, Kenya.

Huxley, P. A. 1999. Tropical Agroforestry.Blackwell Publishers, London.

Huxley, P. A. and Mead, R. 1988. An ecological approach to on-farm experimentation. Working Paper 52. ICRAF, Nairobi, Kenya.

IPCC. 2000. Special Report on Land Use, Land Use Change and Forestry. Summary for Policy Makers.Intergovernmental Panel on Climate Change, Geneva, Switzerland.

IPCC. 2007. Climate Change 2007: Mitigation of Climate Change. Contribution of Working Group III to the Fourth Assessment Report of the Intergovernmental Panel on Climate Change. Cambridge University Press.

IPCC. 2014. Climate Change 2014: Synthesis Report. Contribution of Working Groups I, II and III to the Fifth Assessment Report of the Intergovernmental Panel on Climate Change. IPCC, Geneva, Switzerland.

IUFRO. 2009. Adaptation of forests and people to climate change. A global assessment report. IUFRO World Series Volume 22. Helsinki.

Jaenicke, H. 1999. Good Tree Nursery Practices: Practical Guidelines for Research Nurseries. International Centre for Research in Agroforestry, Nairobi, Kenya.

Jaggi, S., Gupta,V. K. and Sharma,V. K. 2001. Design and analysis of agroforestry experiments: An overview. Indian Journal of Agroforestry 3: 120-129.

Jambulingam, R. and Fernandes, E. C. M. 1986. Multipurpose trees and shrubs on farmlands in Tamil Nadu State, India. Agroforestry Systems 4: 17-23.

Jarecki, M. K. and Lal, R. 2003. Crop Management for Soil Carbon Sequestration. Critical Reviews in Plant Sciences 22: 471-502.

Jarvis, P. G. (Ed.). 1991. Agroforestry: Principles and Practice.Elsevier, Amsterdam, the Netherlands.

Jha, L. K. 1995. Advances in Agroforestry. APH Publishing Company, New Delhi.

Jhon, B. 1988. Litter production and nutrient cycling in coffee plantation with shade. Agroforestry systems 7(2): 103-114.

Johnsen, K. H., Wear, D., Oren, R., *et al.* 2001. Carbon sequestration and southern pine forests. Journal of Forestry 99: 14-21.

Kala, C. P. 2010. Status of an indigenous agroforestry system in changing climate: A case study of the middle Himalayan region of Tehri Garhwal, India. Journal of Forest Science 56(8): 373-380

Kalloo, K. 2003. Research and extension activities on organic agriculture in India. Organic farming in horticulture for sustainable production, 29-30 August, CISH, Lucknow,1p.

Kang, B. T., Reynolds, L. and Atta-Krah, A. N. 1990. Alley farming. Advances in Agronomy 43: 315-359.

Kang, B. T. and Akinnifesi, F. K. 2000. Agroforestry as alternative land-use production system for the tropics. Natural Resources Forum 24: 137–151.

Kang, B. T. and Reynolds, L. 1989. (Eds.). Alley Farming in the Humid and Subhumid Tropics. Proceedings of International Workshop, IITA, 10-14 March 1986, Ibadan, Nigeria.

Kareemulla, K. 2002. Institutional credit flow to agriculture and farm indebtedness in Uttar Pradesh. Ind. Jn. of Agri. Econ. 60(3):372.

Kaur, B., Gupta, S. R. and Singh, G. 2002. Carbon storage and nitrogen cycling in silvopastoral systems on a sodic soil in northwestern India. Agroforestry Systems 54: 21-29.

Kessler, J. J. and Breman, H. 1991. The potential of agroforestry to increase primary production in the Sahelian and Sudanian zones of West Africa. AgroforestrySystems 13(1): 41-62.

Kilewe, A. M., Kealey, K. M. and Kebaara, K. K. 1989. Agroforestry development in Kenya. ICRAF, Nairobi.

King, K. F. S. 1968. Agrisilviculture: The Taungya System, Bulletin No. 1. Department of Forestry, University of Ibadan, Nigeria.

King, K. F. S. 1979. Agroforestry and the utilization of fragile ecosystems. Forest Ecology and Management 2:161-168.

King, K.F.S. and Chandler, M.T. 1978. The Wasted Lands. ICRAF, Nairobi

Kirby, K. R. and Potvin, C. 2007. Variation in carbon storage among tree species: implications for the management of a small-scale carbon sink project. Forest Ecology Management 246:208-221.

Kronick, J. 1984. Temporal analysis of agroforestry systems for rural development. Agroforestry Systems 2:165-176.

Kumar, B. M., Long, J. N. and Kumar, P. 1995. A density management diagram for teak plantations of Kerala in peninsular India. Forest Ecology and Management 74: 125-131.

Kumar, B. M., Nair and P. K. 2004. The enigma of tropical homegardens. Agroforestry System 61:135-152

Kumar, B. M. 2011. Quarter century of agroforestry research in Kerala: an overview. Journal of Tropical Agriculture 49: 1-18.

Kunhamu, T. K. 2015. Indian homegardens as a sustainable land use practice: Prospects and challenges. Agroforestry: Present Status and Way Forward: Biotech Books, New Delhi,PP- 1-41.

Kurtz, W. B., Thurman, S. E., Monson, M. J. and Garrett, H. E. 1991. The use of agroforestry to control erosion -financial aspects. Forestry Chronicle 67: 254-257.

Lal, R. 1989. Agroforestry systems and soil surface management of a tropical Alfisol. Advances in Agronomy 42: 85-197.

Lal, R. 2010. Managing soils and ecosystems for mitigating anthropogenic carbon emissions and advancing global food security. Bioscience 60:708-721.

Lal, R. and Greenland, D. J. (Eds.). 1979. Soil Physical Properties and Crop Production in the Tropics. Wiley, Chichester, UK.

Langton, S. 1990. Avoiding edge effects in agroforestry experiments; the use of neighbour-balanced designs and guard areas. Agroforestry Systems 12: 173-185.

Larue, T. A. and Patterson, T. G. 1981. How much nitrogen do legumes fix? Advances of Agronomy 34: 15-38.

Lavigne, M. B. 1991. Effects of thinning on the allocation of growth and respiration in young stands of balsam fir. The Canadian Journal of Forest Research 21: 186-192.

Leakey, R. R. B. 1996. Definition of agroforestry revisited. Agroforestry Today 8(1): 5–7.

Leakey, R. R. B. 1999. Potential for novel food products from agroforestry trees. Food Chemistry 64: 1-14.

Lee-Smith, D. 2010. Cities feeding people: An up-date on urban agriculture in equatorial Africa. Environment and Urbanization 22(2): 483-499.

Lefroy, E. C., Hobbs, R. J., O'Connor, M. H. and J. S., Pate (Eds.). 1999. Agriculture as a Mimic of Natural Ecosystems. Kluwer Academic Publishers, Dordrecht, the Netherlands.

Lin, B. 2011. Agroforestry management as an adaptive strategy against potential microclimate extremes in crop production. Agricultural and Forest Meteorology 151(9): 1260-1273.

Lin, B. B., Senthilkumar, K., Singh, R. K., Sathish, S. B. A. and Ravichandran, G. 2008. Agroforestry and conservation agriculture: Complementary practices for sustainable development. International Journal of Agricultural Sustainability 6(2), 93-107.

Longman, K. A. 2002. Tropical Trees: A Practical Manual for Growing Good Nursery Stock. Blaketon Hall Ltd., Devon, U. K.

Lorenz, K. and Lal, R. 2010. Carbon Sequestration in Forest Ecosystems. Springer, Dordrecht, The Netherlands.

Lovell, S. 2020. Urban agroforestry and its potential integration into city planning efforts. UrbanAgriculture and Regional Food Systems 5: 1-3.

Lundgren, B. andRaintree, J.B. 1983, Sustained agroforestry In B. Nestel, ed. Agricultural research for development: potentials and challenges in Asia. The Hague, ISNAR.

Lundgren, B. O. 1982. Agroforestry in Development: A Report of the Panel of Experts on Agroforestry. Stockholm: Swedish Agency for Research Cooperation with Developing Countries (SAREC).

MacDicken, K. G., Wolf, G. V. and Briscoe, C. B. 1991. Standard Research Methods for Multipurpose Trees and Shrubs. Winrock International, Arlington, USA.

MacDicken, K. G. and Vergara, N. T. (Eds.). 1990. Agroforestry: Classification and Management. John Wiley, New York, USA.

Mahapatra, A. K. and Panda, P. K. 2003. EthnoPharmacological Knowledge of Juang and MundaTribes of Eastern India. International Journal of Sustainable Development and World Ecology 9(2): 151–158.

Maji, A. K., Obi Reddy, G. P. and Sarkar, D. 2010. Degraded and Wastelands of India: Status and Spatial Distribution. Indian Council of Agricultural Research, New Delhi.

Mann, S. 2014. Urban agroforestry: Connecting agroecology, permaculture, urban forestry and urban agriculture with agroforestry. University of Missouri, Columbia.20 pp.

Markewitz, D. 2006. Fossil fuel carbon emissions from silviculture: Impacts on net carbon sequestration in forests. Forest Ecology and Management 236: 153-161.

Martin, F. B. 1973. Beehive designs for observing variety competition. Biometrics 25: 397-402.

Mazarolli, D. M. 2011. The benefits of tropical homegardens.The Overstory239.Permanent Agriculture Resources, Holualoa, Hawaii.

Mead, R. 1991. Designing experiments for agroforestry research.In: Avery, M. E., Cannell, M. G. R., and Ong, C. K. (Eds.) Biophysical Research for Asian Agroforestry. Winrock International, Arlington, USA.

Mercer, D. E. 1992. The economics of agroforestry.In: Burch, W. R. and Parker, J. K. (Eds.) Social science applications in Asian agroforestry. Winrock International, USA.

Mercer, D. E. and Miller, R. P. 1997. Socioeconomic research in agroforestry: progress, prospects, priorities. Agroforestry Systems 38: 177-193.

Miller, R. W. 1997. Urban forestry: Planning and managing urban green spaces. 2nd edition, Prentice Hall, New Jersey.

Mock, T. 2004. Building a sustainable urban forest. Urban Agriculture Magazine 13: 29-30.

Montagnini, F. and Nair, P. K. R. 2004. Carbon sequestration: an underexploited environmental benefit of agroforestry systems. Agroforestry Systems 61:281-295.

Mosquera-Losada, M. R., Santiago-Freijanes, J. J., Lawson, G., Balaguer, F., Vaets, N., Burgess, P. andRigueiro-Rodríguez, A. 2016. Agroforestry as tool to mitigate and adapt to climate under LULUCF accounting. 3rd European Agroforestry Conference – Montpellier, 23-25 May 2016 Agroforestry and climate change (oral) Abstract number 136, pp-163-165.

Nair, M. A. and Sreedharan, C. 1986. Agroforestry farming systems in the homesteads of Kerala, southern India. Agroforestry Systems 4: 339-363.

Nair, P. K. R. (Ed.). 1989. Agroforestry systems in the tropics. Kluwer Academic Publishers, Dordrecht, the Netherlands.

Nair, P. K. R., 1997. Directions in tropical agroforestry research: Past, present and future. Agroforestry Systems 38: 223-245.

Nair, P. K. R., Fernandes, E. C. M. and Wambugu, P. N. 1984.Multipurpose leguminous trees and shrubs for agroforestry. Pesquisa Agropecuaria Brasileira 19: 295-313.

Nair, P. K. R., Kumar, B. M., and Nair, V. D. 2009. Agroforestry as a strategy for carbon sequestration. Journal of Plant Nutrition and Soil Science 172(1): 10-23.

Nair, P. K. R., Nair, V. D., Kumar, B. M. and Haile, S. G. 2009. Soil carbon sequestration in tropical agroforestry systems: A feasibility appraisal. Environmental Science and Policy 12: 1099-1111.

Nair, P. K. R., Nair, V. D., Kumar, B. M. and Showalter, J. M. 2010. Carbon sequestration in agroforestry systems. Advances in Agronomy 108: 237-307.

Nair, P. K. R., Saha, S. K., Nair, V. D. and Haile, S. G. 2010. Potential for greenhouse gas emissions from soil carbon stock following biofuel cultivation on degraded land. Land Degrad. Develop.(in press)

Nair, P. K. R. 1979. Agroforestry Research: A retrospective and prospective appraisal. Proceedings of International Conference on International Cooperation in Agroforestry. ICRAF, Nairobi, Kenya.

Nair, P. K. R. 1984. Soil Productivity Aspects of Agroforestry. ICRAF, Nairobi, Kenya.

Nair, P. K. R. 1987. Agroforestry systems in major ecological zones of the tropics and subtropics. ICRAF Working Paper No. 25. ICRAF, Nairobi.

Nair, P. K. R. 1987. Agroforestry systems inventory. Agroforestry Systems 5: 301-317.

Nair, P. K. R. 1993. An introduction to agroforestry. Dordrecht: Kluwer Academic Publishers.

Nair, P. K. R. 2008. An Introduction to Agroforestry. Springer (India) Pvt. Ltd., New Delhi.

Nair, P. K. R. 2012. Carbon sequestration studies in agroforestry systems: a reality-check. Agroforestry Systems 86: 243-253.

Nair, P. K. R. and Kumar, B. M. 2004.The enigma of tropical homegardens. Agroforestry Systems 61: 135-152.

Nair, P. K. R. and Muschler, R. G. 1993. Agroforestry.In: Pancel L. (Ed.) Tropical Forestry Handbook. Springer, Berlin.

NCA. 1976. Report of the National Commission on Agriculture: Part V, IX and Abridged. Ministry of Agriculture and Irrigation, Government of India, New Delhi.

Negi, M. S., Tandon, V. N. and Rawat, H. S. 1995. Biomass and nutrient distribution in young teak (Tectona grandis) plantation in Tarai Region of Uttar Pradesh. Indian Forester 121(6): 455-463.

Negi, S. S. 1986. A Hand Book of Social Forestry.International Book Distributers, Dehradun, India.

Negi, S. S. 1999. Agroforestry Handbook. International Book Distributers, Dehradun, India.

Neupane, R. P., Sharma, K. R. and Thapa, G. B. 2002. Adoption of agroforestry in the hills of Nepal: a logistic regression analysis. Agricultural Systems 72: 177–196.

Newaj, R. and Dhyani, S. K. 2008. Agroforestry for carbon sequestration: Scope and present status. Indian Journal of Agroforestry 10: 1-9.

Niez, V. 1987. Household gardens: Theoretical and policy considerations. Agricultural System 23: 167-186.

NRCAF. 2007 Vision 2025. National Research Centre for Agroforestry, Jhansi.

NRCAF. 2013 Vision 2050. National Research Centre for Agroforestry, Jhansi.

NRSA. 2005. Wasteland Atlas of India. Ministry of Rural Development and National Remote Sensing Agency Publication, NRSA, Hyderabad.

Nye, P. H. 1961. Organic matter and nutrient cycles under a moist tropical forest. Plant and Soil 13: 333-346.

Okoji, M. A. and Moses, J. 1998. Adoption of agroforestry for soil conservation in Akwa Ibom State, Nigeria. Journal of Sustainable Agriculture 12: 5-13.

Ong, C. K., Corlett, J. E., Singh, R. P. and Black, C. R. 1991. Above- and below-ground interactions in agroforestry systems. Forest Ecology and Management 45: 45-57.

Ong, C. K. and Huxley, P. (Eds.). 1996. Tree-crop Interactions: A Physiological Approach. CAB International, Wallingford, U. K.

Owino, F. 1992. Improving multipurpose tree and shrub species for agroforestry systems. Agroforestry Systems 19: 131-137.

Palm, C. A. 1995. Contribution of agroforestry trees to nutrient requirements of intercropped plants. Agroforestry Systems 30: 105-124.

Pandey, D. N. 2002. Global climate change and carbon management in multifunctional forests. Current Science 83: 593-602.

Pandey, D. N. 2007. Multifunctional agroforestry systems in India. Current Science 92: 455-463.

Pannell, D. J. 1999. Social and economic challenges in the development of complex farming systems. Agroforestry Systems 45: 393-409.

Pardon, P., Baret, S., Deleuze, C., Clermont-Dauphin, C., Fontaine, S., Nold, F., and Le Bissonnais, Y. 2018. Agroforestry associating fruit trees, vines or legumes with cereals or maize, and the impact on crop productivity, soil fertility and climate regulation - A review. Agronomy for Sustainable Development 38(4): 44.

Pathak, P. S., Pateria, H. M., Solanki, K. R. (Eds.). 2000. Agroforestry systems in India. NRCAF, Jhansi, India.

Pathak, P. S., Solanki, K. R., Rai, P., Handa, A. K. and Pateria, H. M. 2009. Agroforestry.In: Handbook of Agriculture, 6th Edition. Indian Council of Agricultural Research, New Delhi.

Patra, A. K. 2011. Agroforestry for food and wood. Science Horizon 1(7): 24-27.

Patra, A. K. and Mohapatra, A. K., Gantayat, B. P. and Das, S. 2011. Evaluation of Acacia nilotica as bund plantation in coastal Odisha.Journal of Research, OUAT Special Issue 1(1): 81-85.

Patra, A. K. and Mohapatra, A. K. 2011. Agroforestry for climate change adaptation and mitigation. Agron Activities at a Glance 1(1): 14-16.

Paul, S. P. 2000. Sloping agroforestry land technology.Proceedings of National Seminar on BFRI Clients Technology Transfer.Bangladesh Forest Research Institute, Chittagong, Bangladesh.

Pauleit, S., Jones, N., Nyhuus, S., Pirnat, J.andSalbitano, F. 2005.In: Konijnendijk, C., Nilsson, K., Randrup, T. B. and Schipperijn, J. (Eds.) Urban Forest resources in European cities. In: Urban forests and trees. Springer, Berlin. pp. 49-79.

Pinney, A. 1991. Farmer-augmented designs for participatory agroforestry research. Agroforestry Systems 15: 259-274.

Pragason, A. and Karthik, A. 2013. Carbon stock sequestered by tree plantation in university campus at Coimbatore, India. International Journal of Environmental Sciences 3(5): 700-710.

Prasad, R., Shivay, Y. S., Kumar, D., & Kumar, S. 2012. Carbon sequestration and soil carbon pools in agroforestry systems: A review. Agriculture, Ecosystems & Environment 163: 139-151.

Prasad, R. and Bhatnagar, P. 1995. Social forestry experience over a decade. International Book Distributers, Dehradun, India.

Prasad, V. N. 1985. Principles of social-cum-community forestry.International Book Distributers, Dehradun, India.

Puri, S. and Khosla, P. K. 1993. Nursery Technology for Agroforestry- Application in semiarid region. Oxford and IBH Publishing Company, New Delhi.

Putnam, A. R. and Tang, C. S. 1986. The Science of Allelopathy.Wiley Inter-Science, New York, USA.

Rai, P., Solanki, K. R. and Singh, U. P. 2000. Growth and biomass production of multipurpose tree species in natural grassland under semi-arid condition. Indian Journal of Agroforestry 2: 101-103.

Rai, P. and Handa, A. K. (Eds.). 2004. Methodologies in Agroforestry Research. AICRP on Agroforestry, NRCAF, Jhansi.

Raintree, J. B. 1983. Strategies for enhancing the adoptability of agroforestry innovations. Agroforestry Systems 1: 173-187.

Raintree, J. B. 1987. D & D users' manual. ICRAF, Nairobi, Kenya.

Raintree, J.B. 1987. Proposal for a working group on a user-friendly information management system for agroforestry research and extension projects. ICRAF. Nairobi.

Raintree, J. B. and Warner, K. 1986. Agroforestry pathways for the intensification of shifting cultivation. Agroforestry Systems 4: 39-54.

Raj, A., Chakrabarti, B., Pathak, H., Singh, S. D., and Pratap, D. 2015. Impact of elevated temperature on iron and zinc uptake in rice crop. Int. J. Agric. Environ. Biotechnol 8: 691–697.

Ram Newaj, S. K., Singh, R. K. and Shukla, R. 2008. Carbon sequestration potential of agroforestry systems in the sub-humid tropical region of India. Agroforestry Systems 74(2): 115-127.

Rao, K. P. C., Verchot, L. V. and Larman, J. 2007. Adaptation to climate change through sustainable management and development of agroforestry systems. SAT e-Journal 4:1-30.

Rao, L. G. G., Joseph, B. and Sreemannarayana, B. 2000. Growth and biomass production of some important multipurpose tree species on rainfed sandy loam soils. Indian Forester 126(7): 772-781.

Rao, M. R., Nair, P. K. R. and Ong, C. K. 1998. Biophysical interactions in tropical agroforestry systems. Agroforestry Systems 38: 3-50.

Rao, M. R., Sharma, M. M. and Ong, C. K. 1990. A study of potential of hedgerow intercropping in semiarid India using a two-way systematic design. Agroforestry Systems 11:243-258.

Rao, M. R., Sharma, M. M. and Ong, C. K. 1991. A tree/crop interface design and its use for evaluating the potential of hedgerow intercropping. Agroforestry Systems 13: 143-158.

Rao, M. R. and Coe, R. D. 1991. Measuring crop yields in on-farm agroforestry studies. Agroforestry Systems 15: 275-289.

Redondo-Brenes, A. 2007.Growth, carbon sequestration, and management of native tree plantations in humid regions of Costa Rica. New Forests 34: 253-268.

Redwood, M. 2009. Agriculture in urban planning: generating livelihoods and food security. Earthscan, International Development Research Centre, London. 248 pp.

Reed, M. S. 2007. Participatory technology development for agroforestry extension: an innovation-decision approach. African Journal of Agricultural Research 2: 334-341.

Reid, R. 2002. The Principles and Practice of Pruning. Australian Forest Grower Magazine 25(2): 12-16.

Reifsnyder, W. S. and Darnhofer, T. O. 1989. Meteorology and Agroforestry.ICRAF, Nairobi, Kenya.

Ripple, W. J., Wolf, C., Newsome, T. M., Galetti, M., Alamgir, M., Crist, E. and Laurance, W. F. 2019. World Scientists' Warning of a Climate Emergency. BioScience 70(1): 8-12.

Rizvi, R. H., Dhyani, S. K., Yadav, R. S., Singh, R. 2011. Biomass production and carbon stock of poplar agroforestry systems in Yamunanagar and Saharnpur districts of north-western India. Current Science 100: 736-742.

Rizvi, R. H., Hanif, M. K. and Aziz, I. 2012. Potential of agroforestry in carbon sequestration and biofuel production in Pakistan. Pakistan Journal of Botany 44: 339-344.

Rocheleau, D. 1991. Participatory research in agroforestry: Learning from experience and expanding our repertoire. Agroforestry Systems 15: 111-137.

Roger, J. H. and Rao, M. R. 1990. Agroforestry field experiments: Discovering the hard facts Part I, Statistical considerations. Agroforestry Today 2: 4-7.

Saha, S., Nair, P. K. R., Nair, V. D. and Kumar, B. M. 2009. Soil carbon stocks in relation to plant diversity of home gardens in Kerala, India. Agroforestry Systems 76: 53-65.

Sanchez, P. A., 1995. Science in agroforestry. Agroforestry Systems 30: 5-55.

Sang, H. 1988.Project Evaluation: Techniques and Practices for Developing Countries.Wilson Press, New York, USA.

Sanginga, N., Vanlauwe, B. and Danso, S. K. A. 1995. Management of biological N_2 fixation in alley cropping systems: Estimation and contribution to N balance. Plant and Soil 174: 119-141.

Saxena, N.C. 1990. Marketing Constraints for Eucalyptus from farm lands in India. Agroforestry Systems in India 13: 73-85.

Scherr, S. J. 1991. On-farm research: The challenges of agroforestry. Agroforestry Systems 15: 95-110.

Schlesinger, W. H. 2000. Carbon sequestration in soils: Some cautions amidst optimism. Agriculture, Ecosystems and Environment 82: 121-127.

Schoeneberger, M., Bendrub, G., Gooijer, H., Soolanayakanahal, R., Tom, S., Brandle, J., Zhan, X. and Den, C. 2012. Branching out: agroforestry as a climate change mitigation and adaptation for agriculture. Journal of Soil and Water Conservation 67(5): 128-136.

Schroeder, P. 1994. Carbon storage benefits of agroforestry systems. Agroforestry Systems 27: 89-97.

Schroth, G. 1995. Tree root characteristics as criteria for species selection and systems-design in agroforestry. Agroforestry Systems 30: 125-143.

Schroth, G. 1999. A review of belowground interactions in agroforestry, focusing on mechanisms and management options. Agroforestry Systems 43: 5-34.

Schultz, R. C., Isenhart, T. M., Simpkins, W. W., and Colletti, J. P. 2009. Riparian buffers restore hydrologic and ecological functions of channelized agricultural streams. Journal of Soil and Water Conservation 64(6), 427-435.

Shah, S. A. 1988.Forestry for people. India Council of Agricultural Research, New Delhi.

Shankarnarayan, K. A., Harsh, L. N. and Kathju, S. 1987. Agroforestry in arid zones of India. Agroforestry Systems 5: 69-88.

Sharma, A. K. 2000. Addition in tradition: agroforestry in the arid zone of India. LEISA-India 2: 19-20.

Sharma S. K. and Gupta S. K. 2010. Cow dung as a Source of Energy and Fertilizer in India. Renewable and Sustainable Energy Reviews 14(9): 2900-2908.

Shyam Sunder, S. 1985. Urban Tree Planting: Foresters Efforts in Bangalore, India. Proceedings of the IX World Forestry Conference, Mexico. pp. 684-690.

Simons, A. J. and Leakey, R. R. B. 2004. Tree domestication in tropical agroforestry. Agroforestry Systems 61: 167-181.

Sinclair, F. L. 1999. A general classification of agroforestry practice. Agroforestry Systems 46: 161-180.

Singh, M. P. and Tewari, D. N. 1996. Agroforestry and wastelands. Anmol Publications Pvt. Ltd., New Delhi.

Singh, R. P., Ong, C. K. and Saharan, N. 1989. Above and below ground interactions in alley cropping in semiarid India. Agroforestry Systems 9: 259-274.

Singh, S. P. 2006. Handbook of Agroforestry, 2nd Edition. Agrotech Publishing Academy, Udaipur, India.

Sitaram Rao, M. 1994. Introduction to Social Forestry, 2nd Edition. Oxford and IBH Publishing Company Pvt. Ltd., New Delhi.

Somarriba, E. 1992. Revisiting the past: an essay on agroforestry definition. Agroforestry Systems 19:233-240.

Steffan-Dewenter, I., Kessler, M., Barkmann, J., Bos, M. M., Buchori, D., Erasmi, S., and Tscharntke, T. 2007. Tradeoffs between income, biodiversity, and ecosystem functioning during tropical rainforest conversion and agroforestry intensification. Proceedings of the National Academy of Sciences 104(12): 4973-4978.

Steppler, H. A. and Lundgren, B. O. 1988. Agroforestry: now and in the future. Outlook on Agriculture 17: 146-152.

Steppler, H. A. and Nair, P. K. R. (Eds.) 1987. Agroforestry: a decade of development. ICRAF, Nairobi, Kenya.

Streed, E. 1999. Identifying benefits in agroforestry extension. The Agroforestry Advantage 2: 3-7.

Suresh, K. K. and Rai, R. S. V. 1987. Studies on the allelopathic effects of some agroforestry tree crops. International Tree Crops Journal 4: 109-115.

Swami, A. and Puri, S. 2005. Carbon sequestration potential of agroforestry systems in India. Journal of Tropical Agriculture, 43(1-2): 53-56.

Swamy, S. L. and Puri, S. 2005. Biomass production and Csequestration of Gmelina arborea in plantation and agroforestry system in India. Agroforestry Systems 64: 181-195.

Szott, L. T., Palm, C. A. and Sanchez, P. A. 1991.Agroforestry in acid soils of the humid tropics. Advances in Agronomy 45: 275-301.

Tejwani, K. G.,Druvanarayan, V. V. and Satyanarayan, T. 1960. Control of gully erosion in ravine land of Gujarat. Journal of Soil and Water Conservation 11: 20-21.

Tejwani, K. G. 1994. Agroforestry in India. Oxford and IBH Publishing Co. Pvt. Ltd., New Delhi.

Tengnas, B. 1994.Agroforestry extension manual for Kenya.International Centre for Research in Agroforestry, Nairobi, Kenya.

Tewari, J. C., Singh, A. K. and Agnihotri, Y. R. 2013. Role of legumes in sustainable agriculture: A review. Agricultural Reviews 34(3): 222-229.

Tewari, J. C. 2007. Intercropping in semi-arid tropics of India: Concepts and opportunities. Journal of Crop Improvement 19(1-2): 37-66.

Thaman, R. R. 1987. The Pacific Islands and Beyond. Unasylva 39(155): 2-13.

Tiwari, K. M. 1983. Social forestry in India. Natraj Publishers, Dehradun, India.

Tiwari, K. M. and Singh, R. V. 1990. Social Forestry Plantations. Oxford and IBH Publishing Company Pvt. Ltd., New Delhi.

Torquebiau, E. 1992. Are tropical agroforestry home gardens sustainable? Agriculture, Ecosystems and Environment 41: 189-207.

Udawatta, R. P., Anderson, S. H., Wander, M. M., Gantzer, C. J. and Kitchen, N. R. 2011. Phosphorus, sediments, and E. coli removal performance of grass riparian buffers in Midwest agroecosystems. Journal of Soil and Water Conservation 66(4): 217-227.

Umrani, R. and Jain, C. K. 2010.Agroforestry systems and practices. Oxford Book Company, Jaipur.

UNFCCC. 2007. Report of the conference of parties on its thirteenth session. Bali, Indonesia. In: United Nations framework convention on climate change. Geneva, Switzerland.

USDA. 2008. Silvipasture: An Agroforestry Practice. USDA National Agroforestry Centre, Blacksburg, Virginia, USA.

USDA. 2011. Agroforestry Strategic Framework, Fiscal Year 2011-2016. United States Department of Agriculture.

Vandermeer, J. 1989. The ecology of intercropping.Cambridge University Press. Cambridge, U.K.

Verchot, L. V., Van Noordwijk, M., Kandji, S., Tomich, T., Ong, C., Albrecht, A., Mackensen, J., Bantilan, C., Anupama, K. V. and Palm, C. 2007. Climate change: linking adaptation and mitigation through agroforestry. Mitigation and Adaptation Strategies for Global Change, 12(5): 901-918.

Verheij, E. 2003. Agroforestry, 3rd edition. Agromisa Foundation, Wageningen, the Netherlands.

Verheij, E. 2004. Propagating and planting Trees. Agromisa Foundation, Wageningen, the Netherlands.

Verma, P., Bijalwan, A., Dobriyal, M. J. R., Swamy, S. L. and Thakur, T. K. 2017. A paradigm shift in agroforestry practices in Uttar Pradesh. Current Science 112(3): 509-516.

VonCarlowitz, P. G. 1987. ICRAF's multipurpose tree and shrub information system. Agroforestry Systems 5: 319-338.

Vonmaydell, H. J. 1991. Agroforestry for tropical rain-forests. Agroforestry Systems 13: 259–267.

Warkentin, M. E., Nair, P. K. R., Ruth, S. R., and Sprague, K. 1990. A knowledge-based expert system for planning and design of agroforestry systems. Agroforestry Systems 11: 71-83.

Westoby, J. 1968. Changing objects of forest management. Address to the IX Commonwealth Forestry Conference, New Delhi.

Westoby, J. 1989. Introduction to World Forestry: People and their Trees.Basil Blackwell, Oxford, U. K.

Wiersum, K. F. 2004. Social and Community Forestry. In: Burley, J., Evans, J. and Youngquist, J. A. (Eds.) Encyclopaedia of Forest Sciences. Elsevier Academic Press, Oxford, U.K. pp. 1136-1143.

Young, A. 1984. Evaluation of agroforestry potential in sloping areas.Working Paper No. 17.ICRAF, Nairobi, Kenya.

Young, A. 1988. Agroforestry and its potential to contribute to land development in the tropics. Biogeography Journal 15: 19-30.

Young, A. 1997. Agroforestry for Soil Management. 2nd Ed. C.A.B. International. Wallingford, UK, 320 pp.

Young, A. 1997. The effectiveness of contour hedgerows for soil and water conservation. Agroforestry Forum 8: 2-4.

Web Links

http://davesgarden.com/guides
http://en.wikipedia.org
http://forestry.about.com
http://www.aftaweb.org
http://www.agroforestry.co.uk
http://www.agroforestry.net
http://www.agrometeorology.org
http://www.americanforests.org/discover-forests/forest-facts
http://www.bioversityinternational.org
http://www.centerforagroforestry.org/practices
http://www.climatecrisis.net
http://www.conference.ifas.ufl.edu
http://www.earo.org.et/communication
http://www.ejournal.icrisat.org
http://www.eoearth.org/article/agroforestry
http://www.face-thefuture.com
http://www.fao.org/forestry
http://www.greenbeltmovement.org
http://www.greenbooks.co.uk
http://www.livingtreeeducationalfoundation.org
http://www.livingwithclimate.fi
http://www.overstory.org
http://www.pafern.org.ph
http://www.plant-trees.org
http://www.recoftc.org
http://www.ruaf.org/index.php

http://www.sactree.com

http://www.treesofstrength.org

http://www.unimelb.edu.au/publications

http://www.unl.edu/nac/workingtrees

http://www.winrock.org

http://www.worldagroforestry.org

https://www.downtoearth.org.in/coverage/why-is-fodder-so-expensive-802

https://www.ipcc.ch/sr15/

https://www.irri.org/climate-change-ready-rice

Index

A
Acid metabolism 122
Adoptability 129, 148
Adverse climate 20, 151
Aeration 109, 176
Aesthetic forestry 86
Aesthetic value 48, 140, 160, 187
Afforestation 12, 80, 82, 87, 88, 98, 102, 155, 211, 212, 214, 249, 250, 251, 253, 281, 290, 291, 292, 293, 321
Aftercare 178
Agrihorticulture 9, 11, 12, 242
Agrisilviculture 9, 59, 213, 242
Agroclimatic zone 15, 302
Agroecological region 30, 77
Agroecological zones 59, 76, 189, 244
Alkali soils 74, 211
Allelochemicals 115, 121, 122, 145
Allelopathic 13, 121, 122, 149, 150, 153, 154, 186, 316
Alley cropping 28, 32, 33, 34, 35, 50, 55, 56, 67, 69, 73, 104, 105, 112, 119, 124, 134, 135, 136, 147, 189, 190, 191, 193, 207, 235, 298
Alluvial plains 75, 217
Amelioration 63, 64, 73, 102, 119, 120, 124, 246, 247, 248, 280
Apisilviculture 53, 54
Aquaforestry 53, 61, 198
Autoallelopathy 122

B
Bench terracing 195, 200, 209
Bending 185, 304
Biodiversity 5, 7, 57, 63, 162, 258, 259, 261, 263, 265, 266, 270, 276, 281, 290, 292
Bioenergy 7, 10
Biological nitrogen fixation 64, 100, 106, 119, 306
Biotic pressure 79, 210
Boundary plantations 193
Bushing 185

C
Calorific value 10, 156, 157, 298, 300, 302, 307, 315, 321, 327
Canal banks 82, 94, 247
Carbon neutral 160
Carbon pool 282, 286, 287, 289
Carbon sequestration 5, 11, 65, 140, 244, 265, 267, 280, 281, 282, 286, 287, 288, 289, 290
Carbon sink 5, 160, 290, 291
Cation exchange capacity 20, 64
Charcoal 10, 27, 35, 37, 64, 156, 157, 298, 302, 304, 307, 311, 313, 315, 327
Climate change 5, 241, 257, 258, 259, 263, 264, 265, 270, 271, 275, 276, 277, 278, 279, 280, 281, 289, 290, 291
Climate change mitigation 5, 264, 270, 279, 281
Climatic hazards 150
Coastal wastelands 214

Coincident 55
Cold desert 74
Cold stress 163
Combustion efficiency 157
Commercial agroforestry systems 58
Community forestry 26, 80, 81, 85, 249
Companion crops 76, 147, 148, 149
Competition 28, 32, 33, 35, 43, 47, 56, 66, 67, 68, 102, 103, 115, 116, 117, 118, 121, 124, 125, 126, 130, 135, 147, 148, 151, 164, 165, 168, 169, 176, 178, 179, 181, 182, 185, 187, 188, 190, 194, 209, 213, 235, 268, 287, 296, 304, 306, 309, 310, 312
Competitive interaction 117
Complementary interaction 116
Complete block designs 231, 233
Component of agroforestry 4, 66, 208, 262
Concomitant 56
Contour cultivation 209
Contour furrow 209
Coppicing 33, 35, 50, 68, 69, 73, 135, 155, 172, 184, 191, 207, 315, 320
Cottage business 3
Crop rotation 18, 144, 153, 154, 186, 222, 282, 288, 289
Cuttings 29, 32, 46, 72, 132, 171, 172, 176, 299, 301, 303, 305, 306, 312, 315, 317, 319, 322, 323, 325, 329, 330, 331

D

Damping off 177, 300, 306, 322
Deforestation 5, 189, 192, 199, 213, 216, 217, 265, 275, 290, 291, 292
Degraded pasture 217
Departmental taungya 30, 133
Desertification 5, 250, 257, 258, 277
Diagnosis and design 219, 220, 244
Diagnostic survey 243, 244
Drought tolerant 41, 47, 121
Dryland farming 154

E

Ecosystem 3, 26, 60, 81, 84, 98, 103, 109, 131, 137, 140, 204, 210, 215, 241, 255, 258, 261, 262, 263, 264, 265, 267, 271, 280, 281, 286, 288, 289, 290, 291
Ecozone 219, 220, 221, 223
Edapho-climatic 61
Energy plantation 154, 155, 156, 160, 193, 251
Erodibility 46, 112
Evapotranspiration 59, 101, 189, 191, 192, 195, 206, 278, 279, 286, 290
Ex-ante 223, 224, 225
Extension forestry 83, 84, 249

F

Farm forestry 10, 61, 81, 83, 84, 249, 253
Fixed-lift pruning 184
Food security 4, 47, 84, 258, 260, 262, 268, 278, 290
Foreign exchange 12, 255
Forest policy 244, 245, 246, 247, 248, 249, 250, 252, 253, 271
Frankia 100, 107, 108, 109, 306

G

Genetic make-up 169
Global warming 5, 275, 276, 277, 291
Green manure 18, 26, 32, 34, 46, 50, 63, 67, 81, 91, 119, 124, 134, 136, 191, 193, 207, 222, 242, 314
Greenhouse gases 11, 275, 281, 289

H

Hardening-off 177
Heat stress 48, 120, 162, 163, 166
Hedgerow intercropping 28, 32, 55, 57, 134
High density energy plantations 155

Highland 59, 60, 164, 171
Homegarden 31, 50, 51, 52, 53, 56, 57, 59, 68, 136, 137, 138, 139, 140, 173, 188, 189, 190, 200, 208, 242, 300
Hortipasture 9
Hot desert 73
Humid subtropic 74
Humid temperate 74
Humid tropic 50, 74, 119, 189, 295
Hydromorphic 26, 60, 61

I

In situ conservation 203, 209
Industrial wasteland 211, 214
Integrated production system 1
Intercropping 28, 32, 36, 55, 57, 126, 134, 144, 146, 147, 148, 149, 150, 152, 189, 208, 241, 298
Intermittent 55, 164, 209, 257
Interpolated 55, 56

J

Jhum 28, 132
Joint forest management 249

K

Krishishukti 1

L

Land capability classification 18, 20
Land equivalent ratio 126
Land evaluation 21, 22
Land use system 2, 22, 50, 59, 66, 98, 103, 129, 137, 167, 203, 219, 220, 221, 222, 224, 226, 239, 281, 295
Landowners 259
Lateritic soils 75
Layering 169, 172, 173, 305, 306, 312, 315, 326
Leaf litter 34, 100, 101, 136, 149, 166, 206, 208
Leeward 37, 38, 39, 40, 41, 42, 143, 191
Live fences 48, 49, 56, 64, 70, 109, 120, 188, 278
Local control 231
Lopping 8, 12, 69, 70, 72, 118, 135, 185, 193, 213, 300, 307

M

Marketable products 28
Marshy land 215, 216
Mesomorphic 26, 60
Microclimate 5, 40, 41, 45, 49, 52, 63, 64, 73, 109, 115, 117, 118, 119, 120, 124, 125, 134, 139, 140, 163, 166, 173, 208, 232, 250, 277, 278, 279, 280
Microflora 105
Mixed cropping 126, 137, 144, 150, 151, 152, 301
Mixed forestry 84
Monoculture 2, 6, 27, 36, 140
Mulch 32, 34, 43, 44, 45, 46, 51, 73, 101, 106, 109, 120, 124, 126, 127, 134, 135, 141, 175, 176, 178, 191, 193, 200, 206, 207, 217, 231, 258, 279, 280, 314, 329
Multilayer tree gardens 28, 31, 59
Multipurpose trees 28, 35, 51, 60, 63, 65, 66, 68, 70, 197, 199, 208, 211, 215, 225, 295
Mycorrhizal fungi 71, 110

N

National agroforestry policy 15, 253, 254, 255, 256, 257
National forest policy 244, 245, 246, 247, 248, 249, 250, 252, 253
National forest 244, 245, 246, 247, 248, 249, 250, 252, 253, 273
National rainfed area authority 10
Negative interaction 3, 118, 121, 122, 124
Nitrogen fixing trees 44, 56, 72, 102, 107, 110, 119, 135, 206, 310

Non-timber products 41
Nurse trees 72
Nutrient cycling 72, 97, 98, 100, 101, 103, 104, 118, 119, 130, 253

O
On-farm trials 229, 235, 236, 237
Organic fertiliser 3
Organic matter 23, 34, 48, 64, 73, 97, 98, 99, 100, 101, 102, 104, 105, 106, 109, 110, 111, 112, 119, 134, 136, 149, 153, 177, 189, 192, 193, 195, 199, 205, 206, 212, 253, 258, 279, 280, 282, 286, 287, 288, 289, 290, 321, 323
Orientation 38, 39, 42, 134, 135, 141, 144, 219, 228, 234, 235
Overexploitation 258

P
Panchayat lands 84
Petro-plants 159
Phenology 67, 131
Physiognomic 25, 26, 60
Phytotoxic 103
Planting stock 91, 256, 324
Podu 28, 132
Pollarding 69, 124, 135, 155, 172, 184, 185, 266, 315, 320
Pollutants 65, 102, 195, 259, 280
Polyphenols 99
Pricking out 174, 175, 176, 296, 312
Protection forests 246, 248, 249
Protein banks 48, 56
Prototype 232, 237
Pruning 33, 34, 35, 42, 44, 45, 66, 67, 68, 69, 70, 72, 98, 99, 101, 103, 104, 106, 111, 119, 124, 125, 134, 135, 136, 165, 166, 167, 168, 172, 174, 182, 183, 184, 187, 188, 190, 207, 231, 268, 279, 280, 287, 296, 306, 315, 318, 325, 332

Q
Quadruple cropping 152
Quality planting material 89, 173, 254, 256, 257

R
Rainfed land 9
Randomization 230, 234, 235
Rangelands 35, 48, 59, 214, 273
Ratooning 152
Ravinous land 211, 212
Recreation forestry 83, 86
Red soils 74
Reforestation 11, 12, 30, 83, 86, 186, 244, 265, 281, 291, 292
Regeneration 13, 31, 168, 169, 170, 190, 207, 211, 245, 248, 249, 251, 265, 266, 281, 292, 303, 329
Rehabilitation 71, 85, 211, 214, 216, 217
Relevance of agroforestry 4
Renewable energy 86, 155
Renewable resources 87, 290
Research domain 226, 227
Rhizobium 100, 107, 108, 109, 110, 298
Rhizosphere 102

S
Sand dune 211, 213, 214, 250
Scarification 171, 312
Scenic value 86
Self-reliance 72
Sequential cropping 152, 153
Sericulture 53, 54
Shade-tolerant crops 27, 300
Shifting cultivation 28, 29, 30, 34, 55, 58, 59, 61, 79, 98, 131, 132, 189, 211, 212, 213, 240, 242, 251
Silvipasture 7, 8, 9, 11, 46, 47, 48, 57, 59, 160, 161, 162, 163, 164, 166, 213, 242
Simulation models 127

Social forestry 26, 79, 80, 81, 82, 83, 84, 86, 87, 88, 89, 90, 94, 155, 249, 251, 252
Sodicity 102, 206
Soil carbon 12, 286, 287, 288, 289, 290, 291
Soil conservation 11, 18, 28, 45, 46, 50, 54, 55, 57, 60, 63, 82, 84, 88, 98, 104, 111, 118, 119, 139, 191, 194, 200, 208, 212, 231, 244
Soil fertility 4, 5, 26, 28, 31, 32, 45, 57, 60, 68, 69, 70, 76, 91, 97, 98, 99, 103, 104, 118, 119, 125, 126, 131, 132, 134, 145, 149, 153, 154, 156, 161, 178, 181, 189, 190, 199, 205, 222, 229, 265, 267, 277, 279, 282, 300, 307
Spatial arrangement 2, 55, 124, 148, 220, 226
Species selection 28, 144, 167, 194, 268
Stratification 171
Sub-surface drainage 196
Subtropical semi-arid 74
Subtropical semi-humid 74
Succession 152, 153, 154
Supplementary interaction 116
Surface runoff 5, 196, 205
Sustainable development goals 241, 261, 263, 290
Systematic designs 231, 234, 235

T

Tannins 64, 122, 159, 162, 314, 316, 328
Tarai region 75
Taungya system 28, 29, 30, 31, 56, 132, 133, 190
Temperate semi-arid 74
Temperate semi-humid 74
Temporal arrangement 55
Terracing 195, 196, 200, 209, 211, 258
Thinning 165, 167, 168, 180, 181, 182, 183, 187, 287, 296, 299, 301, 304, 306, 310, 331, 332
Tip pruning 184
Training 65, 79, 87, 88, 182, 194, 240, 243, 269
Transpiration 42, 43, 45, 101, 119, 120, 156, 171, 192, 204, 206, 279, 280
Tree borne oil seeds 11
Tree-animal interface 120, 122
Tree-crop interaction 118,123,125,232
Tree-lands 246, 247
Tropical semi-arid 74
Tropical sub-humid 74

U

Understorey 36, 68, 100, 120, 187, 204, 235, 287, 289
Unmanned aerial vehicles 264
Urban forestry 250

V

Value addition 7, 9
Village forest committee 89, 90
Village forests 210, 246, 247

W

Wastelands 7, 8, 9, 12, 82, 84, 159, 210, 211, 212, 213, 214, 215, 216, 217
Water conservation 12, 64, 118, 120, 134, 189, 209, 217, 243, 250, 279
Water use efficiency 98, 279
Wildings 169, 171, 174, 176
Windbreaks 26, 28, 41, 42, 43, 44, 45, 49, 54, 55, 57, 60, 64, 71, 112, 120, 140, 141, 142, 143, 179, 180, 189, 193, 194, 217, 305, 316
Windward 37, 38, 41, 42, 178, 191, 280
World agroforestry centre 2, 241

X

Xeromorphic 26, 60
Xerophytic 60, 192, 216

Social forestry 26, 74, 76, 81, 82, 85, [illegible]
Societies 182, 200
[illegible]
Spatial arrangement 55, 57 [illegible]
Sub-optimal landscape 74
Subsistence 182, 183, 184
Surface runoff [illegible] 205
Sustainable development goals 244, 250
Systematic designs 231, 232, 233

T

Terminal [illegible]
Terracing 64, 122, 155, 162, 214, 216, 238
Thinning 155, 157, 164, 165, 171, 172, 183, 185, 206, 209, 210, 211
Toposequence 181
Topo-sequence semi-arid 74
Topographic semi-humid 74
Temporal arrangement 55
Terracing 195, 198, 200, 204, 211, 238

W

Water conservation 12, 64, 117, 119, 120, 124, 189, 209, 213, 214, 216, 219
Water use efficiency 84, 235
Wildlings 169, 171, 175, 186
Windbreaks 26, 28, 31, 32, 42, 44, 45, 49, 54, 55, 57, 60, 64, 71, 113, 120, 141, 142, 143, 149, 179, 180, 189, 193, 194, 217, 225, 276
Woodlots 47, 78, 81, 82, 125, 191, 230
World agroforestry centre 2, 41

X

Xeromorphic 26, 60
Xerophytes 50, 192, 216